T0142870

Advances in Intelligent Systems and Computing

Volume 546

Series editor

Janusz Kacprzyk, Polish Academy of Sciences, Warsaw, Poland
e-mail: kacprzyk@ibspan.waw.pl

About this Series

The series "Advances in Intelligent Systems and Computing" contains publications on theory, applications, and design methods of Intelligent Systems and Intelligent Computing. Virtually all disciplines such as engineering, natural sciences, computer and information science, ICT, economics, business, e-commerce, environment, healthcare, life science are covered. The list of topics spans all the areas of modern intelligent systems and computing.

The publications within "Advances in Intelligent Systems and Computing" are primarily textbooks and proceedings of important conferences, symposia and congresses. They cover significant recent developments in the field, both of a foundational and applicable character. An important characteristic feature of the series is the short publication time and world-wide distribution. This permits a rapid and broad dissemination of research results.

Advisory Board

Chairman

Nikhil R. Pal, Indian Statistical Institute, Kolkata, India
e-mail: nikhil@isical.ac.in

Members

Rafael Bello Perez, Universidad Central "Marta Abreu" de Las Villas, Santa Clara, Cuba
e-mail: rbellop@uclv.edu.cu

Emilio S. Corchado, University of Salamanca, Salamanca, Spain
e-mail: escorchado@usal.es

Hani Hagras, University of Essex, Colchester, UK
e-mail: hani@essex.ac.uk

László T. Kóczy, Széchenyi István University, Győr, Hungary
e-mail: koczy@sze.hu

Vladik Kreinovich, University of Texas at El Paso, El Paso, USA
e-mail: vladik@utep.edu

Chin-Teng Lin, National Chiao Tung University, Hsinchu, Taiwan
e-mail: ctlin@mail.nctu.edu.tw

Jie Lu, University of Technology, Sydney, Australia
e-mail: Jie.Lu@uts.edu.au

Patricia Melin, Tijuana Institute of Technology, Tijuana, Mexico
e-mail: epmelin@hafsamx.org

Nadia Nedjah, State University of Rio de Janeiro, Rio de Janeiro, Brazil
e-mail: nadia@eng.uerj.br

Ngoc Thanh Nguyen, Wroclaw University of Technology, Wroclaw, Poland
e-mail: Ngoc-Thanh.Nguyen@pwr.edu.pl

Jun Wang, The Chinese University of Hong Kong, Shatin, Hong Kong
e-mail: jwang@mae.cuhk.edu.hk

More information about this series at http://www.springer.com/series/11156

Kusum Deep · Jagdish Chand Bansal
Kedar Nath Das · Arvind Kumar Lal
Harish Garg · Atulya K. Nagar
Millie Pant
Editors

Proceedings of Sixth International Conference on Soft Computing for Problem Solving

SocProS 2016, Volume 1

 Springer

Editors
Kusum Deep
Department of Mathematics
Indian Institute of Technology Roorkee
Roorkee
India

Jagdish Chand Bansal
Department of Mathematics
South Asian University
New Delhi
India

Kedar Nath Das
Department of Mathematics
National Institute of Technology, Silchar
Silchar, Assam
India

Arvind Kumar Lal
School of Mathematics
Thapar Institute of Engineering and
    Technology University
Patiala, Punjab
India

Harish Garg
School of Mathematics
Thapar University Patiala
Patiala, Punjab
India

Atulya K. Nagar
Department of Mathematics and
    Computer Science
Liverpool Hope University
Liverpool
UK

Millie Pant
Department of Applied Science and
    Engineering
Indian Institute of Technology Roorkee
Roorkee
India

ISSN 2194-5357 ISSN 2194-5365 (electronic)
Advances in Intelligent Systems and Computing
ISBN 978-981-10-3321-6 ISBN 978-981-10-3322-3 (eBook)
DOI 10.1007/978-981-10-3322-3

Library of Congress Control Number: 2017931564

© Springer Nature Singapore Pte Ltd. 2017
This work is subject to copyright. All rights are reserved by the Publisher, whether the whole or part of the material is concerned, specifically the rights of translation, reprinting, reuse of illustrations, recitation, broadcasting, reproduction on microfilms or in any other physical way, and transmission or information storage and retrieval, electronic adaptation, computer software, or by similar or dissimilar methodology now known or hereafter developed.
The use of general descriptive names, registered names, trademarks, service marks, etc. in this publication does not imply, even in the absence of a specific statement, that such names are exempt from the relevant protective laws and regulations and therefore free for general use.
The publisher, the authors and the editors are safe to assume that the advice and information in this book are believed to be true and accurate at the date of publication. Neither the publisher nor the authors or the editors give a warranty, express or implied, with respect to the material contained herein or for any errors or omissions that may have been made. The publisher remains neutral with regard to jurisdictional claims in published maps and institutional affiliations.

Printed on acid-free paper

This Springer imprint is published by Springer Nature
The registered company is Springer Nature Singapore Pte Ltd.
The registered company address is: 152 Beach Road, #21-01/04 Gateway East, Singapore 189721, Singapore

Contents

About the Editors

Prof. Kusum Deep is Professor at the Department of Mathematics, Indian Institute of Technology Roorkee, India. Over the past 25 years, her research has made her a central international figure in the areas of nature-inspired optimization techniques, genetic algorithms and particle swarm optimization.

Dr. Jagdish Chand Bansal is Assistant Professor with South Asian University, New Delhi, India. Holding an excellent academic record and having written several research papers in journals of national and international repute, he is an outstanding researcher in the field of swarm intelligence at both national and international levels.

Dr. Kedar Nath Das is Assistant Professor at the Department of Mathematics, National Institute of Technology, Silchar, Assam, India. Over the past 10 years, he has made substantial contribution to research on 'soft computing'. He has published several research papers in prominent national and international journals. His chief area of interest includes evolutionary and bio-inspired algorithms for optimization.

Dr. Arvind Kumar Lal is currently associated with the School of Mathematics and Computer Applications at Thapar University, Patiala. He received his B.Sc. Honors (mathematics) and M.Sc. (mathematics) from Bihar University, Muzaffarpur in 1984 and 1987, respectively. He completed his Ph.D. (mathematics) at the University of Roorkee (now the IIT, Roorkee) in 1995. Dr. Lal has over 130 publications in journals and conference proceedings to his credit. His research areas include applied mathematics (modeling of stellar structure and pulsations), reliability analysis and numerical analysis.

Dr. Harish Garg is Assistant Professor at the School of Mathematics at Thapar University, Patiala, Punjab, India. He received his B.Sc. (computer applications) and M.Sc. (Mathematics) from Punjabi University, Patiala before completing his Ph.D. (applied mathematics) at the Indian Institute of Technology Roorkee. He is currently teaching undergraduate and postgraduate students and is pursuing innovative and insightful research in the area of reliability theory using evolutionary algorithms and fuzzy set theory with their application in numerous industrial engineering areas. Dr. Garg has produced 62 publications, which include 6 book chapters, 50 journal papers and 6 conference papers.

Prof. Atulya K. Nagar holds the Foundation Chair as Professor of Mathematical Sciences and is Dean of the Faculty of Science at Liverpool Hope University, UK. Professor Nagar is an internationally respected scholar working on the cutting edge of theoretical computer science, applied mathematical analysis, operations research and systems engineering.

Dr. Millie Pant is Associate Professor at the Department of Paper Technology, Indian Institute of Technology Roorkee, India. She has published several research papers in national and international journals and is a prominent figure in the field of swarm intelligence and evolutionary algorithms.

Adaptive Scale Factor Based Differential Evolution Algorithm

Nikky Choudhary$^{(\boxtimes)}$, Harish Sharma, and Nirmala Sharma

Rajasthan Technical University, Kota, India
nikkychoudharys8@gmail.com

Abstract. In DE, exploration and exploitation capabilities depend on two processes, namely mutation and crossover. In these two processes exploration capability and exploitation capability is balanced using the tuning of scale factor F and crossover probability CR. In DE, for a high value of CR and F, there is always enough chance to skip the true solution due to large step size in the solution search space. Therefore in this article, a self-adaptive scale factor strategy is proposed in which scale factor is adaptively decided through iterations. In the proposed strategy, in the early iteration, the value of F is kept high to keep the large step size while in later iterations the value of F is kept small to keep the step size short. The proposed strategy is named as Adaptive Scale Factor based Differential Evolution (ASFDE) Algorithm. Further, to increase the exploration capability of the algorithm, a limit is associated with every solution to count the number of not updating iterations. If this count crosses the pre-defined limit, then the solution is randomly initialized. The proposed algorithm is tested over 12 different benchmark functions and correlate with standard DE, and another swarm intelligence based algorithm, namely artificial bee colony (ABC) algorithm, and particle swam optimization (PSO) algorithm. The obtained results reveal that ASFDE is a competitive variant of DE.

Keywords: Evolutionary Algorithm · Differential Evolution Algorithm · Optimization · Nature inspired algorithms

1 Introduction

Nature-inspired algorithms (NIA) are inspired by the natural behavior and solve various real-world optimization problems [11]. Evolutionary Algorithm (EA) is a method used for searching the optimum value for a problem by yielding a populous of results over numeral generations. Differential Evolution (DE) is a populous based and random probability search technique, comparatively an easy method to search an optimum value to the optimization problems. There are possibilities that populous has not merged to local optima due to which it can't reach the global optimum [7].

Mezura-Montes et al. [12] emulate the different forms of DE for global best and bring out that DE demonstrates a degraded performance and stays

© Springer Nature Singapore Pte Ltd. 2017
K. Deep et al. (eds.), *Proceedings of Sixth International Conference on Soft Computing for Problem Solving*, Advances in Intelligent Systems and Computing 546, DOI 10.1007/978-981-10-3322-3_1

ineffectual in analyzing the search space, particularly for multimodal functions. Price et al. [5] concluded the same. The problems of premature convergence and stagnation have to be considered seriously for developing a comparatively efficient differential evolution algorithm. Research is constantly operational to enhance the premature convergence of DE [3,8,13]. In [9] Some latest variants of DE with remarkable purposes are depicted.

In this paper, modified version is adaptive through a change in iterations which make the step-size adaptive due to which a proper balance is maintained among two mechanism that is exploration and exploitation which helps a leader to remove stagnation and achieve a good convergence speed. Further, to avoid the stagnation, a limit based counter is associated with every solution to check the not updating counts of the solution.

The remaining paper is described as: Sect. 2 covers the Condensed summary of DE. Adaptive Scale Factor based Differential Evolution (ASFDE) Algorithm is presented in Sect. 3. Performance of ASFDE is tested with several benchmark functions in Sect. 4. Last, Sect. 5 comprises a summary and conclusion.

2   Condensed Summary of the DE

Price and Storn propounded DE algorithm [2] in 1995. It is a fast, easy and populous based random probabilistic search technique. $DE/rand/1/bin$ technique is used, rand denotes that parent is elected arbitrarily, 1 represents the counts of differential vectors and bin indicates the binomial crossover. DE comprises of three essential components that are mutation, crossover and selection respectively. At the initial phase, uniformly distributed population is generated randomly. Mutation results in the generation of an experimental vector which further used within crossover to create offspring and then selection is committed to elect the best for next generation [10]. A Dim-dimensional vector $(x_{i1}, x_{i2}, \ldots, x_{iDim})$ is used to represent an Dim-dimensional area and i = 1, 2, ..., S. Here, S is the populous size. Initialization of ith vector in j^{th} component is displayed Eq. 1:

$$X_{i,j} = X_{j,lo} + rand_{i,j}[0,1] * (X_{j,hi} - X_{j,lo}) \tag{1}$$

where, $X_{i,j}$ is a position, lo and hi are lower and upper limits of searching area. $rand_{ij}$ is an evenly dispersed random number in the range of 0 to 1.

2.1   Mutation

For every individual of the current populous, an experimental vector is produced by mutation operator. Experimental vector is created when a parent is altered with a subjective differential which produces an offspring in crossover operation. For generating an experimental vector $u_i(t)$, mutation operation is defined in Eq. 2:

- Parent $x_{i1}(t)$ is elected randomly from initialized populous, as $i \neq i_1$.
- Election of two candidates are done arbitrarily that are x_{i2} and x_{i3}, from populous with a condition that $i \neq i_1 \neq i_2 \neq i_3$.
- After this, experimental vector is computed by mutating the parent using Eq. 2:

$$u_i(t) = x_{i_1}(t) + F \times (x_{i_2}(t) - x_{i_3}(t)) \tag{2}$$

Here, $F \in [0,1]$, which controls differential variation.

2.2   Crossover

Crossover is applied to get offspring $x_i'(t)$ which is produced by crossover of parent $x_i(t)$ and experimental vector $u_i(t)$ depicted in the following Eq. 3:

$$y_{ij}'(t) = \begin{cases} u_{ij}(t), & \text{if } q \in Q \\ x_{ij}(t), & \text{otherwise.} \end{cases} \tag{3}$$

Here Q is the set of crossover points that will go under perturbation, $x_{ij}(t)$ is the jth element of the vector $x_i(t)$. Basically two types of crossover are used in DE. The presented variant ASFDE uses the binomial crossover. Here, $R(1, Dim)$ is a uniformly distributed between 1 and Dim. Crossover point (Q) $\in \{1, 2, ...,$ $Dim\}$ is used to select the crossover points in random fashion. Algorithm 1 shows binomial crossover. CR (Crossover probability) is used to select the crossover points.

Q (Set of crossover points) $=$ empty set, $q^* \sim R(1, Dim)$;
$Q \leftarrow Q \cup q^*$;
for each $q \in 1.......Dim$ (Problem dimension) **do**
   if $R(0,1) < CR$ (Crossover probability) *and* q $\neq q^*$ **then**
      $Q \leftarrow Q \cup q$;
   end if
end for

Algorithm 1. Binomial Crossover.

2.3   Selection

The solution having low-cost value or less objective value is chosen to survive in next generation i.e. group. It elects the better among parent and offspring depending on objective cost for the next group.

$$x_i(t+1) = \begin{cases} y_i'(t), & \text{if } f(y_i'(t)) > f(x_i(t)). \\ x_i(t), & \text{otherwise.} \end{cases} \tag{4}$$

The solution having less objective value will survive in the next generation.

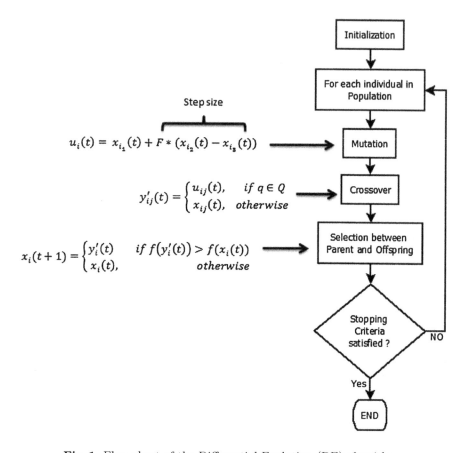

$$u_i(t) = x_{i_1}(t) + F * (x_{i_2}(t) - x_{i_3}(t))$$

Step size

$$y'_{ij}(t) = \begin{cases} u_{ij}(t), & if\ q \in Q \\ x_{ij}(t), & otherwise \end{cases}$$

$$x_i(t+1) = \begin{cases} y'_i(t) & if\ f(y'_i(t)) > f(x_i(t)) \\ x_i(t), & otherwise \end{cases}$$

Fig. 1. Flow chart of the Differential Evolution (DE) algorithm

3 Adaptive Scale Factor Based DE

To avoid premature convergence and stagnation, a modification is done in mutation phase of DE. DE algorithm has two control parameters named as scaling factor F and crossover probability CR. CR controls the perturbation rate of the algorithm and F used to maintain the step-size of individuals. CR is directly proportional to perturbation rate which is directly proportional to the exploration of the search area. When F is greater, then resultant step-size explore the search area, and lesser the F exploitation is performed. In basic DE, F is constant due to which step-size gets affected.

For improving this drawback of DE, concept of adaptive step-size is introduced in the algorithm by which initially step-size is large and later on it decreases gradually. Modified equation for mutation phase is given below:

$$u_i(t) = x_{i_1}(t) + r \times (1 - iter/Max\_iterations) \times (x_{i_2}(t) - x_{i_3}(t)) \tag{5}$$

In above Eq. 5, r is random number between $(0.1,1)$, *iter* is present iteration and $Max\_iterations$ is maximum iterations. From Eq. 5 there is a uniform change in step-size that helps in doing exploration initially and as an increase in iteration number results in exploitation of search space. This modification results in a uniform decrease in step-size by which global optima can't be skipped.

Hence, modified version is adaptive through a change in iterations which make the step-size adaptive due to which a proper balance is maintained among two mechanism that is exploration and exploitation which helps a leader to remove stagnation and achieve a good convergence speed.

Further, to reduce the possibilities of premature convergence, a counter is associated with every solution which is incremented by one in case the solution is not updated itself in each iteration. If a solution gets updated then the counter is initialized to zero. If the counter is reached to a predefined threshold then the associated solution is randomly initialized in the solution search space by considering that the solution has been stuck in a local optima.

Based on above discussion pseudo code of the purposed strategy is shown in Algorithm 2:

Initialize, control parameters, r and CR;
Initialize, the populous, $s(0)$, of S individuals;
while termination criteria(s) **do**
   for each solution, $x_i(t) \in S(t)$ **do**
     Find the Objective value, $f(x_i(t))$;
     Produce the experimental vector depicted in the following Eq. 6:

$$u_i(t) = x_{i_1}(t) + r \times (1 - iter/Max\_iterations) \times (x_{i_2}(t) - x_{i_3}(t)) \quad (6)$$

     Produce offspring, $y_i'(t)$, using crossover;
     if $f(y_i'(t))$ is more fit than $f(x_i(t))$ **then**
       Add $y_i'(t)$ to $P(t+1)$;
     else
       Add $x_i(t)$ to $P(t+1)$;
     end if
   end for
   if Counter associated with x_i **is reached to the threshold then**
     Counter=0 and randomly initialize x_i.
   end if
end while
Return the best solution;
       Algorithm 2. Adaptive Scale Factor based DE.

4   Outcomes and Discussions

4.1   Test Problems Under Consideration

For examining the outcomes of the ASFDE, 12 different benchmark functions (f_1 to f_{12}) are picked and displayed in Table 1.

Table 1. Test function problems

Test Problem (TP)	Objective function (OF)	Search Range (SR)	Optimum Value (OV)	D	Acceptable Error (AE)
Cosine Mixture	$f_1(x) = \sum_{i=1}^{D} x_i^2 - 0.1\left(\sum_{i=1}^{D}\cos 5\pi x_i\right) + 0.1D$	$[-1, 1]$	$f(0) = -D \times 0.1$	30	$1.0E-05$
Step function	$f_2(x) = \sum_{i=1}^{D} (\lfloor x_i + 0.5 \rfloor)^2$	$[-100, 100]$	$f(-0.5 \le x \le 0.5) = 0$	30	$1.0E-05$
Inverted cosine wave	$f_3(x) = -\sum_{i=1}^{D-1}\left(\exp\left(\frac{-(x_i^2+x_{i+1}^2+0.5x_ix_{i+1})}{8}\right)\times I\right)$ where, $I = \cos\left(4\sqrt{x_i^2+x_{i+1}^2+0.5x_ix_{i+1}}\right)$	$[-5, 5]$	$f(0) = -D + 1$	10	$1.0E-05$
Levy montalvo 1	$f_4(x) = \frac{\pi}{D}\left(10\sin^2(\pi y_1) + \sum_{i=1}^{D-1}(y_i-1)^2 \times (1+10\sin^2(\pi y_{i+1})) + (y_D-1)^2\right)$, where $y_i = 1 + \frac{1}{4}(x_i+1)$	$[-10, 10]$	$f(-1) = 0$	30	$1.0E-05$
Colville	$f_5(x) = 100|x_2 - x_2^2|^2 + (1-x_1)^2 + 90(x_4 - x_3^2)^2 + (1-x_3)^2 + 10.1|(x_2-1)^2 + (x_4-1)^2| + 19.8(x_2-1)(x_4-1)$	$[-10, 10]$	$f(1) = 0$	4	$1.0E-05$
Kowalik	$f_6(x) = \sum_{i=1}^{11}\left[a_i - \frac{x_1(b_i^2+b_ix_2)}{b_i^2+b_ix_3+x_4}\right]^2$	$[-5, 5]$	$f(0.192833, 0.190836, 0.123117, 0.135766) = 0.000307486$	4	$1.0E-05$
2D Tripod	$f_7(x) = p(x_2)\{1 + p(x_1)\} + |(x_1 + 50p(x_2)(1-2p(x_1)))| + |(x_2 + 50(1-2p(x_2)))|$	$[-100, 100]$	$f(0, -50) = 0$	2	$1.0E-04$
Shifted Rosenbrock	$f_8(x) = \sum_{i=1}^{D-1}(100(z_i^2 - z_{i+1})^2 + (z_i-1)^2) + f_{bias}, z = x - o + 1, x = [x_1, x_2,....x_D], o = [o_1, o_2,...o_D]$	$[-100, 100]$	$f(0) = f_{bias} = 390$	10	$1.0E-01$
Goldstein-Price	$f_9(x) = (1+(x_1+x_2+1)^2 \cdot (19 - 14x_1 + 3x_1^2 - 14x_2 + 6x_1x_2 + 3x_2^2)) \cdot (30 + (2x_1 - 3x_2)^2 \cdot (18 - 32x_1 + 12x_1^2 + 48x_2 - 36x_1x_2 + 27x_2^2))$	$[-2, 2]$	$f(0, -1) = 3$	2	$1.0E-14$
Hosaki Problem	$f_{10} = (1 - 8x_1 + 7x_1^2 - 7/3x_1^3 + 1/4x_1^4)x_2^2\exp(-x_2)$, subject to $0 \le x_1 \le 5, 0 \le x_2 \le 6$	$[0, 5], [0, 6]$	-2.3458	2	$1.0E-05$
Meyer and Roth Problem	$f_{11}(x) = \sum_{i=1}^{5}\left(\frac{x_1x_3t_i}{1+x_1t_i+x_2v_i} - y_i\right)^2$	$[-10, 10]$	$f(3.13, 15.16, 0.78) = 0.4E-04$	3	$1.0E-03$
Sinusoidal Problem	$f_{12}(x) = -[A\prod_{i=1}^{n}\sin(x_i - z) + \prod_{i=1}^{n}\sin(B(x_i - z))], A = 2.5, B = 5, z = 30$	$[0, 180]$	$f(90 + z) = -(A + 1)$	10	$1.0E-02$

Table 2. Comparison of the results of test function problems

TF	Algorithm	SD	ME	AFE	SR
f_1	DE	4.72E−02	1.33E−02	37386.0	92
	ASFDE	8.08E−07	8.99E−06	39007.5	100
	ABC	2.30E−06	7.20E−06	22897.5	100
	PSO	7.05E−02	3.70E−02	77107.5	77
f_2	DE	2.92E−01	7.00E−02	26625.5	94
	ASFDE	0.00E+00	0.00E+00	25508.5	100
	ABC	0.00E+00	0.00E+00	11615.0	100
	PSO	9.95E−02	1.00E−02	38549.5	99
f_3	DE	6.59E−01	9.83E−01	173894.5	18
	ASFDE	7.03E−01	8.38E−01	170203.0	30
	ABC	1.18E−01	2.36E−02	91146.9	93
	PSO	6.95E−01	1.37E+00	195749.5	7
f_4	DE	1.03E−02	1.05E−03	21515.0	99
	ASFDE	1.00E−06	8.80E−06	34098.0	100
	ABC	2.23E−06	7.52E−06	19553.0	100
	PSO	6.71E−07	9.31E−06	33939.0	100
f_5	DE	3.66E−01	8.49E−02	32264.5	86
	ASFDE	2.59E−03	6.71E−03	7264.9	100
	ABC	1.04E−01	1.48E−01	200022.3	0
	PSO	2.18E−04	8.01E−04	49955.0	100
f_6	DE	2.00E−03	4.39E−04	55793.0	74
	ASFDE	2.71E−04	1.90E−04	34975.0	88
	ABC	7.54E−05	1.87E−04	184167.6	18
	PSO	1.02E−05	9.20E−05	35835.0	100
f_7	DE	3.67E−01	1.60E−01	34831.0	84
	ASFDE	2.55E−01	7.01E−02	17044.8	93
	ABC	2.24E−07	6.56E−07	12356.6	100
	PSO	3.57E−01	1.51E−01	46957.0	84
f_8	DE	1.78E+00	2.46E+00	194758.5	3
	ASFDE	2.12E−03	9.83E−02	66069.7	100
	ABC	1.08E+00	7.85E−01	172996.3	20
	PSO	2.98E+00	4.88E−01	190951.5	60
f_9	DE	4.00E−15	4.29E−15	3806.0	100
	ASFDE	4.32E−15	4.88E−15	3875.6	100
	ABC	2.96E−06	5.85E−07	111670.7	65
	PSO	3.01E−15	5.27E−15	9759.5	100

Table 2. *(Continued)*

TF	Algorithm	SD	ME	AFE	SR
f_{10}	DE	6.37E−06	5.86E−06	34787.0	83
	ASFDE	6.50E−06	5.82E−06	16845.1	92
	ABC	6.52E−06	5.96E−06	657.0	100
	PSO	3.36E−06	5.85E−06	23302.5	89
f_{11}	DE	8.84E−05	1.96E−03	7751.5	97
	ASFDE	2.85E−06	1.95E−03	1981.5	100
	ABC	3.02E−06	1.95E−03	25993.6	100
	PSO	2.50E−06	1.95E−03	3401.5	100
f_{12}	DE	2.37E−01	5.07E−01	199314.5	1
	ASFDE	1.89E−01	1.64E−01	169981.6	37
	ABC	2.11E−03	7.66E−03	57215.8	100
	PSO	3.05E−01	4.01E−01	177590.5	25

4.2   Trial Settings

For analyzing the performance of the developed algorithm ASFDE, a comparison is done among ASFDE, DE, ABC [6] and PSO [4]. Following trial setting is limited to test the algorithm DE, ASFDE, ABC and PSO over the considered test problem.

– The number of run $=100$,
– Population $S = 50$,
– $r = U$ [0.1, 1],
– Settings for ABC [6] and PSO [4] are taken from their elementary papers.

4.3   Outcomes

Outcomes of algorithms are displayed in Table 2 in a form of standard deviation (SD), mean error (ME), average number of function evaluations (AFE) and success rate (SR). Results in Table 2 replicates, many times ASFDE exceeds by other algorithms in terms of reliability, efficiency, and accuracy.

Further, for comparison of examined algorithms, in a form of consolidated achievement boxplots [1] study of AFE is carried out. Boxplot study presents the empirical circulation of results graphically. The boxplots for DE, ASFDE, ABC and PSO are depicted in Fig. 2. The outcomes clearly display that interquartile span and median of ASFDE is relatively low.

Further, Mann-Whitney U rank sum test [8] is performed between ASFDE - DE, ASFDE - ABC and ASFDE - PSO. Table 3 display the compared outcomes of mean function evaluation and Mann-Whitney test for 100 simulations. In Mann-Whitney test, we observe the remarkable difference between two data set. If an outstanding difference is not seen then = symbol appears, and when a

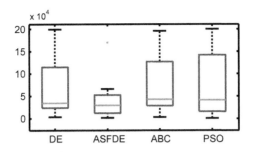

Fig. 2. Boxplot for AFE

remarkable difference is observed then, a comparison is performed regarding the AFEs. And we use + and − symbol, + represent the ASFDE is superior to the examined algorithms and − represent the algorithm is inferior. The total number of + sign in the last line of Table 3, authorize the excellence of ASFDE over chosen algorithms.

Further, all examined algorithms are analyzed regarding ME, SR and AFE by performance indices (PI) [1] graph that are computed for DE, ASFDE, ABC, and PSO respectively and shown in Fig. 3.

Table 3. Evaluations of outcomes in Table 2

Test problems	ASFDE Vs DE	ASFDE Vs ABC	ASFDE Vs PSO
f_1	=	−	+
f_2	+	−	+
f_3	=	+	+
f_4	+	−	=
f_5	+	+	+
f_6	+	+	+
f_7	+	−	+
f_8	+	+	+
f_9	=	+	+
f_{10}	+	−	+
f_{11}	+	+	+
f_{12}	+	−	+
Total number of + sign	09	06	11

It is evident from Fig. 3 that *PI* of ASFDE algorithm is superior as compared to others. At every phase, ASFDE performs better as compared to other established algorithms.

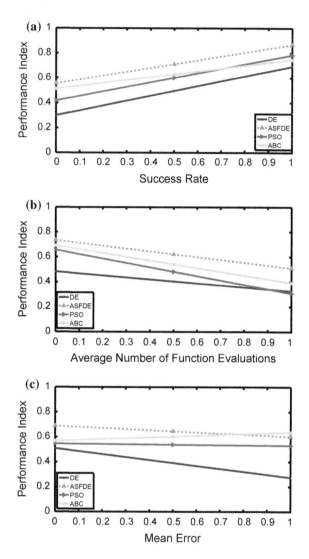

Fig. 3. Performance index; (a) SR, (b) AFE and (c) ME.

5   Conclusion

This paper presents a variant of DE algorithm, known as Adaptive Scale Factor based DE (ASFDE). In ASFDE, modified version is adaptive through a change in iterations which make the step-size adaptive due to which a proper balance is maintained among two mechanism that is exploration and exploitation which helps a leader to remove stagnation and achieve a good convergence speed. Further, to avoid the stagnation, a limit based counter is associated with every solution to check the not updating counts of the solution. The proposed algorithm is compared with DE, ABC, and PSO over different benchmark functions. The obtained results state that ASFDE is a competitive variant of DE and also a good choice for solving the continuous optimization problems. In future, the newly developed algorithm may be used to solve various real-world optimization problems of continuous nature.

References

1. Bansal, J.C., Sharma, H., Arya, K.V., Nagar, A.: Memetic search in artificial bee colony algorithm. Soft Comput. **17**(10), 1911–1928 (2013)
2. Das, S., Mullick, S.S., Suganthan, P.N.: Recent advances in differential evolution-an updated survey. Swarm Evol. Comput. **27**, 1–30 (2016)
3. Das, S., Suganthan, P.N.: Differential evolution: a survey of the state-of-the-art. IEEE Trans. Evol. Comput. **15**(1), 4–31 (2011)
4. Eberhart, R.C., Kennedy, J., et al.: A new optimizer using particle swarm theory. In: Proceedings of the Sixth International Symposium on Micro Machine and Human Science, New York, NY, vol. 1, pp. 39–43 (1995)
5. Engelbrecht, A.P.: Computational Intelligence: An Introduction. Wiley, New York (2007)
6. Karaboga, D., Basturk, B.: A powerful and efficient algorithm for numerical function optimization: artificial bee colony (ABC) algorithm. J. Glob. Optim. **39**(3), 459–471 (2007)
7. Lampinen, J., Zelinka, I.: On stagnation of the differential evolution algorithm (2000)
8. Neri, F., Tirronen, V.: Scale factor local search in differential evolution. Memetic Comput. **1**(2), 153–171 (2009)
9. Panigrahi, B.K., Suganthan, P.N., Das, S.: Swarm, Evolutionary, and Memetic Computing. LNCS, vol. 8947. Springer, Cham (2015)
10. Price, K.V.: Differential evolution: a fast and simple numerical optimizer. In: 1996 Biennial Conference of the North American Fuzzy Information Processing Society, NAFIPS 1996, pp. 524–527. IEEE (1996)
11. Storn, R., Price, K.: Differential evolution-a simple and efficient adaptive scheme for global optimization over continuous spaces, vol. 3. ICSI, Berkeley (1995)
12. Yan, J.-Y., Ling, Q., Sun, D.: A differential evolution with simulated annealing updating method. In: 2006 International Conference on Machine Learning and Cybernetics, pp. 2103–2106. IEEE (2006)
13. Yang, X.-S.: Nature-Inspired Metaheuristic Algorithms. Luniver Press, Bristol (2010)

Adaptive Balance Factor in Particle Swarm Optimization

Siddhi Kumari Sharma and R. S. Sharma$^{(\boxtimes)}$

Rajasthan Technical University, Kota, India
rssharma@rtu.ac.in

Abstract. Particle Swarm Optimization (PSO) is a refined optimization method, that has drawn interest of researchers in different areas because of its simplicity and efficiency. In standard PSO, particles roam over the search area with the help of two accelerating parameters. The proposed algorithm is tested over 12 benchmark test functions and compared with basic PSO and two other algorithms known as Gravitational search algorithm (GSA) and Biogeography based Optimization (BBO). The result reveals that ABF-PSO will be a competitive variant of PSO.

Keywords: Meta-heuristic optimization techniques · Particle swarm optimization algorithm · Swarm intelligence · Acceleration coefficients · Nature inspired algorithm

1 Introduction

Generally, real-world optimization problems are very difficult to solve. Optimization tools are used to solve these kind of problems, though there is no surety to get optimal solution always. So, by using different optimization methods several problems are solved by trial and errors [8]. Development of Swarm intelligence and bio-inspired algorithms make a new subject, inspired by nature. Based on the origins of motivation, these kind of meta-heuristic algorithms can be known as swarm-intelligence-based, bio-inspired-based algorithm [6]. Particle Swarm Optimization (PSO) is a refined optimization method, that has drawn interest of researchers in different areas because of its simplicity and efficiency. Different versions of PSO have been suggested already. PSO is a swarm - based, modifying search development facility firstly suggested by James Kennedy and Russell Eberhart (1995). This algorithm is inspired by mimicking the collective behavior of natural swarm's like fishes and birds and even human common routine etc. [9]. In standard PSO (SPSO), particles roam over the search area with the help of two accelerating parameters. One parameter, known as the cognitive parameter, controls the local exploration of the particles, while the second parameter, known as the social parameter, guides the global search capability of the particles. Generally, diversification and intensification properties are managed by these two parameter. Various researchers have

© Springer Nature Singapore Pte Ltd. 2017
K. Deep et al. (eds.), *Proceedings of Sixth International Conference on Soft Computing for Problem Solving*, Advances in Intelligent Systems and Computing 546, DOI 10.1007/978-981-10-3322-3_2

found that, in the SPSO, particles immediately get a fine local solution, though get stuck to that solution for rest of the iterations without a further improvement [7,12,13].

In order to increase the convergence speed as well as the exploration capabilities of PSO algorithms, an acceleration parameter PSO strategy has been presented in this paper. In this paper a time differing acceleration parameter scheme is introduced to efficiently manage the universal search and convergence to the universal best solution. The primary attention of this modification is to neglect overearly convergence in the initial phases and to enhance convergence to the universal optimal solution in the later phases [14].

The rest of the paper is structured as follows: In Sect. 2, the particle swarm optimization algorithm is discussed. In Sect. 3, introduction of scheduled algorithm and the quality of the scheduled algorithm is tested with several benchmark datasets in Sect. 4. To show the quality of the scheduled strategy, a provisional study is carried out among scheduled strategy, basic PSO and other algo. namely Gravitational search algorithm (GSA) [10] and Biogeography based Optimization (BBO) [15]. The simulation results reveals that the scheduled strategy outperforms among the aforementioned algorithms. At last, Sect. 5 presents a summary and the conclusion of the work.

2 Particle Swarm Optimization Algorithm (PSO)

The particle swarm optimization (PSO) is a swarm- based, modifying search development facility firstly suggested by James Kennedy and Russell Eberhart (1995). As the name suggests it is a swarm intelligence algorithm. This algorithm is inspired by mimicking the collective behavior of natural swarm's like fishes, birds and even human common routine. It can be executed and practiced simply to clarify distinct function optimization issues and the problems which could be converted to function optimization issues [9].

To search the optimal solution, swarms are distributed to spot the food site. When the swarms are seeking for food here and there, there is ever a swarm which may scent the food positively, i.e., the swarm is detectable of the site where it can get the food, keeping the superior food site knowledge. When they seeking the food site, they are transferring the knowledge, specially the fine knowledge at each time, forwarded by the fine knowledge, the swarms will finally went to the food site [2]. Every particle modifies its flying based on its individual flying experience and its colleagues flying experience. Y. Shi and Russell Eberhart (1998) termed the initial as the cognition section and the later the social section. For the social section, James Kennedy and Russell Eberhart (1995) introduced U_{best} (Universal best, experience based on all swarms) and L_{best} (Local best, experience based on individual swarm) components [5,18].

The primary PSO algorithm contains a group of swarms roaming in an n-dimensional, real valued search area of feasible problem solutions. Generally, a convinced fitness aspect is described for swarms to analyze distinct problem solutions. Each swarm i at the time t has the subsequent attributes:

$p_i(t)$ is the position vector;
$s_i(t)$ is the speed (velocity) vector;
$L_i(t)$ is the limited memory saving its individual finest position seen earlier;
$U_i(t)$ is the overall finest position.

Following are some points in PSO process [11]:

At first point, the population (swarms) size is assumed as N. The value of N should be moderate that would give different positions to get optimum solution.

At second point, initial population p is evaluated with arbitrary order to get the p_1, p_2, p_3, ... p_n. Objective function evaluation for every swarm is given by f $[p_1(0)]$, f $[p_2(0)]$, f $[p_3(0)]$, ... f$[$ $p_n(0)]$.

At third point, update the speed for each swarm. The swarms roam facing the optimum solution with a speed. At the initial point, speed of all the swarms is taken as 0. Set iteration t =1. Now, at t^{th} iteration, find some necessary components for every swarm j such as:

- The values of local best (L_{best}) and universal best (U_{best}).
- When the speed is upgraded then the swarm is positioned to a latest position. The latest position is easily computed with the addition of the earlier position and the latest speed:

$$p_i(t+1) = p_i(t) + s_i(t+1) \qquad (1)$$

- Where the speed upgrading is computed as following relation:

$$s_i(t+1) = ws_i(t) + a_1r_1(L_{best}(t) - p_i(t)) + a_2r_2(U_{best}(t) - p_i(t)) \qquad (2)$$

where, w represents inertia weight constant, r_1 and r_2 are random numbers. a_1 and a_2 represent constant values and p represents the position of the swarm.

At the final point, check if the latest solution is convergent. If yes, then stop the iteration, otherwise, repeat last phase by doing $t = t+1$ and compute the values of local best (L_{best}) and universal best (U_{best}).

> **for** every swarm **do**
>> Evaluate fitness value. If the current fitness value is better than the previous best fitness value (Lbest) then set current value as the latest L_{best}.
>> Select the swarm that contains the best fitness value among all swarms as the U_{best}.
> **for** every swarm **do**
>> Evaluate latest speed:
>> $s_i(t+1) = ws_i(t) + a_1r_1(L_{best}(t)\text{-}p_i(t)) + a_2r_2(U_{best}(t)\text{-}p_i(t))$
>> Upgrade position of the swarm:
>> $p_i(t+1) = p_i(t) + s_i(t+1)$
>> Until stopping pattern is found.
> **end for**
> **end for**

Algorithm 1. PSO algorithm

3 Adaptive Balance Factor in Particle Swarm Optimization (ABF-PSO)

In populous-based search algorithms, exploration and exploitation are the two key properties of any Nature Inspired Algorithm (NIA). A proper balance between these two properties are required to find the global optima. Exploration identifies the promising regions by searching the given search space while exploitation helps in finding the optimum solution in the promising search regions. From the explanation of Particle swarm optimization, it is noted that, the search process is managed with the help of two acceleration parameters (the cognitive parameter and the social parameter). So, suitable discipline of those two parameters is really necessary to get the optimal solution precisely as well as effectively.

Mostly, in swarm-related optimization strategies, it is necessary to inspire the swarms to roam over the full search area, beyond gathering around local optimum, during the initial stages of the optimization. While, during the later stages, it is necessary to increase convergence speed, to search the optimal solution effectively.

Suganthan [16] evaluated a form of linear diminishing both acceleration parameters with the time, but realized that the established acceleration parameters at value 2 develop superior solutions. Anyhow, over experimental exercises he offered that the acceleration parameters should not be equivalent to value 2 always. In PSO, for a high value of these two parameter, there is always enough chance to skip the true solution due to large step size in the solution search space.

Since those involvement, in this paper, time differing acceleration parameter scheme is introduced for the PSO technique. The purpose of this expansion is to increase the step size as well as the universal search in the initial stage, and to decrease the step size of the swarms to motivate them to converge toward the global optimum at the last stage of the search. In this scheme, the social parameter is modified with the iterations to manage the step size. In this paper, a balance between cognitive section and the social section is presented, to get the optimal result. The proposed speed updating strategy is shown in following equation:

$$s_i(t+1) = ws_i(t) + a_1 * r_1(L_{best}(t) - p_i(t)) + a_0 * (1 - (it/MaxIt)) * r_2(U_{best}(t) - p_i(t)) \tag{3}$$

where, w is inertia constant, r_1 and r_2 are random values, a_1 has constant value, $a_2 = a_0 * (1 - (it/MaxIt))$, $a_0 = 2.5$, it is the latest iteration no, $MaxIt$ is the total no of iterations and p is particle position.

It is clear from above equation that initially, the social parameter will be large so the step size will also be large and it will help in exploration of the search area. Further, at later stage as the iteration increases the value of social parameter will decrease so that the step size will also gradually decrease, this will help in exploitation.

Based on the above explanation the algorithmic representation is as shown:

for every swarm **do**
    Evaluate fitness value. If the current fitness value is better than the previous best fitness value (Lbest) then set current value as the latest L_{best}.
    Select the swarm that contains the best fitness value among all swarms as the U_{best}.
    for every swarm **do**
        Evaluate latest speed:
        $s_i(t+1) = ws_i(t) + a_1 * r_1(L_{best}(t)\text{-}p_i(t)) + 2.5 * (1 - (it/MaxIt)) * r_2(U_{best}(t)\text{-}p_i(t))$
        Upgrade position of the swarm:
        $p_i(t+1) = p_i(t) + s_i(t+1)$
        Until stopping pattern is found.
    end for
end for

Algorithm 2. ABF-PSO algorithm

4   Experimental Results

4.1   Test Problems Under Consideration

To study the quality of the scheduled algorithm ABF-PSO, 12 distinct universal optimization issues(f_1–f_{12}) are selected as indexed in Table 1. All the issues are continuous optimization issues and having distinct rates of difficulty. Test problems f_1 to f_{12} are yielded from [1,17] with their correlated offset values.

4.2   Experimental Setting

To verify the quality of the scheduled algorithm ABF-PSO, a performance analysis is carried out among ABF-PSO, basic PSO and other algorithms namely Gravitational search algorithm (GSA) [10], Biogeography based Optimization (BBO) [15]. To analyze ABF-PSO, PSO, GSA and BBO over the specified problems, consecutive empirical setting is used:

- The number of simulations/run = 30,
- Population size $nPop = 100$ and Number of food sources $SN = nPop/2$,
- $r_{ij} = rand[0, 1]$,
- $a1 = 1.5$ and $a2 = 2.5$ in PSO update equation.

4.3   Results Comparison

Numeral outcomes according to the empirical setting are given in Table 2. This table represents the outcomes of the scheduled and other considered algorithms in terms of standard deviation (SD), mean error (ME), average number of function

Table 1. Test problems

S. No.	Test problem	Objective function	Search space	Objective value	Dimension	Acceptable error
1	Ackley	$f_1(x) = -20 + e +$ $exp(-\frac{0.2}{D}\sqrt{\sum_{i=1}^{D} x_i^3})$	[−1 1]	$f(0) = 0$	30	$1.0E-05$
2	Alpine	$f_2(x) = \sum_{i=1}^{n} \|x_i \sin x_i + 0.1 x_i\|$	[−10 10]	$f(0) = 0$	30	$1.0E-05$
3	Michalewicz	$f_3(x) =$ $-\sum_{i=1}^{D} \sin x_i (\sin (\frac{i.x_i^2}{\pi})^{20})$	[0 π]	$f_{min} =$ -9.66015	10	$1.0E-05$
4	Cosine Mixture	$f_4(x) = \sum_{i=1}^{D} x_i^2 -$ $0.1(\sum_{i=1}^{D} \cos 5\pi x_i) + 0.1D$	[−1 1]	$f(0) =$ $-D \times 0.1$	30	$1.0E-05$
5	Schewel	$f_5(x) =$ $\sum_{i=1}^{D} \|x_i\| + \prod_{i=1}^{D} \|x_i\|$	[−10 10]	$f(0) = 0$	30	$1.0E-05$
6	Salomon Problem	$f_6(x) = 1 - cos(2\pi\sqrt{\sum_{i=1}^{D} x_i^2}) +$ $0.1(\sqrt{\sum_{i=1}^{D} x_i^2})$	[−100 100]	$f(0) = 0$	30	$1.0E-01$
7	Levy montalvo 1	$f_7(x) = \frac{\Pi}{D}(10\sin^2(\Pi y_1) +$ $\sum_{i=1}^{D-1}(y_i - 1)^2 \times (1 +$ $10\sin^2(\Pi y_{i+1})) + (y_D - 1)^2),$ where $y_i = 1 + \frac{1}{4}(x_i + 1)$	[−10 10]	$f(-1) = 0$	30	$1.0E-05$
8	Levy montalvo 2	$f_8(x) =$ $0.1(\sin^2(3\Pi x_1) + \sum_{i=1}^{D-1}(x_i -$ $1)^2 \times (1 + \sin^2(3\Pi x_{i+1})) +$ $(x_D - 1)^2(1 + \sin^2(2\Pi x_D))$	[−5 5]	$f(1) = 0$	30	$1.0E-05$
9	*Braninss function*	$f_9(x) = a(x_2 - bx_1^2 + cx_1 - d)^2 +$ $e(1 - f)\cos x_1 + e$	$x_1 \in [-510]$, $x_2 \in [015]$	f(0)=0.3979	2	$1.0E-05$
10	Kowalik	$f_{10}(x) =$ $\sum_{i=1}^{11}[a_i - \frac{x_1(b_i^2+b_i x_2)}{b_i^2+b_i x_3+x_4}]^2$	[−5 5]	$f(0.192833,$ $0.190836,$ $0.123117,$ $0.135766) =$ 0.000307486	4	$1.0E-05$
11	Shifted Rastrigin	$f_{11}(x) = \sum_{i=1}^{D}(z_i^2 -$ $10\cos(2\pi z_i) + 10) + f_{bias}$ z=(x-o), x=$(x_1,x_2,........x_D)$, o=$(o_1,o_2,........o_D)$	[−5 5]	$f(0) =$ $f_{bias} =$ -330	10	$1.0E-02$
12	Six-hump camel back	$f_{12}(x) = (4 - 2.1x_1^2 + x_1^4/3)x_1^2 +$ $x_1 x_2 + (-4 + 4x_2^2)x_2^2$	[−5 5]	$f(-0.0898,$ $0.7126) =$ -1.0316	2	$1.0E-05$

evaluations (AFE), and success rate (SR). According to outcomes in Table 2 maximum time ABF-PSO shows best outcomes in terms of performance, accuracy as well as efficiency from the considered algorithm like PSO, GSA and BBO.

Moreover, boxplots study of *AFE* is accomplished for comparing the studied algorithms in terms of centralized quality, still it can simply illustrate the empirical distribution of statistic data graphically. The box plots for ABF-PSO, PSO,

Table 2. Comparison of the results of test problems

Test problem	Algorithm	SD	ME	AFE	SR
f_1	PSO	8.26E−01	7.79E−01	107183.33	15
	ABF-PSO	4.46E−07	9.39E−06	38833.33	30
	GSA	5.83E−07	9.37E−06	161030	30
	BBO	1.27E−06	8.61E−06	60383.33	30
f_2	PSO	5.51E−04	1.40E−04	86393.33	19
	ABF-PSO	1.41E−06	9.12E−06	33290	30
	GSA	5.59E−07	9.29E−06	154615	30
	BBO	5.78E−03	1.04E−02	200000	00
f_3	PSO	9.42E−01	1.69E+00	200000	00
	ABF-PSO	4.78E−01	5.79E−01	191080	02
	GSA	2.35E−01	4.75E−01	197296.67	01
	BBO	2.81E−01	6.20E−01	200000	00
f_4	PSO	4.28E−01	9.95E−01	200000	00
	ABF-PSO	7.32E−02	3.45E−02	69750	24
	GSA	8.13E−07	8.66E−06	111176.67	30
	BBO	1.97E−01	2.36E−01	156776.67	07
f_5	PSO	4.47E−02	1.32E−02	158420	07
	ABF-PSO	6.05E−07	9.26E−06	33406.67	30
	GSA	4.94E−07	9.42E−06	181821.67	30
	BBO	9.51E−07	9.09E−06	48673.33	30
f_6	PSO	6.53E−02	3.20E−01	175233.33	04
	ABF-PSO	4.42E−02	2.27E−01	84983.33	22
	GSA	5.82E−02	8.00E−01	200000	00
	BBO	1.12E−01	6.73E−01	200000	00
f_7	PSO	1.85E−01	1.35E−01	116980	13
	ABF-PSO	3.63E−01	2.73E−01	101920	17
	GSA	8.82E−07	8.81E−06	90630	30
	BBO	8.52E−01	1.17E+00	182070	03
f_8	PSO	2.48E−02	6.64E−03	20616.67	28
	ABF-PSO	8.29E−07	9.11E−06	26570	30
	GSA	6.12E−07	9.00E−06	95498.33	30
	BBO	1.93E−06	8.66E−06	18816.67	30
f_9	PSO	3.18E−05	3.43E−05	1146.67 0	30
	ABF-PSO	2.84E−05	2.65E−05	903.33	30
	GSA	3.29E−05	4.93E−05	37113.33	30
	BBO	1.87E−05	7.40E−05	54626.67	30
f_{10}	PSO	7.39E−05	2.82E−04	303.33	30
	ABF-PSO	7.28E−05	2.83E−04	303.33	30
	GSA	1.07E−04	2.23E−04	75	30
	BBO	5.51E−05	2.75E−04	110	30
f_{11}	PSO	7.07E+00	1.39E+01	200000	00
	ABF-PSO	1.84E+00	3.32E+00	188313.33	02
	GSA	1.56E+00	5.14E+00	200000	00
	BBO	3.35E+00	8.56E+00	200000	00
f_{12}	PSO	9.83E−06	1.44E−05	1263.33	30
	ABF-PSO	1.15E−05	1.59E−05	956.67	30
	GSA	1.16E−05	1.17E−05	49801.67	30
	BBO	3.74E−01	2.45E−01	76745	21

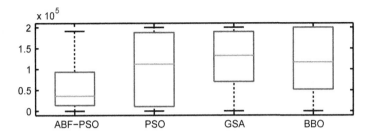

Fig. 1. Box plot graphs (average function evaluation)

GSA and BBO are displayed in Fig. 1. The outcomes declares that interquartile dimensions and medians of ABF-PSO are relatively small.

Next, all studied algorithms are also observed by offering sufficient priority to the AFE, ME, and SR. This observation is calculated using the quality basis that is detailed in [3,4]. The concluded values of PI for the ABF-PSO, PSO, GSA and BBO are measured and subsequent performance index (PIs) graphs are exhibited in Fig. 2.

The graphs relating to any of the cases i.e. offering sufficient priority to the AFE, SR and ME (as described in [3,4]) are displayed in Fig. 2[a], [b] and [c] subsequently. In the mentioned diagrams, parallel axis shows the priority and perpendicular axis specifies the PI.

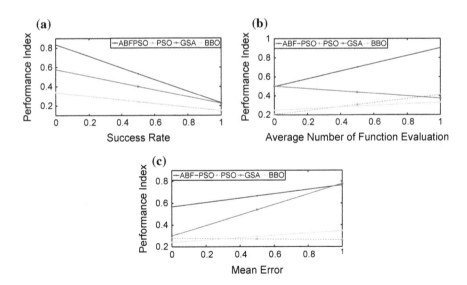

Fig. 2. Performance index for test problems; [a] for SR, [b] for AFE and [c] for ME.

It is clear from Fig. 2 that PI of ABF-PSO are better than the studied algo. in every case. i.e. performance of ABF-PSO is superior on the studied test issues as compared to the PSO, GSA and BBO.

5   Conclusion

To balance the step size of the swarms during the solution search process, a new variant of PSO is presented namely, Adaptive Balance Factor PSO (ABF-PSO) algorithm. In the ABF-PSO, the social parameter is modified such that in initial iterations, the step size of the swarms will be high whereas in later iterations, it will be low. Therefore, by managing the step size, an effort is made to balance the diversification and intensification properties of the PSO algorithm. The proposed ABF-PSO algorithm is practiced on the 12 standard functions and compared with PSO, GSA and BBO algorithms. Through intensive analysis of outcomes, it can be state that the ABF-PSO is an efficient varient of PSO and it can be applied to solve the real world complex optimization problems.

References

1. Montaz Ali, M., Khompatraporn, C., Zabinsky, Z.B.: A numerical evaluation of several stochastic algorithms on selected continuous global optimization test problems. J. Global Optimization **31**(4), 635–672 (2005)
2. Aote, S.S., Raghuwanshi, M.M., Malik, L.: A brief review on particle swarm optimization: limitations & future directions. Intl. J. Comput. Sci. Eng. (IJCSE) **14**, 196–200 (2013)
3. Bansal, J.C., Sharma, H.: Cognitive learning in differential evolution and its application to model order reduction problem for single-input single-output systems. Memetic Comput. **4**(3), 209–229 (2012)
4. Bansal, J.C., Sharma, H., Arya, K.V., Nagar, A.: Memetic search in artificial bee colony algorithm. Soft Comput. **17**(10), 1911–1928 (2013)
5. Eberhart, R.C., Kennedy, J., et al.: A new optimizer using particle swarm theory. In: Proceedings of the Sixth International Symposium on Micro Machine and Human Science, vol. 1, New York, NY, pp. 39–43 (1995)
6. Fister Jr., I., Yang, X.-S., Fister, I., Brest, J., Fister, D.: A brief review of nature-inspired algorithms for optimization. arXiv preprint, arXiv:1307.4186 (2013)
7. Jadon, S.S., Sharma, H., Bansal, J.C., Tiwari, R.: Self adaptive acceleration factor in particle swarm optimization. In: Proceedings of Seventh International Conference on Bio-Inspired Computing: Theories and Applications (BIC-TA 2012), pp. 325–340. Springer (2013)
8. Kennedy, J.: How it works: collaborative trial and error. Intl. J. Comput. Intell. Res. **4**(2), 71–78 (2008)
9. Kennedy, J.: Particle swarm optimization. In: Encyclopedia of Machine Learning, pp. 760–766. Springer, New York (2011)
10. Rashedi, E., Nezamabadi-Pour, H., Saryazdi, S.: GSA: a gravitational search algorithm. Inf. Sci. **179**(13), 2232–2248 (2009)
11. Rini, D.P., Shamsuddin, S.M., Yuhaniz, S.S.: Particle swarm optimization: technique, system and challenges. Intl. J. Comput. Appl. **14**(1), 19–26 (2011)

12. Sharma, K., Chhamunya, V., Gupta, P.C., Sharma, H., Bansal, J.C.: Fitness based particle swarm optimization. Intl. J. Syst. Assur. Eng. Manage. **6**(3), 319–329 (2015)
13. Shi, Y., Eberhart, R.: A modified particle swarm optimizer. In: The 1998 IEEE International Conference on Evolutionary Computation Proceedings, 1998, IEEE World Congress on Computational Intelligence, pp. 69–73. IEEE (1998)
14. Shi, Y., Eberhart, R.C.: Parameter selection in particle swarm optimization. In: Porto, V.W., Saravanan, N., Waagen, D., Eiben, A.E. (eds.) EP 1998. LNCS, vol. 1447, pp. 591–600. Springer, Heidelberg (1998). doi:10.1007/BFb0040810
15. Simon, D.: Biogeography-based optimization. IEEE Trans. Evol. Comput. **12**(6), 702–713 (2008)
16. Suganthan, P.N.: Particle swarm optimiser with neighbourhood operator. In: Proceedings of the 1999 Congress on Evolutionary Computation, 1999, CEC 99, vol. 3. IEEE (1999)
17. Suganthan, P.N., Hansen, N., Liang, J.J., Deb, K., Chen, Y.-P., Auger, A., Tiwari, S.: Problem definitions and evaluation criteria for the CEC 2005 special session on real-parameter optimization. KanGAL report, 2005005 (2005)
18. Tchomté, S.K., Gourgand, M.: Particle swarm optimization: a study of particle displacement for solving continuous and combinatorial optimization problems. Intl. J. Prod. Econ. **121**(1), 57–67 (2009)

Community Detection in Complex Networks: A Novel Approach Based on Ant Lion Optimizer

Maninder Kaur[✉] and Abhay Mahajan

Computer Science and Engineering Department,
Thapar University, Patiala 147004, India
manindersohal@thapar.edu

Abstract. The problem of community detection in complex networks has established an increased amount of interest since the past decade. Community detection is a way to discover the structure of network by assembling the nodes into communities. The grouping performed for the communities encompasses denser interconnection between the nodes than community's intra-connections. In this paper a novel nature-inspired algorithmic approach based on Ant Lion Optimizer for efficiently discovering the communities in large networks is proposed. The proposed algorithm optimizes modularity function and is able to recognize densely linked clusters of nodes having sparse interconnects. The work is tested on Zachary's Karate Club, Bottlenose Dolphins, Books about US politics and American college football network benchmarks and results are compared with the Ant Colony Optimization (ACO) and Enhanced Firefly algorithm (EFF) approaches. The proposed approach outperforms EFF and ACO for Zachary and Books about US politics and produces results better than ACO for Dolphins and EFF for American Football Club. The conclusion drawn from experimental results illustrates the potential of the methodology to effectively identify the network structure.

Keywords: AntLion optimization · Community detection · Modularity · Social networks

1 Introduction

In the past decade, the research on complex networks has become more eye catching in the fields of mathematics, sociology, physics, biology (Ferrara and Fiumara 2011; Newman 2003; Clauset et al. 2004). The topological structure of complex systems can simply be represented as a complex network with connected nodes. The existing networks of well-known social media and online social networking websites like Twitter, Facebook, and Google+ (Ferrara and Fiumara 2011), characterize the system by means of links and nodes. The nodes signify the systems and links represent the relationship between the connecting or interrelating nodes. The network links in different type of areas represent different types of relationship e.g. animal's physical proximity, interconnectivity of infrastructures, human friendship, organizational structures, web hyperlinks and abstract relationships like similarity between data points.

© Springer Nature Singapore Pte Ltd. 2017
K. Deep et al. (eds.), *Proceedings of Sixth International Conference on Soft Computing for Problem Solving*, Advances in Intelligent Systems and Computing 546, DOI 10.1007/978-981-10-3322-3_3

The presence of communities shows the structure of the networks existing in nature. Communities, named as modules/clusters, are the groups of relatively connected nodes, and are said to be inherent arrangement of the networks present in nature (Newman 2003). Nodes of the same community or cluster typically share common interesting features such as a function, purpose and interest. For this reason, one of the most crucial problems in network analysis is community detection.

Various contributions have been anticipated by the researchers in the field of community detection in recent years to detect communities in complex networks, with each methodology being classified according to its algorithm type. Many authors dug into the field of community detection by proposing various analytical approaches. The authors (Newman and Girvan 2004) proposed one of the first successful betweenness based divisive algorithms for community detection. The proposed approach determined the communities but it could not determine the strength of communities formed. Later on the author (Newman 2004) proposed an algorithm based on agglomerative clustering which used the modularity function determining the strength of communities. This algorithm was efficient in case of speed but in practice, the modularity produced by this algorithm was not high. The authors (Clauset et al. 2004) observed that the Newman's approach was not efficient for sparse networks as well as it was also inefficient with respect to time and memory. They improved the Newman's original algorithm with the help of max-heap. Their algorithm was the first algorithm used for analyzing large networks (about 10^6 nodes). The algorithm has a drawback that it might form large communities in the early phase at the cost of existing small communities. The author (Newman 2006) proposed a new method for optimizing the fitness function with the help of associated matrix eigenvectors and eigenvalues. The authors (Schuetz and Callisch 2008) proposed a variation of (Clauset et al. 2004) algorithm by using the "Touched-Community-Exclusion-Rule" (TCER) in the implementation. The algorithm had same complexity as (Clauset et al. 2004) algorithm but instead of creation and maintenance of the max-heap, it required the computation of pairwise gains that increases the computational cost. The authors (Lambiotte et al. 2008) proposed a greedy hierarchical clustering algorithm named as Louvain method considering modularity as objective function. The proposed algorithm is fast but on multiple core architecture the sequential corrections make it slow and it is also inefficient to be applied for very large networks. The authors (Le Martelot and Hankin 2013) proposed a new method based on local and global criteria, with global criteria algorithm almost similar to Louvain method but with an advantage of application for multiscale detection of communities. The local criterion has its advantage that it can be used for overlapping communities.

Community detection being an NP hard problem (Fortunato 2010) various heuristic approaches have been anticipated that assist in detecting communities of the complex networks. There are the widely used properties to calculate the quality factors of the clustered structure of networks. Amongst them, the most well-liked method is reliant on the optimization of the profit function recognized as "modularity" over the feasible partitions of a network. It is one approach to community detection that has been set primarily competent. A growing number of evolutionary approaches (Pizzuti 2008; Honghao et al. 2013; Shang et al. 2013; Hafez et al. 2014; Ma et al. 2014) dependent on modularity optimization in detecting communities have been published in past few years. The author (Pizzuti 2008) presented a (GA) algorithm for uncovering communities in large networks named GA-Net. They introduced the concept named community score. The author also introduced the Safe individual criteria in GA to avoid useless computation making the algorithm efficient. The algorithm showed effective results in finding the communities in networks for synthetic as well as real world data set. The author (Pizzuti 2012) further modified his work by presenting a new GA to discover optimal communities in complex networks named MOGA-Net i.e. multiobjective GA basically a Non-dominated Sorting GA (NSGA-II). The algorithm used modularity criteria as objective function and Normalized Mutual Information (NMI) for measuring the performance of algorithm. The algorithm produced promising results in implementation for synthetic as well as real world data set. The authors (Amiri et al. 2013) proposed a multicriteria optimization approach that utilizes EFF algorithm Fuzzy- based grouping and mutation techniques for the detection of network communities. The implementation and testing of EFA on real world and other synthetic data networks showed its efficiency in finding different communities in large networks. The authors (Shang et al. 2013) proposed a community detection approach named MIGA on the basis of modularity function and an improved GA. The authors also used prior information regarding the number of detecting communities. The results shown a lesser computational complexity of MIGA in comparison to memetic algorithm (ME) and GA for both real-world and computer-generated data networks. The authors (Honghao et al. 2013) proposed an ACO method for detecting communities in networks using max-min ant system method for community detection. The algorithm was tested on four real-life network and LFR bench mark and results showed the great potential of algorithm in finding communities in networks. The authors (Hafez et al. 2014) used Artificial bee colony (ABC) optimization procedure to solve the community detection problem. The algorithm has advantage of automatically detecting the count of communities. The algorithm shown efficient results in terms of accuracy and detection of communities when applied on real-world data as well as online social network. The authors (Ma et al. 2014) proposed fast multi-level memetic algorithm for community detection named MLCD that uses GA with multi-level learning strategies. The results showed a lesser computation time in comparison with original memetic algorithm.

Various evolutionary approaches aforementioned have been applied in the field of community detection for identification of network communities. There is still a gap between the results obtained by existing evolutionary approaches for solving community detection in comparison to original ones. The current work focuses on development of a novel nature-inspired algorithmic approach based on Ant Lion Optimizer (Mirjalili 2015) that is aimed at maximizing the benefit function modularity to produce good partitions of a network into different set of communities through exploration and exploitation, by searching through the possible candidates for ones with high modularity.

2   Problem Statement

Given a graph $G(V, E)$ with a set of n nodes/vertices $V = \{v1, v2, \ldots..vn\}$ and a set of m interconnections/links $E = \{e1, e2, \ldots.em\}$, the graph reflecting the social structure corresponding to the given community is represented with an adjacency matrix A of size $V \times V$ such that for any pair of vertices i and j

$$A_{i,j} = \begin{cases} 1 & \text{if an edge between i and j} \\ 0 & \text{Otherwise} \end{cases} \tag{1}$$

Where the adjacency matrix A is symmetric as shown in Eq. (1) (Assuming the graph as undirected graph).

The problem of community detection relies on finding the subgraphs i.e. partitioning the graph G into n subgraphs $G_1, G_2, \ldots \ldots G_n$ and $V = G_1 \cup G_2 \cup \ldots \ldots G_n$ such that all $G_i \forall i \in n$ correspond to the communities of densely linked nodes, with the nodes belonging to different communities being only sparingly connected based on the criteria of optimizing modularity value. The modularity function evaluates the quality of cluster signifying the extent to which a given community partition is distinguished by high number of intra-community connectivity in comparison to inter-community ones (Newman and Girvan 2004).

The Modularity (Q) value can be mathematically stated as in Eq. (2)

$$Q = 1/2m \sum_{i,j} \left(A_{ij} - \frac{d_i d_j}{2m} \right) \delta(i,j) \tag{2}$$

Where, $A_{i,j}$ is the adjacency matrix, m is the number of edges in network, d_i, d_j are the degree (or strength) of nodes i, j and $\delta(i,j)$ is the function which return 1 when both i, j are in same community, 0 otherwise. The modularity value of a community ranges from -1 to 1 that computes the degree of cohesiveness within community as compared to interconnections between communities (Newman 2004; Pizzuti 2012). More the modularity value better is the quality of the communities detected.

3   The Proposed Algorithm

The work presents a novel nature-inspired algorithmic approach based on Ant Lion Optimizer for solving the Community Detection named ALOCD approach. The underlying algorithm proposed by Seyedali Mirjalili (Mirjalili 2015) utilizes the unique hunting behavior of antlion. The algorithm is inspired by of hunting behavior of victim such as random walk of ants, constructing traps, entrapment of ants inside traps, catching preys, and re- constructing traps. These features allow the antlions to take the positions of ants, making the antlions move towards better fit ants to achieve optimal solution. The main components of the proposed algorithm are as follows:

3.1   Solution Representation

In the proposed approach each solution/antlion in the population is encoded as a collection of n antlion positions, $s = [a_1, a_2, \ldots a_n]$ such that each value $a_i \in [1, n]$ interprets the community to which the ith node belongs. The node i and node j belong

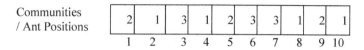

Fig. 1. Solution representation

to same cluster if the value a_i equals a_j for any set of i and j nodes. One antlion stands for one solution that divides given community structure into sub-community partitions. Figure 1 illustrates a solution representation of a social network with 10 nodes using array data structure with set of nodes {2,4,8,10} belonging to the 1st sub-community/cluster and set {1,5,9}, {3,6,7} belonging to 2nd and 3rd sub-community respectively.

3.2   Random Walk of Ants

In each iteration, the position of each ant is updated with respect to elite (best antlion obtained so far) and a selected antlion based on roulette wheel selection operator. This updation of position is performed with the help of two random walks i.e. random walk on the basis of roulette selected antlion and elite antlion.

For a given community of size say 'n', pick value at random position [1:n] from candidate solution(elite/roulette selection) which represents sub-community number to which that indexed vertex belongs to. Figure 2 shows the representation of Candidate Solution before random walk.

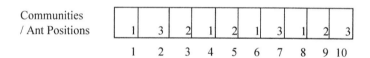

Fig. 2. Candidate solution (before random walk)

Figure 3 shows one step in random walk procedure by taking a random position say 4th in candidate solution and generates new solution after merging 1st and 3rd communities. If the newly generated solution gives better modularity than candidate solution, the candidate solution is updated. This step of random walk is applied recursively depending upon the size of the trap. For assuring exploitation of search space, the radius of updating ant's positions is contracted. This step is modeled by decreasing the random walk rate as the iteration value approaches Num_of_gen value.

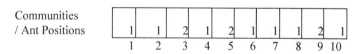

Fig. 3. Solution after merging 1st and 3rd communities

3.3   Updating the Position of Ants

During each iteration of the algorithm, the movements of all the ants are affected by elite as well as selected antlion as every ant randomly walks around a selected antlion by the roulette wheel and the elite concurrently. This concurrent effect of both selected and elite antlion on the movement of ant is modeled using Update_Ant_pos() procedure as shown in Fig. 4. The steps shown in Fig. 4 are repeated for all the possible communities while retaining better solutions in every repetition.

The ALOCD approach is illustrated by the pseudo code as shown in Table 1. A solution is encoded as a random permutation of n positions of ant/antlion representing the communities of given input net list file pertaining to the social network. The initial solutions are preprocessed by randomly picking a vertex Vi and finding its adjacent vertex say Vj. The community value of Vi is allocated to that of Vj. This process is repeated for rest of vertices until sufficient number of solutions for initial population are generated. After initialization, the best antlion (elite) is chosen from the generated population on the basis of modularity value of antlion. More modularity

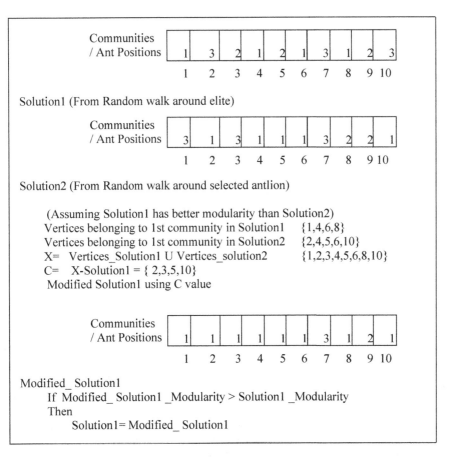

Fig. 4. One step of *Update_Ant_pos()* procedure

Table 1. Pseudocode of proposed ALOCD algorithm

Community_AntLionDetector (Input_File, Num_Nodes, Num_nets, Numclass , class, Pop_size, Num_of_gen)

Input : Read the Benchmark files of communities
Variables: pop_ant, pop_antlion - Array of structures of solution of size (Pop_size X Num_Nodes), Merge_solution -Array of structures of solution of size ((2 X Pop_size) X Num_Nodes)
Class- True community of community structure, Numclass -Total no .of true communiites
Output: Set of Best Communiites

Begin
[adj_array]=Create Netlist(Input_File, Num_Nodes, Num_nets)
[pop_ant]=Initalise(Pop_size,Num_Nodes,adj_array)
[pop_antlion]=Initalise(Pop_size ,Num_Nodes,,adj_array)
Set iteration:=1
 While (iteration <= Num_of_gen)
          [Elite]=Findbest(pop_antlion,Pop_size,Num_Nodes)
           For every ant i=1:Pop_size
             For j=1:Pop_size
                                 w(j)=pop_antlion(j).fit
                    EndFor
                    choice = RouletteWheelSelection(1./w,Pop_size)
                 If choice==-1
                                 choice=1
                    Endif
                 roulete_antlion=pop_antlion(choice)
                 RA = Random_walk (Elite,Num_Nodes,adj_array)
                        RB = Random_walk (roulete_antlion,Num_Nodes,adj_array)
                        Cross_R= Update_Ant_pos(adj_array,RA,RB,Num_Nodes)
                        pop_ant(i)=Cross_R
                        If pop_ant(i).fit>Elite.fit
                                      Elite= pop_ant(i)
                        EndIf
          EndFor
          Merge_solution =Merge_Sort(pop_ant,pop_antlion, Pop_size,Num_Nodes)
          [pop_antlion]=Update_solution(Merge_solution, Pop_size,Num_Nodes)
 Endwhile
[C]= pop_antlion(1).bit
[NMI] = Compute_NMI(class, C)
Print: 'nmi ', NMI
Print: 'Optimized Community ', pop_antlion(1).bit
End

---

contributes to better antlion. The quality of solutions/antlions is improved through random walk. After random walk, if the new ants have high fitness than antlions, then antlion new positions will become positions of the ants for imitating the process of catching the prey, the antlion is required to change its position to the most recent position of the hunted ant to boost its chance of catching new prey. In each iteration, the antlion with highest fitness is substituted as best antlion (elite). The process is repeated until optimal solution is obtained. At final stage, the NMI value of the first solution of the population with best modularity value is calculated.

4   Simulation Results

The proposed ALOCD approach is implemented using matlab 7.11.0 (R2012a) on intel core i5 processor, with 4 GB RAM under 64-bit Operating System. The work is tested on Zachary's Karate Club, Bottlenose Dolphins, American college football and Books about US politics network (Newman 2009) benchmarks and results are compared with the ACO and EFF (Amiri et al. 2013; Honghao et al. 2013). The characteristics of these benchmarks are shown in Table 2.

Table 2. Shows the basic features of the real world networks and their true number of community structures

Benchmark networks	Nodes	Edges	True communities
Zachary's Karate Club	34	78	2,4
Bottlenose Dolphins	62	159	2,4
Books about US politics	105	441	3
American College Football	115	613	12

The performance of ALOCD approach is evaluated using Normalized mutual information (NMI) (Le Martelot and Hankin 2013; Pizzuti 2008; Pizzuti 2012; Amiri et al. 2013) that quantifies the similarity between the detected and true community structure. NMI denoted as I (X,Y) is calculated using the following formula as shown in Eq. (3).

$$I(X,Y) = -\frac{2\sum_{i=1}^{C_X}\sum_{j=1}^{C_Y} C_{ij}\log\left(\frac{C_{ij}N}{C_i}\cdot C_j\right)}{\sum_{i=1}^{C_X} C_i \cdot \log(C_i/N) + \sum_{j=1}^{C_Y} C_{.j}\log(C_{.j}/N)} \tag{3}$$

Where X and Y denotes two network structures, C - the confusion matrix; C_{ij} - the count of nodes present in community i of X as well as in community j of Y; C_X, C_Y - the number of classes in part X and Y; $C_i, C_{.j}$ - the count of elements in row i and column j of C; N - the total count of nodes in networks.

The larger value of NMI reflects more similarity between true and detected communities leading to better solution quality. For both communities being same, the NMI value equals 1 and NMI = 0 signifies different communities.

Table 3 shows the results of implementation of the proposed approach over 10 different runs for a given set of benchmark networks. The algorithm computes the NMI value and determines the total number of communities in each run, as shown in Table 3.

From the tabulated values as given in Table 3, it is concluded that the proposed approach is competent to identify 100% community structure for Zachary's karate network (Newman 2009). Figure 5 reveals NMI value equal to 1 at 5th and 8th run of the program execution with the number of communities same as that of true community structure for Zachary's karate network.

Figures 6, 7 and 8 shows the NMI values of ten runs of ALOCD on Bottlenose Dolphins, Books about US politics and American Football Club benchmark networks

Table 3. Represents the NMI values and number of communities for multiple runs of ALOCD approach

Benchmark networks		Number of runs									
		1st	2nd	3rd	4th	5th	6th	7th	8th	9th	10th
Zachary's Karate Club	NMI	0.732	0.732	0.592	0.494	**1.000**	0.732	0.516	1.000	0.732	0.732
	NC	2	2	4	4	**2**	2	8	2	2	2
Bottlenose Dolphins	NMI	0.502	0.534	0.699	0.798	0.698	0.835	**0.887**	0.802	0.513	0.756
	NC	5	3	2	3	4	2	**2**	3	6	4
Books about US politics	NMI	0.696	0.508	0.449	0.506	0.726	**0.733**	0.431	0.668	0.705	0.534
	NC	4	3	10	15	4	**3**	7	5	3	9
American College Football	NMI	0.537	0.608	0.503	0.804	0.775	0.795	0.665	**0.850**	0.611	0.759
	NC	17	16	14	13	12	13	22	**12**	20	14

NC-represents the number of Communities and bold faced values represent best results of ALOCD approach for respective benchmark networks.

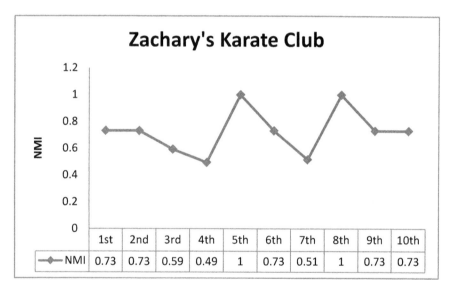

Fig. 5. NMI values of ten runs of ALOCD on Zachary's Karate Club

respectively. In 7th run of the implementation of ALOCD approach, the no. of communities found for bottlenose dolphins is equal to true number of community structures with NMI value of 0.887 as shown in Fig. 6.

From Fig. 7 it is clear that the number of communities found for Books about US politics is equal to the number of true communities with NMI value of 0.733 at 6th run of program execution. Figure 8 depicts same number of communities as that of actual one at 8th run of implementation for American Football with NMI value of 0.850.

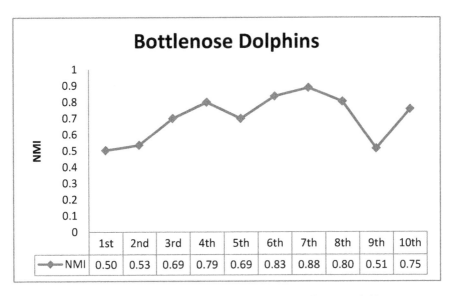

Fig. 6. NMI values of ten runs of ALOCD on Bottlenose Dolphins

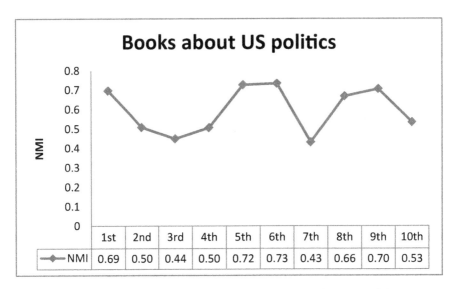

Fig. 7. NMI values of ten runs of ALOCD on books about US politics

Table 4 depicts excellent performance of ALOCD approach for Zachary's Karate Club Benchmark Networks. The solutions found by ALOCD approach split the communities into 2, 3, 12 clusters with 1, 6, 13 nodes misplaced for Bottlenose Dolphins, Books about US politics, American College Football respectively.

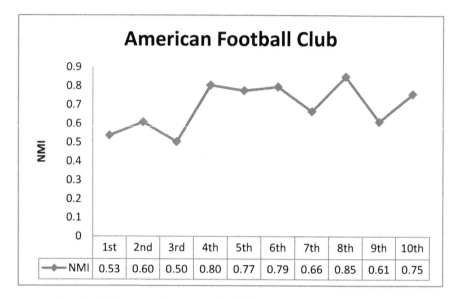

Fig. 8. NMI values of ten runs of ALOCD on American Football Club

Table 4. Shows the count and nodes of wrong communities along with best NMI of all networks

Benchmark networks	NMI	Wrong communities	Nodes in wrong communities
Zachary's Karate Club	1.000	–	–
Bottlenose Dolphins	0.887	1	[20]
Books about US politics	0.733	6	[8, 15, 28, 46, 54, 89]
American College Football	0.850	13	[10, 18, 23, 32, 39, 47, 52, 54,65, 78, 82, 86, 101]

Table 5. Shows the comparison of best NMI values of ALOCD, ACO and EFF algorithm and also shows at best NMI the respective communities found in all networks

Benchmark networks	NMI and no. of communities	ALOCD	ACO	EFF
Zachary's Karate Club	NMI	1.000	0.687	0.998
	Communities	2	2	4
Bottlenose Dolphins	NMI	0.887	0.587	0.988
	Communities	2	2	4
Books about US politics	NMI	0.733	0.560	0.599
	Communities	3	2	4
American College Football	NMI	0.850	0.890	0.798
	Communities	12	12	11

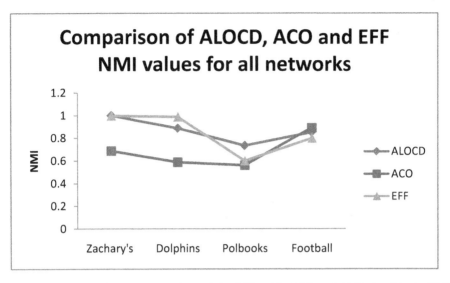

Fig. 9. Comparison of best NMI values of ALOCD with ACO and EFF algorithm's NMI Values for four real world networks

The results of the proposed approach are compared with ACO and EFF in terms of best NMI values obtained. From Table 5 and Fig. 9, it is concluded that the proposed ALOCD approach detects 100% true community structure statics for Zachary's karate network. For Dolphins networks, the algorithm obtained best normalized mutual information of 0.887 closer to EFF best NMI value, while the NMI of ACO was 0.798. For Books about US politics, the proposed approach has highest NMI value of 0.733 in comparison with other approaches. On the American College Football network, ALOCD obtained best normalized mutual information of 0.850 closer to ACO's best NMI value, while the NMI of EFF was 0.798. Consequently the proposed approach is able to detect the true count of communities with NMI value close enough to NMI value of true community structure.

5 Conclusion

The proposed approach optimizes the modularity function and is able to recognize densely connected clusters of nodes bearing sparse interconnections. The performance of the algorithm is measured in terms of Normalized Mutual Information (NMI) function. The algorithm is tested on real-world networks i.e. Bottlenose Dolphins, Zachary's Karate Club, Books about US politics and American Football Club. The experimental results show 100% community structure statics detection for Zachary's Karate club. The approach also gives promising results for other networks by finding the true number of communities with NMI values close to true community structures. The ALOCD approach is also compared with ACO and EFF over given set of benchmarks. The approach outperforms EFF and ACO for Zachary and Books about

US politics and produces results better than ACO and comparable to EFF for Dolphins and also produces results better than EFF and comparable to ACO for American Football Club. The work can be further extended for other bigger real world networks like Facebook, Twitter etc. This nature inspired approach can be improved to solve the problem of community detection for overlapping communities and dynamic community detection.

References

Amiri, B., Hossain, L., Crawford, J.W., Wigand, R.T.: Community detection in complex networks: multi-objective enhanced firefly algorithm. Knowl.-Based Syst. **46**, 1–11 (2013)

Clauset, A., Newman, M.E., Moore, C.: Finding community structure in very large networks. Phys. Rev. E **70**(6), 1–6 (2004)

Ferrara, E., Fiumara, G.: Topological features of online social networks. Commun. Appl. Indus. Math. **2**(2), 1–20 (2011)

Fortunato, S.: Community detection in graphs. Phys. Rep. **486**(3), 75–174 (2010)

Hafez, A.I., Zawbaa, H.M., Hassanien, A.E., Fahmy, A.A.: Networks community detection using artificial bee colony swarm optimization. In: Proceedings of the Fifth International Conference on Innovations in Bio-Inspired Computing and Applications, IBICA 2014, pp. 229–239 (2014)

Honghao, C., Zuren, F., Zhigang, R.: Community detection using ant colony optimization. In: IEEE Congress on Evolutionary Computation (CEC), pp. 3072–3078 (2013)

Lambiotte, R., Lefebvre, E., Guillaume, J.L., Blondel, V.D.: Fast unfolding of communities in large networks. J. Stat. Mech: Theory Exp. **2008**(10), P10008 (2008)

Le Martelot, E., Hankin, C.: Fast multi-scale detection of relevant communities in large-scale networks. Comput. J. **56**(9), 1136–1150 (2013)

Ma, L., Gong, M., Liu, J., Cai, Q., Jiao, L.: Multi-level learning based memetic algorithm for community detection. Appl. Soft Comput. **19**, 121–133 (2014)

Mirjalili, S.: The Ant lion optimizer. Adv. Eng. Softw. **83**, 80–98 (2015)

Newman, M.E.: The structure and function of complex networks. SIAM Rev. **45**(2), 167–256 (2003)

Newman, M.E.: Fast algorithm for detecting community structure in networks. Phys. Rev. E **69**(6), 1–5 (2004)

Newman, M.E.: Finding community structure in networks using the eigenvectors of matrices. Phys. Rev. E **74**(3), 036104 (2006)

Newman, M.E.: (2009). http://www-personal.umich.edu/mejn/netdata/

Newman, M.E., Girvan, M.: Finding and evaluating community structure in networks. Phys. Rev. E **69**(2), 1–16 (2004)

Pizzuti, C.: GA-Net: a genetic algorithm for community detection in social networks. In: Rudolph, G., Jansen, T., Beume, N., Lucas, S., Poloni, C. (eds.) PPSN 2008. LNCS, vol. 5199, pp. 1081–1090. Springer, Heidelberg (2008). doi:10.1007/978-3-540-87700-4_107

Pizzuti, C.: A multiobjective genetic algorithm to find communities in complex networks. IEEE Trans. Evol. Comput. **16**(3), 418–430 (2012)

Schuetz, P., Caflisch, A.: Efficient modularity optimization by multistep greedy algorithm and vertex mover refinement. Phys. Rev. E **77**(4), 046112 (2008)

Shang, R., Bai, J., Jiao, L., Jin, C.: Community detection based on modularity and an improved genetic algorithm. Physica A Stat. Mech. Appl. **392**(5), 1215–1231 (2013)

Hybrid SOMA: A Tool for Optimizing TMD Parameters

Shilpa Pal[1], Dipti Singh[2(✉)], and Varun Kumar[3]

[1] Department of Civil Engineering, Gautam Buddha University,
Greater Noida, India
[2] Department of Applied Mathematics, Gautam Buddha University,
Greater Noida, India
diptipma@rediffmail.com
[3] School of Engineering, Gautam Buddha University,
Greater Noida, India

Abstract. Tuned Mass Damper are widely used in the engineering community for reduction in response of the structure during the hazardous earthquake excitations or for other uses such as vibration control in slender and tall buildings. But it is not necessary that the TMD used is reducing the response of the structure effectively for the parameters set for it during the application. So for the TMD to work at its best, the optimal parameters have to be found. The work discusses the optimum parameters of Tuned Mass Damper for seismically excited structures. The Hybrid Self Organizing Migrating Genetic Algorithm (SOMGA) and Self Organizing Migrating algorithm with Quadratic Interpolation (SOMAQI) are used to find the Optimum values of TMD parameters. All parameters of TMD are searched in order to find the best results. TMD parameters are checked under different excitations and the present approach is also compared with other published results.

Keywords: Tuned Mass Damper · SOMGA · SOMAQI · Displacement response · TMD Parameter Optimization

1 Introduction

Earthquake phenomenon is not new in India and tremors of earthquake from nearby countries are often felt which causes threats to public safety and damage to property. Based on geometric location quiet near to the seismic fault line of Himalayan range, any structure has chances to fail especially high rise buildings. Damage induced due to earthquake has increased over the last few years and some devastating examples are from Gujarat earthquake in 2001, Nepal earthquake in 2015 and earthquake in Sumatra in 2004 which left India and other South Asian countries shocked. Collapse of engineered and non-engineered structure is the major contributor in loss to human life.

Over the past few decades progress has been made in making the structural control system such as vibration control a practicable technology for improving structure functionality and safety against natural hazards like earthquakes.

© Springer Nature Singapore Pte Ltd. 2017
K. Deep et al. (eds.), *Proceedings of Sixth International Conference
on Soft Computing for Problem Solving*, Advances in Intelligent Systems
and Computing 546, DOI 10.1007/978-981-10-3322-3_4

The effectiveness of the Tuned Mass Damper (TMD) depends on the right standardization of the characteristics of TMD in a structure. Hybrid Self Organizing Migrating Algorithm is used to optimize the TMD so that structural response can be improved.

Frahm [2] in 1909 was the first to work and implement the concept of a Tuned Mass Damper (TMD), to reduce vibrations and motion in ships. Later Den Hartog [3] developed theory for a single degree of freedom system. A detailed discussion about optimal parameters of Tuned Mass Damper was provided by them later in 1940. Sadek et al. [4] suggested that TMD performs efficiently when first two modal damping modes are equal, as earlier formulations by other authors don't show the equality in the damping of the first two modes. Bekdaş and Nigdeli [1] optimized the damper properties using metaheuristic technique known as Harmony search.

2 Simplification of a Structure to a SDOF System

A structure is actually not a single degree, but rather it is an infinite degree of freedom system. It is however, mathematically impossible to simplify the structure's model, an approximation requires to use the lumped mass model, where the model goes from an infinite to a multiple (finite) degree of freedom system. This is achieved by considering each floor as a Single Degree of Freedom (SDOF), where the mass is the total mass of the considered floor, and the stiffness and damping are calculated using equivalence formulas - which depend on the properties of elements (E, l, C), the fixations at the end.

Equation of motion for a Multi Degree of Freedom (MDOF) system is given by

$$[m]\{\ddot{x}(t)\} + [c]\{\dot{x}(t)\} + [k]\{x(t)\} = [m]\{r\}\ddot{x}_g(t) \tag{1}$$

Where m, c, k are mass matrix, damping matrix and stiffness matrix and $\{r\}$ is influence coefficient vector (nX1) with $\ddot{x}(t)$ denoting relative acceleration vector $\dot{x}(t)$ relative velocity vector and $x(t)$ relative displacement vector, $\ddot{x}_{g(t)} =$ EQ ground acceleration or excitation.

The undamped eigen values and eigen vectors of MDOF are found for the characteristic equation for n stories by:

$$\{k - \omega_i^2 [m]\}\emptyset_i = 0;$$
$$\det|\{k - \omega_i^2 [m]\}| = 0;$$

Where; i = 1 to n where n is total number of stories.

Where displacement response of MDOF is expressed as $x(t) = [\emptyset]\{y(t)\}$;

Where y(t) represent modal displacement vector and $[\emptyset]$ is the mode shape matrix given by $[\emptyset] = [\emptyset_1 \emptyset_2 \emptyset_3 \emptyset_4 \ldots \emptyset_n]$.

Substituting $\{x\} = [\emptyset]\{y\}$ in Eq. 1 and pre multiplying by transpose of $[\emptyset]$ i.e. $[\emptyset]^T$

$$[\emptyset]^T [m][\emptyset]\{\ddot{y}(t)\} + [\emptyset]^T [c][\emptyset]\{\dot{y}(t)\} + [\emptyset]^T [k][\emptyset]\{y(t)\}$$
$$= [\emptyset]^T [m]\{r\}\ddot{x}_g(t)$$

$[\varnothing]^T[m][\varnothing] = [M]$; Generalized mass matrix

$[\varnothing]^T[c][\varnothing] = [C]$; Generalized damping matrix

$[\varnothing]^T[k][\varnothing] = [K]$; Generalized stiffness matrix

and equation becomes

$$\ddot{y}_i(t) + 2\varepsilon_i \omega_i \dot{y}_i(t) + \omega_i^2 y_i(t) = \tau_i \ddot{x}_g(t) \tag{2}$$

Where; $y_i(t)$ = modal displacement response in i^{th} mode

ε_i = modal damping ratio in i^{th} mode

τ_i = modal participation factor for i^{th} mode expressed by

$$\tau_i = \left([\varnothing]^T[m]\{r\}\ddot{x}_g\right) / \left([\varnothing]^T[m][\varnothing]\right)$$

Equation (2) can also be written as

$$m\ddot{x}(t) + c\dot{x}(t) + kx(t) = -m\ddot{x}_g(t)$$

Where; $\omega_\circ = \sqrt{k/m}$ and $\varepsilon = c/(2m\omega_\circ)$

2.1 Optimization Procedure

The primary objective of optimization is to reduce the vibration of structure subjected to seismic loading under different earthquake excitations recorded in the past. The fundamental parameter considered is the maximum displacement of the structure. To increase the effectiveness of the TMD, the following parameters are optimized: Mass ratio (μ) i.e., ratio of mass of TMD (m_d) to the mass of the structure (M)• Stiffness of the Tuned Mass Damper (k_d)• Damping coefficient of the TMD (c_d)• The methodology aims at optimizing the parameters for the first mode of vibration of the structure. In order to optimize the parameters of TMD, the structure was idealized as lumped spring–mass-damper single degree of freedom system (SDOF). The mass of the SDOF was considered as the modal mass participated in the first mode of the structure. The TMD was attached to the structure as spring-mass-damper system and the SDOF structure becomes 2DOF system. The parameters of the TMD are found as follows,

- Mass of TMD (m_d)

  $m_d = \mu M$

- Stiffness of TMD (k_d)

  $k_d = \omega_o^2 * \alpha * 2m$

- Damping of the TMD (c_d)

  $c_d = 2 m \varepsilon_d \omega_d$

The equation of motion of system equipped with a TMD is given in Eq. 3, where the parameters m_d, c_d, k_d were optimized by setting the range of the TMD parameters of the structure and then finding the best optimum solution of TMD parameters with the help of SOMGA and SOMAQI.

$$(m + m_d)\ddot{u} + c\ddot{u} + \text{k u} + m_d \ddot{u}_d + c_d \dot{u}_d + k_d u_d = \text{p} \qquad (3)$$

Where;

m = mass of the structure alone
c = damping of the structure
k = stiffness of the building alone
m_d = mass of damper
c_d = damping of the tuned mass damper
k_d = stiffness of the damper
p = loading applied on the building
u = displacement of primary mass
u_d = displacement of damper

The range is as follows:

$$m_d = 13850 \, to \, 69250$$
$$k_d = 1.3375 \times 10^5 \, to \, 6.68759 \times 10^5$$
$$c_d = 8.608 \times 10^3 \, to \, 43.04 \times 10^3$$

3   Case Studies

Two examples from the existing literature are selected and the use of Hybrid SOMA method is applied on them for finding the optimum TMD parameters for them.

3.1   Case Study-1

Case Study 1 is on example taken from Singh et al. [5] in which all floors have the same properties i.e. same mass, stiffness and damping coefficient as 360t, 650 MN/m, and 6.2 MNs/m for all the ten floors. After performing the analyses for optimization, optimum TMD parameters are found as m_d = 70.312t, c_d = 93.6 kNs/m and k_d = 2973.34 kN/m using SOMGA Technique and from SOMAQI technique parameters are found as m_d = 180 t, c_d = 46.8 kNs/m and k_d = 7612.02 kN/m. These values for TMD parameters are smaller than the values obtained by previous studies and this example was analyzed under El Centro NS (1940) excitation for comparison with previous results. The maximum displacement response obtained for first story, top story and TMD under different earthquakes are presented in Table 1 and for the comparison, response of all the stories under El Centro NS (1940) with displacement responses from previous studies are presented in Table 2.

Table 2 shows the maximum displacement response reduction for the structure stories in terms of percentage for the present study which are written against the values

Table 1. Maximum displacements respect to ground under different earthquakes for Case Study-1

Story	Maximum absolute displacement under ground excitation in m								
	El Centro			El Centro NS			Tabas		
	Without TMD	With TMD SOMAQI	With TMD SOMGA	Without TMD	With TMD SOMAQI	With TMD SOMGA	Without TMD	With TMD SOMAQI	With TMD SOMGA
First	0.025	0.010	0.010	0.031	0.015	0.013	0.039	0.032	0.027
Top	0.173	0.078	0.078	0.189	0.114	0.089	0.264	0.226	0.205
TMD	–	0.268	0.389	–	0.393	0.450	–	0.684	0.709

Table 2. Maximum displacements respect to ground under EL Centro NS excitation for Case Study-1

Story	Maximum absolute displacement respect to ground (m)					% of reduction			
	Without TMD	With TMD (GA)	Without TMD (HS)	With TMD SOMAQI	With TMD SOMGA	GA	HS	SOMAQI	SOMGA
1	0.031	0.019	0.016	0.015	0.013	38.71	48.93	51.61	58.06
2	0.060	0.037	0.031	0.030	0.025	38.33	48.33	50.00	58.33
3	0.087	0.058	0.044	0.044	0.036	33.33	49.43	49.43	58.62
4	0.112	0.068	0.057	0.058	0.047	39.29	49.11	48.21	58.03
5	0.133	0.082	0.068	0.071	0.057	38.35	48.87	46.62	57.14
6	0.151	0.094	0.078	0.082	0.067	37.75	48.34	45.69	55.63
7	0.166	0.104	0.087	0.093	0.075	37.35	47.59	43.98	54.82
8	0.177	0.113	0.094	0.101	0.081	36.61	46.89	42.94	54.24
9	0.184	0.119	0.099	0.108	0.086	35.35	46.20	41.30	53.26
10	0.188	0.122	0.102	0.113	0.090	35.11	45.74	39.89	52.13
TMD	–	0.358	0.395	0.393	0.450	–	–	–	–

of optimized result from the past studies of Harmony search and Genetic algorithm optimization. From the study, it is found that the maximum top floor displacement of the building is reduced by 52.13% under EL Centro NS (1940) excitation with SOMGA and 39.89% by SOMAQI, whereas GA showed 35.11% reduction and Harmony Search algorithm showed reduction of 45.74%.

3.2   Case Study-2

The second example in Case Study 2 is also a ten story building which was optimized before by Sadek et al. [4] and the structure has different values of parameters for all floors as seen in Table 3 below.

In this study the damping matrix is taken proportional to the stiffness matrix i.e. $C = 0.0129 K$ for the second example as it was given in the paper that it can be taken proportional to the stiffness matrix or the mass matrix. Optimum TMD parameters for the Case Study-2 are found as $m_d = 20.789$ t, $c_d = 17.25$ kNs/m and $k_d = 198.536$ kN/m

Table 3. Structure properties of Case Study 2

Story	10	9	8	7	6	5	4	3	2	1
Mass (t)	98	107	116	125	134	143	152	161	171	179
Stiffness (MN/m)	34.31	37.43	40.55	43.67	46.79	49.91	53.02	56.14	52.26	62.47

using SOMGA Technique and TMD parameters from SOMAQI are found as m_d = 69.25 t, c_d = 43.04 kNs/m and k_d = 668.759 kN/m. The maximum displacement response obtained for first story, top story and TMD under different earthquakes are presented in Table 4 and for the comparison, response of all the stories under El Centro NS (1940) with displacement responses from previous studies are presented in Table 5.

Table 4. Maximum displacements respect to ground under different earthquakes for Case Study-2

Story	Maximum absolute displacement under ground excitation in m								
	El Centro			El Centro NS			Tabas		
	Without TMD	With TMD SOMAQI	With TMD SOMGA	Without TMD	With TMD SOMAQI	With TMD SOMGA	Without TMD	With TMD SOMAQI	With TMD SOMGA
First	0.036	0.021	0.015	0.041	0.022	0.017	0.132	0.0460	0.043
Top	0.285	0.196	0.134	0.327	0.205	0.151	0.821	0.431	0.378
TMD	–	0.663	0.801	–	0.693	0.899	–	1.459	2.250

Table 5. Maximum displacements respect to ground under EL Centro NS excitation for Case Study-2

Story	Maximum absolute displacement respect to ground from different Optimization methods (m)						
	Without TMD	Den Hartog	Sadek et al.	Hadi and Arfiadi	Gebrail and Sinan	SOMAQI	SOMGA
1	0.041	0.034	0.036	0.034	0.027	0.021	0.017
2	0.088	0.074	0.077	0.072	0.058	0.047	0.037
3	0.129	0.106	0.113	0.105	0.083	0.071	0.055
4	0.166	0.136	0.145	0.134	0.105	0.093	0.073
5	0.197	0.163	0.172	0.160	0.124	0.115	0.090
6	0.222	0.187	0.194	0.184	0.140	0.137	0.106
7	0.252	0.213	0.219	0.210	0.157	0.157	0.120
8	0.286	0.239	0.245	0.236	0.177	0.175	0.133
9	0.313	0.261	0.266	0.258	0.195	0.191	0.143
10	0.327	0.276	0.281	0.272	0.205	0.205	0.151
TMD	–	0.602	0.456	0.635	0.449	0.693	0.899

From Tables 4 and 5 it can be seen that both the soft computing techniques which are used for optimization of the TMD parameters are working fine, as for the values of TMD parameters, significant reduction in maximum absolute displacement of the structure storey can be seen for Example 2 under different earthquake records. The values of displacement obtained from the present study are better than the values obtained from previous studies. While the results obtained from the TMD parameters derived from SOMGA are performing in a better way as compared to the results obtained from SOMAQI.

It is found that the maximum displacement of the top story is reduced by a reduction of 53.82% by SOMGA and 37.31% by SOMAQI under El Centro NS (1940) excitation, while the top story displacement is reduced with a reduction of 15.6% by Den Hartog 14.07% by Sadek et al. and 37.31% by Gebrail-Sinan Harmony Search method and 16.82% by Hadi-Arfiadi methods. TMD parameters obtained by SOMGA are numerically smaller than various other methods which were compared and efficient in reducing the displacement response, while the parameters obtained by SOMAQI are similar to the parameters obtained by other studies in reducing the maximum absolute response of the structure with respect to the ground under different earthquake excitations.

4 Conclusions

In this paper two case studies have been considered for analysis. In case study 1, all the floors have the same properties i.e. same mass, stiffness and damping coefficient for all the ten floors and in Case study 2, different values of mass, stiffness and damping coefficient has been taken for all floors. In both cases, the parameter of TMD has been optimized by two soft computing techniques. Both the techniques are performing better as the displacement is reduced as compared to previously published results. Results showed that among these two techniques, SOMGA is showing better reduction of displacement of the top and bottom storey as compared to SOMAQI. Hence it can be concluded that displacement responses can be achieved for any structure incorporated with TMD with the help of Hybrid Self Organizing Migration Algorithm.

References

1. Bekdaş, G., Nigdeli, S.: Estimating optimum parameters of tuned mass dampers using harmony search. Eng. Struct. **33**(9), 2716–2723 (2011)
2. Frahm, H.: Device for damping of bodies. US Patent No 989, 958 (1911)
3. Den Hartog, J.P.: Mechanical Vibrations, 3rd edn. McGraw-Hill, New York (1947)
4. Sadek, F., Mohraz, B., Taylor, A., Chung, R.: A method of estimating the parameters of tuned mass dampers for seismic applications. Earthq. Eng. Struct. Dyn. **26**(6), 617–635 (1997)
5. Singh, M., Singh, S., Moreschi, L.: Tuned mass dampers for response control of torsional buildings. Earthq. Eng. Struct. Dyn. **31**(4), 749 (2002)

Fast Convergent Spider Monkey Optimization Algorithm

Neetu Agarwal$^{(\boxtimes)}$ and S.C. Jain

Rajasthan Technical University, Kota, India
neetuagarwal974@gmail.com

Abstract. Spider Monkey Optimization (SMO) is a recent optimization method, which has drawn interest of researchers in different areas because of its simplicity and efficiency. This paper presents an effort to modify Spider Monkey Optimization Algorithm with higher exploitation capabilities. A new acceleration coefficient based strategy is proposed in the basic version of SMO. The proposed algorithm is named as Fast Convergent Spider Monkey Optimization Algorithm (FCSMO). FCSMO is tested over 14 benchmark test functions and compared with basic SMO. The result reveals that FCSMO will surely become a good variant of SMO.

Keywords: Meta-heuristic optimization techniques · Swarm intelligence · Acceleration coefficient

1 Introduction

Optimization works to unfolds all potential outputs meeting some stated constraints. From past years NIA have proposed various methods to explain NP-Hard and NP- complete optimization problems of real world [7]. Nature inspired algorithm, inspired by nature, is a stochastic approach wherein an individual or a neighbor's interacts with each other intellectually to explain complicated preexisting mechanisms in an efficient manner. NIA is focused mainly on evolutionary based algorithm and swarm based algorithm. Evolutionary algorithm is a computational standard motivated by Darwinian Evolution [9]. Swarm intelligence assets in unlocking optimization problems considering collaborative nature of self-sustaining creatures like bees, ants, monkeys whose food-gathering capabilities and civilized characteristics have been examined and simulated [5,6,8]. SMO is a subclass of swarm intelligence, proposed by Jagdish Chand Bansal et al., in the year 2014 [4]. SMO is a food foraging based algorithm, considering nature and social frame work of spider monkeys. Fission-Fusion social system relates to social configuration of spider monkey. Many researchers have been studied that SMO algorithm is good at exploration and exploitation but there is possibilities of further improvements.

© Springer Nature Singapore Pte Ltd. 2017
K. Deep et al. (eds.), *Proceedings of Sixth International Conference on Soft Computing for Problem Solving*, Advances in Intelligent Systems and Computing 546, DOI 10.1007/978-981-10-3322-3_5

To improve the convergence speed, a variant of SMO is proposed, i.e. name as Fast Convergent Spider Monkey Optimization Algorithm. In the proposed modification acceleration coefficients based strategy is incorporated in the basic version of SMO.

The rest of the paper is structured as follows: In Sect. 2, SMO is described. Fast Convergent Spider Monkey Optimization Algorithm (FCSMO) is proposed in Sect. 3. In Sect. 4, performance of FCSMO is tested with several benchmark functions. Finally, Sect. 5 includes a summary and conclude the work.

2   Overview of Spider Monkey Optimization (SMO) Technique

A distinct class of NIA proposed by JC Bansal et al. [4], by trivial behavior of monkeys i.e. Spider Monkey Optimization (SMO) technique. Spider Monkey optimization, a Fission-Fusion mode is an extension of above discussed predicament. Here, a populous, consistently dictated by a female, is fragmentized into tiny clusters for seeking, chiefly food and they are buddy up to 40 to 50 singular who rift into small groups in search of food who again are headed by a female. In case she fails to meet the objective (food finding), further subdivides, again succeeded by a female, replicating the process until reach the food. For recent updates in their positions, various steps are undertaken: inspection of probing of wide search space and picking or electing of superlative practical results [10].

2.1   Steps of SMO Technique

SMO technique is based on population repetitive methodology. It consists of seven steps. Each step is described below in a detailed manner:

1. **Initialization of Population**: Originally a population comprised of N spider monkeys signifying a D-dimensional range M_i where i=1,2,...N and i represents i^{th} spider monkey. Each spider monkey (M) exhibits possible results of the problem under consider. Each M_i is initialized as below:

$$M_{ij} = M_{minj} + R(0,1) \times (M_{maxj} - M_{minj}) \qquad (1)$$

   Here M_{minj} and M_{maxj} are limits of M_i in j^{th} vector and R(0,1) is a random number (0,1).

2. **Local Leader Phase (LLP)**: This phase relies on the observation of local leader and group mates, M renew its current position yielding a fitness value. If the fitness measure of the current location is larger than that of the former location, then M modifies his location with the latest one. Hence i^{th} M that also exists in k^{th} local group modify its position.

$$M_{newij} = M_{ij} + R(0,1) \times (LL_{kj} - M_{ij}) + R(-1,1) \times (M_{rj} - M_{ij}) \qquad (2)$$

   Here M_{ij} define i^{th} M in j^{th} dimension, LL_{kj} correlate to the kth leader of local assembly location in j^{th} dimension. M_{rj} defines r^{th} M which is randomly picked from k^{th} troop such that $r \neq i$ in j^{th} dimension.

3. **Global Leader Phase (GLP)**: This following phase initiates just after accomplishing LLP. Depending upon the observation of global leader and mates of local troop, M updates their location. The position upgrade equation for GLP phase is as follows:

$$M_{newij} = M_{ij} + R(0,1) \times (GL_j - M_{ij}) + R(-1,1) \times (M_{rj} - M_{ij}) \quad (3)$$

Here GL_j poises for global leader's location in j^{th} dimension and j=1,2,3,...,D defines an arbitrarily chosen index. M_i modify their locus considering probabilities $Pr_i's$. Fitness is used to calculate probability of a specific solution, with various methods such as

$$Pr_i = 0.1 + (\frac{fitness_i}{fitness_{max}}) \times 0.9 \quad (4)$$

4. **Global Leader Learning (GLL) Phase**: Here greedy selection strategy is applied on the population which modifies the locus of global leader i.e. the location of M which has best fitness in the group is chosen as the modified global leader location. Also its is verified that global leader location is modifying or not and in case not then GlobalLimitCount(GLC) is increased by 1.
5. **Local Leader Learning (LLL) Phase**: Here, local leader locus is modified by implement greedy selection in that population i.e. the location of M which has best fitness among the entire group is chosen as the latest location of local leader. Afterwards, this modified local leader location and old values are compared and LocalLimitCount (LLC) is increment by 1.
6. **Local Leader Decision (LLD) Phase**: Here, updating of local leader location is done in two ways i.e. by arbitrary initialization or by mixing information obtained via global and local leader, if local leader location is not modified up to a precalculated limit named as LocalLeaderLimit through equation based on perturbation rate (p).

$$M_{newij} = M_{ij} + R(0,1) \times (GL_j - M_{ij}) + R(0,1) \times (M_{ij} - LL_{kj}) \quad (5)$$

Clearly, it is seen in equation that modified dimension of this M is fascinated towards global leader and oppose local leader. Moreover, modified M's fitness is determined.
7. **Global Leader Decision (GLD) Phase**: Here, global leader location is examine and if modification is not done up to precalculated iterations limit named as GlobalLeaderLimit then division of population in small group is done by local leader. Primarily population division is done in two classes and further three, four and so on until the upper bound called groups of maximum number (GM) is reached. Meanwhile, local leaders are selected using LL method for newly formed subclasses.

The pseudo-code of the SMO algorithm is as follows:-

(1) Define Population, LocalLeaderLimit, GlobalLeaderLimit, Perturbation rate.
(2) Determine fitness (each individual distance from sources of food)
(3) Apply greedy selection to choose global and local leaders.
while *Termination condition is not met* **do**
    (i) To hit target, new locations for group population is formulated with the help of self experience as well as local and group population experience, using Local Leader Phase (LLP).
    (ii) Relied on fitness value of group members, employ greedy selection strategy.
    (iii) Assess probabilities Pr_i for all companions using equation (4).
    (iv) Generate new locations for each group companions, chosen by Pr_i, by self experience, global leader experience also consider experience of group member using Global Leader Phase (GLP).
    (v) Greedy selection method is applied to modify global and local leaders locations of entire groups.
    (vi) Any local leader of a group, if fails to modify her locus within LocalLeaderLimit then deflect that specific group companions for further foraging using Local Leader Decision (LLD) Phase.
    (vii) Any global leader if fails to modify her locus within GlobalLeaderLimit then she diversifies group into subgroups by Global Leader Decision Phase with the minimum threshold of each groups size being 4
end

Algorithm 1. Spider Monkey Optimization (*SMO*)

3 Fast Convergent Spider Monkey Optimization Algorithm

In population repetitive methodology, exploration and exploitation are the two basic properties of NIA. A convenient balance between both these two properties are required. Exploration describe the promising regions by searching the given search space while exploitation helps in finding the optimal solution in the promising search regions. In the basic SMO, it is good at exploration and exploitation but there is possibilities of further improvements. So to improve the basic SMO, on new variant named Fast Convergent Spider Monkey Optimization Algorithm (FCSMO) is designed.

From the results of search process of basic SMO that it will get higher opportunity for advancement in various iteration using two ways: (1) Global Leader Phase (GLP) and (2) Local Leader Decision (LLD) phase. Exploration and exploitation capacity should be managed in an effective way. The details of these two steps of FCSMO implementation are explained below:

1. **Global Leader Phase (GLP):** In the GLP phase, depending upon the observation of Global leader and mates of local troop, M updates their location. In the GLP phase of iteration, solutions in search space are having

large step-size resulting in exploration. In later iterations, there is a gradual decrease in step-size by moving slowly iteration by iteration due to which solution exploits the search space well and resulting good convergence. The position upgrade equation for this phase is as follow:

$$M_{newij} = M_{ij} + R(0,1) \times (GL_j - M_{ij}) + (M_{rj} - M_{ij}) \times [1 - (\frac{iter}{Maxiteration})] \times c$$
(6)

Here GL_j poises for global leader's location in j^{th} dimension and j=1,2,3,...,D defines an arbitrarily chosen index and iter and Max iteration show the present iteration and the maximum iteration number, respectively. c is the random number. Its value is 2. In this acceleration coefficient is added with random member. The position upgrade method of GLP phase is exhibited in following algorithm.

```
count=0;
while count < group do
    for each member M_i ∈ group do
        if R(0,1) < Pr_i then
            count=count+1;
            Randomly select j ∈ 1...D
            Randomly select M_r ∈ group s.t r ≠ i
            M_newij = M_ij + R(0,1) × (GL_j − M_ij) + (M_rj − M_ij) × [1 − (iter/Maxiteration)] × c
        end if
    end for
end
```

Algorithm 2. Global Leader Phase (GLD)

2. **Local Leader Decision (LLD) Phase**: Here, updating of Local Leader location is done in two ways i.e. by arbitrary initialization or by mixing information obtained via global and local leader, if local leader location is not modified up to a precalculated limit named as LocalLeaderLimit through equation based on p. In the LLD phase of iteration, solutions in search space are having large step-size resulting in exploration. In later iterations, there is a gradual decrease in step-size by moving slowly iteration by iteration due to which solution exploits the search space well and resulting good convergence. The position upgrade equation for this phase is as follow:

$$M_{newij} = M_{ij} + (GL_j - M_{ij}) \times (1 - (\frac{iter}{Maxiteration})) + R(0,1) \times (M_{ij} - LL_{kj})$$
(7)

Clearly, it is seen in equation that modified dimension of this M is fascinated towards global leader and oppose local leader. Moreover, modified M's fitness is determined and iter and Max iteration show the present iteration and the maximum iteration number, respectively. In this acceleration coefficient is added with global leader. The position upgrade method of LLD phase is exhibited in following algorithm.

if LocalLimitCount > LocalLeaderLimit **then**
   LocalLimitCount=0
   for each j ∈ 1...D **do**
     if R(0,1) ≥ p then **then**
       $M_{new_{ij}} = M_{ij} + R0,1 \times (M_{maxj} - M_{minj})$
     else
       $M_{newij} = M_{ij} + (GL_j - M_{ij}) \times (1 - (\frac{iter}{Maxiteration})) + R(0,1) \times (M_{ij} - LL_{kj})$
     end if
   end for
end if

Algorithm 3. Local Leader Decision (LLD) Phase

The pseudo-code of the FCSMO algorithm is as follows:-
(1) Define Population, LocalLeaderLimit, GlobalLeaderLimit, Perturbation rate.
(2) Determine fitness (each individual distance from sources of food)
(3) Apply greedy selection to choose global and local leaders.
while *Termination condition is not met* **do**
  (i) To hit target, new locations for group population is formulated with the help of self experience as well as local and group population experience, using Local Leader Phase (LLP).
  (ii) Relied on fitness value of group members, employ greedy selection strategy.
  (iii) Assess probabilities Pr_i for all companions using equation (4).
  (iv) Generate new locations for each group companions, chosen by Pr_i, by self experience, global leader experience also consider experience of group member using algorithm 2.
  (v) Greedy selection method is applied to modify global and local leaders locations of entire groups.
  (vi) Any local leader of a group, if fails to modify her locus within LocalLeaderLimit then deflect that specific group companions for further foraging using algorithm 3.
  (vii) Any global leader if fails to modify her locus within GlobalLeaderLimit then she diversifies group into subgroups by Global Leader Decision Phase with the minimum threshold of each groups size being 4
end

Algorithm 4. Fast Convergent Spider Monkey Optimization ($FCSMO$)

4   Experimental Results

4.1   Test Problems Under Consideration

To evaluate the quality of proposed FCSMO algorithm, 14 opposed global optimization issue (f_1 - f_{14}) are selected as presented in Table 1. All the issues are continuous optimization issues and having various rates of complexity. Test problems (f_1 - f_{14})) are yield from [1,11] with the correlated offset values.

Table 1. Test problems

Test Problem	Objective function	Search Space	Objective Value	Dimension	Acceptable Error
Rastrigin	$f_1(x) = 10D + \sum_{i=1}^{D}[x_i^2 - 10\cos(2\Pi x_i)]$	[-5.12,5.12]	$f(0)=0$	30	1.00E-05
Ackley	$f_2(x) = -20 + e + exp(-\frac{0.2}{D}\sqrt{\sum_{i=1}^{D} x_i^3})$	[-30,30]	$f(0)=0$	30	1.00E-05
Michalewicz	$f_3(x) = -\sum_{i=1}^{D}\sin x_i(\sin(\frac{i \cdot x_i^2}{\pi})^{20})$	$[0,\pi]$	$f_{min} = -9.66015$	10	1.00E-05
Cosin Mixture	$f_4(x) = \sum_{i=1}^{D} x_i^2 - 0.1(\sum_{i=1}^{D}\cos 5\pi x_i) + 0.1D$	[-1,1]	$f(0) = -D \times 0.1$	30	1.00E-05
Step Function	$f_5(x) = \sum_{i=1}^{D}(\lfloor x_i + 0.5\rfloor)^2$	[-100,100]	$f(-0.5 \le x \le 0.5) = 0$	30	1.00E-05
Neumaier 3 Problem (NF3)	$f_6(x) = \sum_{i=1}^{D}(x_i - 1)^2 - \sum_{i=2}^{D} x_i x_{i-1}$	[-900,900]	$f(0 = -(D*(D+4)*(D-1))/6.0)$	10	1.00E-05
Levy montalvo 1	$f_7(x) = \frac{\Pi}{D}(10\sin^2(\Pi y_1) + \sum_{i=1}^{D-1}(y_i-1)^2 \times (1 + 10\sin^2(\Pi y_{i+1})) + (y_D - 1)^2)$, where $y_i = 1 + \frac{1}{4}(x_i+1)$	[-10,10]	$f(-1)=0$	30	1.00E-05
Levy montalvo 2	$f_8(x) = 0.1(\sin^2(3\Pi x_1) + \sum_{i=1}^{D-1}(x_i-1)^2 \times (1 + \sin^2(3\Pi x_{i+1})) + (x_D-1)^2(1+\sin^2(2\Pi x_D))$	[-5,5]	$f(1)=0$	30	1.00E-05
Shifted Griewank	$f_9(x) = \sum_{i=1}^{D}\frac{z_i^2}{4000} - \prod_{i=1}^{D}\cos(\frac{z_i}{\sqrt{i}}) + 1 + f_{bias}, z = (x-o), x = [x_1, x_2, ...x_D], o = [o_1, o_2, ...o_D]$	[-600,600]	$f(o) = f_{bias} = -180$	10	1.00E-05
Goldstein-Price Function	$f_{10}(x) = (1 + (x_1+x_2+1)^2 \cdot (19 - 14x_1 + 3x_1^2 - 14x_2 + 6x_1x_2 + 3x_2^2)) \cdot (30 + (2x_1 - 3x_2)^2 \cdot (18 - 32x_1 + 12x_1^2 + 48x_2 - 36x_1x_2 + 27x_2^2))$	[-2,2]	$f(0,-1)=3$	2	1.0E-14
Six-Hump Camel back Function	$f_{11}(x) = (4 - 2.1x_1^2 + x_1^4/3)x_1^2 + x_1x_2 + (-4 + 4x_2^2)x_2^2$	[-5,5]	$f(-0.0898, 0.7126) = -1.0316$	2	1.0E-05
Hosaki Problem (HSK)	$f_{12} = (1 - 8x_1 + 7x_1^2 - 7/3x_1^3 + 1/4x_1^4)x_2^2\exp(-x_2)$	$x_1 \in [0,5], x_2 \in [0,6]$	$f(0) = -2.3458$	2	1.00E-06
Sinusoidal Problem (SIN)	$f_{13}(x) = -[A\prod_{i=1}^{D}\sin(x_i - z) + \prod_{i=1}^{D}\sin(B(x_i - z))], A = 2.5, B = 5, z = 30$	[0,180]	$f(90+z) = -(A+1)$	10	1.00E-02
Pressure Vessel	$f_{14}(x) = 0.6224 \times x_1 \times x_3 \times x_4 + 1.7781 \times x_2 \times x_3^2 + 3.1611 \times x_1^2 \times x_4 + 19.84 \times x_1^2 \times x_3$	$x_1 \in [1.125, 112.5], x_2 \in [0.625, 12.5], x_3 \in [1.0E-8, 240], x_4 \in [1.0E-8, 240]$	$f(0) = 7197.729$	4	1.00E-05

4.2   Experimental Setting

To verify the efficiency of proposed algorithm FCSMO, a relative study is taken between FCSMO and SMO. To analysis FCSMO and basic SMO, over the examine testing issues, subsequent observational setting is emulated:

– The number of simulations/run =100,
– Population size (Monkeys) $NP = 50$
– $R = rand[0, 1]$
– GlobalLeaderLimit $\in [N/2, 2\times N]$ [4]
– LocalLeaderLimit= $D\times N$ [4]
– Perturbation rate (p) $\epsilon[0.1, 0.8]$

4.3   Results Comparison

Table 2 represent the observational results of relative algorithm. Following Table 2 gives a information about Standard Deviation (SD), Mean Error (ME),

Table 2. Comparison of the results of test functions, TP: Test Problem

TP	Algorithm	SD	ME	AFE	SR
f_1	FCSMO	1.81E-06	7.99E-06	88298.07	100
	SMO	1.56E-06	8.24E-06	96073.45	100
f_2	FCSMO	4.79E-07	9.36E-06	23416.47	100
	SMO	9.32E-07	9.26E-06	32438.70	100
f_3	FCSMO	3.50E-06	5.52E-06	32438.20	100
	SMO	3.72E-06	4.98E-06	52153.75	100
f_4	FCSMO	4.01E-02	1.18E-02	32651.24	92
	SMO	4.80E-02	1.77E-02	62144.30	88
f_5	FCSMO	0.00E+00	0.00E+00	10792.21	100
	SMO	0.00E+00	0.00E+00	14261.81	100
f_6	FCSMO	4.38E-07	9.80E-06	124902.53	100
	SMO	1.15E-05	1.23E-05	169082.28	90
f_7	FCSMO	8.70E-07	9.02E-06	12917.52	100
	SMO	1.45E-02	2.08E-03	20795.99	98
f_8	FCSMO	1.09E-06	8.80E-06	12867.18	100
	SMO	1.53E-03	2.28E-04	17814.27	98
f_9	FCSMO	1.23E-03	1.39E-04	96079.51	97
	SMO	4.27E-03	1.64E-03	134399.12	81
f_{10}	FCSMO	4.70E-14	6.18E-14	130131.22	38
	SMO	4.31E-14	7.05E-14	150535.16	28
f_{11}	FCSMO	1.48E-05	1.68E-05	104691.82	50
	SMO	1.42E-05	1.72E-05	106712.17	49
f_{12}	FCSMO	4.11E-06	9.90E-06	177221.60	15
	SMO	3.39E-06	1.05E-05	187621.00	10
f_{13}	FCSMO	5.05E-03	8.29E-03	55146.68	97
	SMO	5.38E-03	1.11E-02	154825.33	62
f_{14}	FCSMO	1.07E-03	1.35E-04	89666.40	67
	SMO	3.63E-05	3.39E-05	111436.91	57

Average Number of Function valuations (AFE) and Success Rate (SR). According-ing to Results of Table 2, at maximum time FCSMO shows best results from SMO, in terms of performance, efficiency and accuracy.

Moreover, boxplots evaluation of AFE is taken for comparing the relevant algorithms in scheme of consolidated quality, so it can simply show the observed distribution of statistic graphically. The boxplots for FCSMO and SMO are presented in Fig. 1. The results declares that interquartile scope and medians of FCSMO are comparatively low. Further, all relevant algorithms are studied by allowing entire attention to the SR, AFE and ME. This study is determined

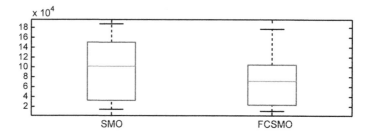

Fig. 1. Boxplots graph for average function evaluation

using the quality basis i.e. represented in $[2,3]$. The evaluated values of PI for the FCSMO and SMO are calculated and consecutive PIs graphs are represented in Fig. 2. The graphs analogous to several cases i.e. allowing entire attention to SR, AFE and ME (as explained in $[2,3]$) are represent in Figs. 2(a), (b), and (c) respectively. In these diagram, horizontal axis means the weights and vertical axis means the PI. It is clear from Fig. 2 that PI of FCSMO are superior than the other studied algorithms in various case. i.e. FCSMO observe better on the studied testing issues as compare to the SMO.

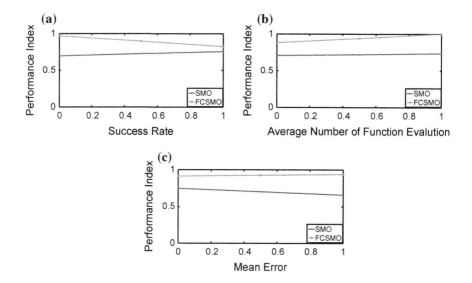

Fig. 2. Performance index for test problems; (a) for weighted importance to SR, (b) for weighted importance to AFE and (c) for weighted importance to ME.

5   Conclusion

This paper presents a variant of SMO algorithm, known as Fast Convergent Spider Monkey Optimization Algorithm (FCSMO). In FCSMO, an acceleration coefficient based strategy is proposed in which step size is decreased through iteration. To evaluate the proposed algorithm, it is tested over 14 benchmark function. The results collected by the FCSMO is better than the basic SMO algorithm. In future, newly developed algorithm may be used to solve various real-world optimization problems of continuous nature.

References

1. Ali, M.M., Khompatraporn, C., Zabinsky, Z.B.: A numerical evaluation of several stochastic algorithms on selected continuous global optimization test problems. J. Global Optim. **31**(4), 635–672 (2005)
2. Bansal, J.C., Sharma, H.: Cognitive learning in differential evolution and its application to model order reduction problem for single-input single-output systems. Memetic Comput. **4**(3), 209–229 (2012)
3. Bansal, J.C., Sharma, H., Arya, K.V., Nagar, A.: Memetic search in artificial bee colony algorithm. Soft. Comput. **17**(10), 1911–1928 (2013)
4. Bansal, J.C., Sharma, H., Jadon, S.S., Clerc, M.: Spider monkey optimization algorithm for numerical optimization. Memetic Comput. **6**(1), 31–47 (2014)
5. Bonabeau, E., Dorigo, M., Theraulaz, G.: Swarm Intelligence: From Natural to Artificial Systems, Number 1. Oxford University Press, New York (1999)
6. Colorni, A., Dorigo, M., Maniezzo, V., et al. Distributed optimization by ant colonies. In: Proceedings of the first European Conference on Artificial Life, vol. 142, pp. 134–142. Paris, France (1991)
7. Fister Jr., I., Yang, X.-S., Fister, I., Brest, J., Fister, D.: A brief review of nature-inspired algorithms for optimization. arXiv preprint arXiv:1307.4186 (2013)
8. Karaboga, D.: An idea based on honey bee swarm for numerical optimization. Technical report, Technical report-tr06, Erciyes University, Engineering Faculty, Computer Engineering Department (2005)
9. Price, K.V.: Differential evolution: a fast and simple numerical optimizer. In: Fuzzy Information Processing Society, 1996. NAFIPS., 1996 Biennial Conference of the North American, pp. 524–527. IEEE (1996)
10. Sharma, A., Sharma, H., Bhargava, A., Sharma, N.: Power law-based local search in spider monkey optimisation for lower order system modelling. Int. J. Syst. Sci. 1–11 (2016)
11. Suganthan, P.N., Hansen, N., Liang, J.J., Deb, K., Chen, Y.-P., Auger, A., Tiwari, S.: Problem definitions and evaluation criteria for the CEC 2005 special session on real-parameter optimization. KanGAL report, 2005005:2005 (2005)

Bi-level Problem and SMD Assessment Delinquent for Single Impartial Bi-level Optimization

Srinivas Vadali[1](✉), Deekshitulu G.V.S.R.[2](✉), and Murthy J.V.R.[3](✉)

[1] Research Scholar, Department of CSE, Jawaharlal Nehru Technological University Kakinada (JNTUK), Kakinada, India
vadalisrinivas16@gmail.com
[2] Department of Mathematics, University College of Engineering Kakinada (UCEK), Jawaharlal Nehru Technological University Kakinada (JNTUK), Kakinada, India
dixitgvsr@hotmail.com
[3] Department of CSE, University College of Engineering Kakinada (UCEK), Jawaharlal Nehru Technological University Kakinada (JNTUK), Kakinada, India
mjonnalagedda@gmail.com

Abstract. In this paper, the proposed strategy is versatile such that the paper moreover gives a test suite of twelve test problems, which includes eight unconstrained and four compelled problems. The test suite comprises problems with adaptable variables and necessities, which can be used to survey the limit of the calculations in dealing with bi-levelproblems. To give design results, we have handled the proposed test problems using a settled bi-level transformative calculation. The results can be used for examination, while evaluating the execution of some other bi-level streamlining calculation.

Keywords: Bi-level optimization · Test problem development system · Baseline arrangement · SMD problems

1 Introduction

Bi-level enhancement constitutes a testing class of headway problems, where one change errand is settled within the other. A broad number of studies have been coordinated in the field of bi-level programming [7, 9–13], and on its practical applications [2]. Customary procedures commonly used to handle bi-levelproblems fuse the Karush-Kuhn-Tucker approach [2, 4], Branch-and-bound frameworks [3] and the use of censure limits [1]. Different studies have been performed towards using formative calculations [7–9, 11, 13] for dealing with bi-levelproblems. In any case, the examination on formative calculations for bi-levelproblems is still in early stage, and foremost change in the present approaches is required.

Past studies [3] on bi-level upgrade have displayed different fundamental test problems. In any case, the levels of inconvenience can't be controlled in these test problems. In most of the studies, the problems are either immediate [4], or quadratic [5, 6], or

© Springer Nature Singapore Pte Ltd. 2017
K. Deep et al. (eds.), *Proceedings of Sixth International Conference on Soft Computing for Problem Solving*, Advances in Intelligent Systems and Computing 546, DOI 10.1007/978-981-10-3322-3_6

non bi-scalable with settled number of decision variables. Application problems in transportation (framework diagram, perfect assessing), monetary angles (Stackelberg entertainments, crucial administrators problem, charge gathering, procedure decisions), organization (framework office region, coordination of multi-divisional firms), building (perfect design, perfect substance equilibria) et cetera [10] have also been used to display the efficiency of calculations. For most of these problems, the veritable perfect course of action is dark. Thusly, it is hard to recognize, whether a particular course of action got using a present procedure is close to the optima. Under these vulnerabilities, it is doubtful to proficiently evaluate course of action frameworks on realistic problems. These drawbacks stance hindrances in calculation headway, as the execution of the calculations can't be surveyed on various inconvenience unsettled areas.

In this paper, we perceive the troubles normally experienced in bi-level streamlining problems. A test problem improvement technique is proposed, which mimics these difficulties controllably. Using the advancement technique, we propose a social event of bi-level test problems flexible in regards to variables and objectives. The proposed arrangement grants to control the difficulties at the two levels self-governingly of each other.

The paper is made as takes after. In the accompanying zone, we clear up the structure of a general bi-level advancement problem and present the documentation that is used all through the paper. Area 3 presents our structure for building adaptable test problems for bi-level programming. Starting there, taking after the principles of the advancement procedure, we prescribe a course of action of twelve versatile test problems in Sect. 4. To make a benchmark for surveying distinctive arrangement calculations, the problems are handled using an essential settled bi-level developmental calculation which is a settled arrangement depicted in Sect. 5. The outcomes for the standard calculation are inspected in Sect. 6.

2   Portrayal of a Bi-level Problem

A bi-level change problem incorporates two levels of progression errands, where one level is settled within the other. The outer progression undertaking is commonly called as upper level upgrade errand, and the internal change task is called as lower level streamlining assignment. The problem contains two sorts of variables; particularly the upper level variables x_u, and the lower level variables x_l. In the going with, we give two equivalent definitions to a general bi-level upgrade problem with one focus at both levels:

Definition 1 (Bi-level Optimization Problem). Let $X = X_U \times X_L$ show the product of the upper-level decision space X_U and the lower-level decision space X_L, i.e. $x = (x_u, x_i) \in X$, if $x_u \in X_U$ and $x_l \in X_L$. For upper-level target limit F: $X \rightarrow$ R and lower-level target limit f: $X \rightarrow$ R, a general bi-level change problem is given by

$$
\begin{aligned}
&\text{Minimize } F(x), \, x \in X \\
&\text{s.t. } x_l \in \operatorname{argmin}\{f(x) \mid g_i(x) \geq 0, i \in I\}, \\
&x_l \in X_L \\
&G_j(x) \geq 0, j \in J.
\end{aligned}
\tag{1}
$$

where the limits $g_i: X \to R$, $i \in I$, address lower-level restrictions and $G_j: X \to R$, $j \in J$, is the aggregation of upper-level requirements.

In the above definition, a vector x is seen as functional at

(0) the upper level, in case it satisfies all the upper level goals, and vector x_l is perfect

(0) at the lower level for the given x_u. We find in this specifying the lower level problem is a parameterized constraint to the upper-level problem. An indistinguishable specifying of the bi-level change problem is procured by supplanting the lower level streamlining problem with a set worth limit which maps the given upper-level decision vector to the relating set of perfect lower-level game plans.

In the test problem development methodology, the Ψ limit gives an accommodating delineation of the relationship between the upper and lower level problems. Figures 1 and 2 plot two circumstances, where Ψ can be a singular vector regarded or a multi-vector regarded limit independently. In Fig. 1, the lower level problem is had all the earmarks of being a paraboloid with a lone minimum limit regard identifying with the game plan of upper level variables x_u. Figure 2 addresses a circumstance where the lower level limit is a paraboloid cut from the base with an even plane.

3 Test Problem Construction Procedure

The region of an additional streamlining errand within the requirements of the upper headway assignment prompts a gigantic augmentation in multifaceted nature, when appeared differently in relation to any single level improvement problem. The test problems made using the development system are depended upon to be adaptable to the extent number of decision variables and imperatives, such that the execution of the computations can be evaluated

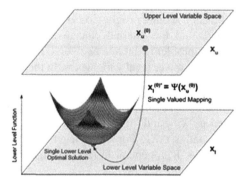

Fig. 1. Relationship amidst upper and lower level variables if there ought to be an event of a single vector regarded mapping. For ease the lower level limit is alive and well of a paraboloid.

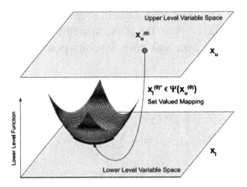

Fig. 2. Relationship amidst upper and lower level variables if there ought to emerge an event of a multi-vector regarded mapping. The lower level limit is seemed alive and well of a paraboloid with the base cut with a plane.

Case 1. Making co-agent communication: A test problem with co-operators association case can be made by picking

$$F_2(x_{l1}) = F_2(x_{l1})$$
$$F_3(x_{u2}, x_{l2}) = F_4(x_{u2}) + f_3(x_{u2}, x_{l2})$$

(4)

where $F_4(x_{u2})$ is any limit of x_{u2} whose base is known.

Case 2. Making clashing communication: A test problem with a conflict between the two levels can be made by basically changing the signs of terms f_2 and f_3 on the right hand side in (4):

$$F_2(x_{l1}) = -f_2(x_{l1})$$
$$F_3(x_{u2}, x_{l2}) = F_4(x_{u2},) - f_3(x_{u2}, x_{l2}).$$

(5)

The choice of F_2 and F_3 prescribed here is an excellent case, and there can be various diverse ways to deal with fulfill battle or co-operation using the two limits.

Case 3. Making blended connection: There may be a condition of both investment and strife if limits F_2 and F_3 are picked with reverse signs as,

$$F_2(x_{l1}) = f_2(x_{l1})$$
$$F_3(x_{u2}, x_{l2}) = F_4(x_{u2}) - f_3(x_{u2}, x_{l2})$$

(6)

Or

$$F_2(x_{l1}) = -f_2(x_{l1})$$
$$F_3(x_{u2}, x_{l2}) = F_4(x_{u2}) + f_3(x_{u2}, x_{l2}).$$

(7)

Illustration 2: Consider a bi-level enhancement problem where the lower level errand is given by Example 1. According to the above strategies, we can make a test problem with a conflict between the upper and lower level by portraying the upper level target limit as takes after:

$$F_1(X_{u1}) = \sum_{i=1}^{U1} (x_{u1}^i)^2$$
$$F_2(x_{l1}) = F_2 = - \sum_{i=1}^{L1} (x_{l1}^i)^2 \qquad (8)$$
$$F_3(x_{u2}, x_{l2}) = - \sum_{i=1}^{U2} (x_{u1}^i - x_{l2}^i)^2$$

The picked definition thinks about to illustration 2, where $F_4(x_{u2}) = 0$. The last perfect course of action of the bi-level problem is $F(F(x_u, x_l).) = 0$ for $(x_u, x_l) = 0$.

Various Global Solutions at Lower Level

In this sub-territory, we discuss creating test problems with lower level limit having various overall responses for a given course of action of upper level variables. To fulfil this, we detail a lower level limit which has various lower level optima for a given $(x_u$, such that $x_l^* \in \Psi(x_u))$. By then, we promise that out of all these possible lower level perfect courses of action, one of them (x_l^{**}) looks at to the best upper level limit regard, i.e.,

$$x_l^{**} \quad \in argmin\{F(X_u, X_l^*) | X_l^* \in (X_u)\} \qquad (9)$$

To go along with this inconvenience in the problem, we pick the second limits at the upper and lower levels. Given that the term $f_2(x_{l1})$ is responsible for realizing complexities exactly at the lower level, we can energetically arrange it such that it has various lower level perfect game plans. From this it generally takes after that the entire lower level limit has different perfect courses of action.

Sample 3: We depict the development methodology by considering a fundamental case, where the cardinalities of the variables are, dim $(x_{u1}) = 2$, dim $(x_{u2}) = 2$, dim $(x_{l1}) = 2$ and dim $(x_{l2}) = 2$, and the lower level limit is described as takes after,

$$f_1(x_{u1}, x_{u2},) = ((x_{u1}^1)^2 + (x_{u1}^2)^2 + (x_{u2}^1)^2 + (x_{u2}^2)^2$$
$$f_2(x_{l1}) = (x_{l1}^1 - x_{l1}^1)^2 \qquad (10)$$
$$f_3(x_{u2}, x_{l1}) = (x_{u2}^1 - x_{l1}^1)^2 + (x_{u2}^2 - x_{l2}^2)^2$$

Here, we watch that $f_2(x_{l1})$ influences different perfect game plans, as its base quality is 0 for all. At the base $f_3(x_{u2}, x_{l1})$ fixes the estimations of and to and independently. Next, we make the upper level limit ensuring that out of the set, one of the plans is best at upper level.

$$F_1(X_{u1}) = (x_{u1}^1)^2 + (x_{u1}^2)^2$$
$$F_2(X_{l1}) = (x_{l1}^1)^2 + (x_{l1}^2)^2 \qquad (11)$$
$$F_2(x_{u2}, x_{l2}) = (x_{u2}^1 - x_{l2}^2)^2 + (x_{u2}^2 - x_{l2}^2)^2$$

The meaning of $F_2(X_{l1})$, as aggregate of squared terms ensures that gives the best course of action at the upper level for any given (X_{u1}, X_{u2}).

Table 1. Composition of the requirement sets at both levels.

Level	Constraint Set	Subsets	Dependence
Upper	$G = \{G_j : j \in J\}$	$G = G_a \cup G_b \cup G_c$	G_a depends on x_u G_b relies on x_l G_c relies on x_u and x_l
Lower	$g = \{g_j : i \in I\}$	$g = g_a \cup g_b \cup g_c$	g_a depends on x_u g_b reliess on x_l g_c relies on x_u and x_l

Inconveniences influenced by limitations. In this subsection, we inspect about the sorts of requirements which can be knowledgeable about a bi-level headway problem. Considering that the bi-level problems have the probability to have limitations at both levels, and each basic could be a component of two different sorts of variables, the constrained set at both levels can be further isolated into smaller subsets as takes after:

In Table 1, G and g mean the plan of confinements at the upper and lower level independently. Each of the basic set can be broken into three more diminutive subsets, as showed up in the table. If the important impediment subset (G_a or g_a) is non-unfilled at both of the two levels, then for any given x_u we should check the achievability of imperatives in the sets Ga and ga, before dealing with the lower level progression problem. If, there is one or more infeasible objectives in g_a, then the lower level headway problem does not contain perfect lower level course of action (X_l^*) for the given X_u. Regardless, in the event that one or more goals are infeasible within, G_b then a lower level perfect course of action (X_l^*) may exist for the given X_u, however the pair (x) will be infeasible for the bi-level problem. In light of this property, a decision can be made, whether it is useful to handle the lower level progression problem at all for a given X_u.

4   SMD Test Problems

By adhering to the layout benchmarks exhibited in the past portion, we now propose a plan of twelve problems which we call as the SMD test problems. Each problem addresses another inconvenience level similarly as meeting at the two levels, versatile nature of coordinated effort between two levels and multi-modalities at each of the levels. The underlying eight problems are unconstrained and the staying four are obliged.

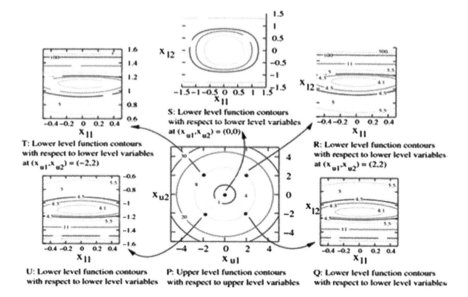

Fig. 3. Upper and lower level four-variable SMD1

SMD1

This is a direct test problem where the lower level problem is a curved streamlining undertaking, and the upper level is raised concerning upper level variables and perfect lower level variables. The two levels partake with each other.

$$F_1 = \sum_{i=1}^{p} (x_{u1}^i)^2$$
$$F_2 = \sum_{i=1}^{q} (x_{l1}^i)^2$$
$$F_3 = \sum_{i=1}^{r} (x_{u2}^i) + \sum_{i=1}^{r} (x_{u2}^i - \tan x_{l2}^i)^2 \quad (12)$$
$$f_1 = \sum_{i=1}^{p} (x_{u1}^i)^2$$
$$f_2 = \sum_{i=1}^{q} (x_{l1}^i)^2$$

$$f_3 = \sum_{i=1}^{r} (x_{u2}^i - \tan x_{l2}^i)^2 \quad (13)$$

The extent of variables is according to the accompanying,

Relationship between upper level variables and lower level perfect variables is given as takes after,

The estimations of the variables at the optima are $x_u = 0$ and xl is gotten by the relationship given above. Both the upper and lower level limits are identical to zero at the optima.

Figure 3 shows the states of the upper and lower level limits with respect to the upper and lower level variables for a four-variable test problem. The problem has two

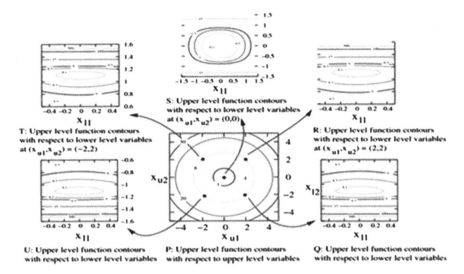

Fig. 4. Upper level for a four-variable SMD1 test.

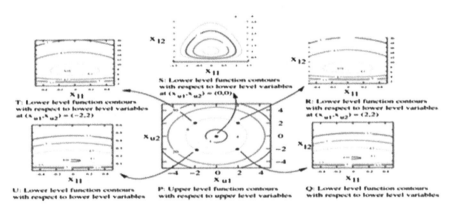

Fig. 5. Upper and lower level limit shapes for a four-variable SMD2 test problem.

upper level variables and two lower level variables, such that the estimations of x_{u1}, x_{u2}, x_{l1} and x_{u2} are each one of the one. Sub-figure P exhibits the upper level limit shapes with respect to the upper level variables, tolerating that the lower level variables are at the optima. Settling the upper level variables (x_{u1}, x_{u2}) at five unmistakable zones, i.e. (2,2), (−2,2), (2, − 2), (−2, − 2) and (0,0), the lower level limit structures are showed up with respect to the lower level variables.

Figure 4 exhibits the types of the upper level limit concerning the upper and lower level variables. Sub-figure P before long shows the upper level limit shapes concerning the upper level variables. In any case, sub-figures Q, R, S, T and V now address the upper level limit frames at different (x_{u1}, x_{u2}), i.e. (2,2), (−2,2), (2, − 2), (−2, − 2) and (0,0).

From sub-figures Q, R, S, T and V, we watch that if the lower level variables move a long way from its optimal zone, the upper level limit regard goes into disappear.

5 Baseline Solution Methodology

Around there, we depict the course of action system used to deal with the constructed test problems. The prescribed strategy is a settled bi-level transformative count, and requires that a lower level progression task unwound for each new course of action of upper level variables made using the innate managers. We have realized a changed adjustment of the procedure [1, 14, 15], which is used to handle the bi-level test problems. The proposed methodology relies on upon a lone target Parent Centric Crossover (PCX) [8]. A directed strategy for the computation is depicted as takes after:

5.1 Upper Level Optimization Procedure

Step 1: Initialization Scheme. Present a sporadic people (N_p) of upper level variables. For each upper level people part execute a lower level improvement system to choose the relating perfect lower level variables. Dole out upper level wellbeing in light of the upper level limit quality and goals

Step 2: Selection of upper level people. Pick 2μ people from the past masses and conduct an opposition decision to choose μ people

Step 3: Evolution at the upper level. Perform a PCX based crossover [8] (Refer Sub-fragment 5.4) and a polynomial change to make λ off-springs. This gives the upper level variables to each successor

Step 4: Lower level progression. Deal with the lower level progression problem (Refer Sub-zone 5.2) for each successor. This gives the lower level variables to each descendant

Step 5: Evaluate off-springs. Join the upper level variables with the relating perfect lower level variables for each successors. Evaluate every one of the off-springs in perspective of upper level limit worth and goals

Step 6: Population upgrade. Pick r self-assertive people from the watchman masses and pool them with the λ off-springs. The best r people from the pool supplant the picked r people from the masses

Step 7: Termination check. Proceed to the front line (Step 2) if the end check (Refer Sub-fragment 5.6) is false

5.2 Lower Level Optimization Procedure

The lower level change system resembles the upper level technique beside the presentation step which fluctuates to some degree. Allow the lower level masses to size be n Step 1, and the upper level part being progressed to be the execution is traded from Step 1 of the upper level optimization x_{0u}.

Assignment then goes to (a) by and large go to (b), an: Initialize n_p lower level part indiscriminately, and apportion lower level health (Fig. 5).

In perspective of the lower level limit worth and constraints. Go to Step 2.

b: Initialize n_p lower level people aimlessly. Choose the part closest to x_{ou} in the upper level people. The lower level perfect variables from the closest upper level part transforms into the part in the lower level people. Allot lower level health in light of the lower level limit quality and objectives. Go to Step 2.

Step 2:   Choose 2μ people discretionarily from the lower level masses. Perform an opposition decision in regards to lower level health to make μ people

Step 3:   Perform half and half and change to make λ off-springs

Step 4:   Evaluate each descendants with respect to lower level limit and requirements

Step 5:   Choose r people discretionarily from the lower level masses and pool them with the λ lower level off-springs. The best r people concerning lower level health supplant the picked r people from the lower level masses

Step 6:   Proceed to the front line (Step 2) if the end check (Refer Sub-region 5.6) is false

6   Results

In this fragment, we give the results gained from dealing with the proposed test problems using the bi-level transformative count. We performed 11 number of continues running for each of the test problems with 5 and 10 estimations. In the event that there ought to be an event of 5 estimations, for SMD1 to SMD5 and SMD7 to SMD12 we pick p = 1, q = 2 and r = 1, and for SMD6 we pick p = 1, q = 0, r = 1 and s = 2. In the event that there ought to be an event of 10 estimations, for SMD1 to SMD5 and SMD7 to SMD12 we pick p = 3, q = 3 and r = 2, and for SMD6 we pick p = 3, q = 1, r = 2 and s = 2. The upper level people size N_p and the lower level masses size were picked as 30 for the five dimensional cases. Both masses sizes were picked as 50 for the 10 dimensional cases.

We consider a test problem settled if the refinement between the limit regard finished by the estimation and the perfect limit worth is near 0.1. Exactly when the range of the test problems is extended to 10. For SMD6 the accomplishment rate was 87%, for SMD7 it was 66% and for SMD8 it was 62%. The lower level problems couldn't be completely disentangled for SMD9 to SMD12, which introduced infeasible people at the upper level.

7   Conclusion

The development method offers the flexibility to control the difficulties at the two levels only and taking all things into account. The development technique has been used to make a demonstrating ground of 12 bi-level change problems, out of which 8 are unconstrained and 4 are constrained. The test-suite contains problems, which are adaptable with respect to number of variables furthermore requirements.

Five and ten-variable events of all the test problems have been comprehended, which display the high computational need of bi-level problems despite for humbler cases. This adequately exhibits the game plan of bi-level problems, even with a constructional figuring, is a trying errand and more thought ought to now be made to become computationally speedier counts.

References

1. Aiyoshi, E., Shimizu, K.: Hierarchical decentralized systems and its new solution by a barrier method. IEEE Trans. Syst. Man Cybern. **11**, 444–449 (1981)
2. Bard, J.F.: Practical Bilevel Optimization: Algorithms and Applications. Kluwer, The Netherlands (1998)
3. Bard, J.F., Falk, J.: An explicit solution to the multi-level programming problem. Comput. Oper. Res. **9**, 77–100 (1982)
4. Bianco, L., Caramia, M., Giordani, S.: A bilevel flow model for hazmat transportation network design. Transp. Res. Part C: Emerg. Technol. **17**(2), 175–196 (2009)
5. Calamai, P.H., Vicente, L.N.: Generating linear and linear-quadratic bilevel programming problems. SIAM J. Sci. Comput. **14**(1), 770–782 (1992)
6. Calamai, P.H., Vicente, L.N.: Generating quadratic bilevel programming test problems. ACM Trans. Math. Softw. **20**(1), 103–119 (1994)
7. Colson, B., Marcotte, P., Savard, G.: An overview of bilevel optimization. Ann. Oper. Res. **153**, 235–256 (2007)
8. Deb, K., Anand, A., Joshi, D.: A computationally efficient evolutionary algorithm for real-parameter optimization. Evolu. Comput. J. **10**(4), 371–395 (2002)
9. Deb, K., Sinha, A.: An efficient and accurate solution methodology for bilevel multi-objective programming problems using a hybrid evolutionary-local-search algorithm. Evolu. Comput. J. **18**(3), 403–449 (2010)
10. Dempe, S.: Annotated bibliography on bilevel programming and mathematical programs with equilibrium constraints. Optimization **52**(3), 339–359 (2003)
11. Garg, H.: Bi-criteria optimization for finding the optimal replacement interval for maintaining the performance of the process industries. In: Modern Optimization Algorithms and Application in Engineering and Economics (2016). doi:10.4018/978-1-4666-9644-0. ch025
12. Garg, H., Rani, M., Sharma, S.P., Vishwakarma, Y.: Bi-objective optimization of the reliability-redundancy allocation problem for series-parallel system. J. Manuf. Syst. **33**(3), 335–347 (2014). Elsevier
13. Garg, H., Rani, M., Sharma, S.P., Vishwakarma, Y.: Intuitionistic fuzzy optimization technique for solving multi-objective reliability optimization problems in interval environment. Expert Syst. Appl. **41**(7), 3157–3167 (2014). Elsevier
14. Rajesh Eswarawaka, S.K., Mahammad, Noor, Eswara Reddy, B.: Genetic annealing with efficient strategies to improve the performance for the NP-hard and routing problems. J. Exper. Theor. Artif. Intel. **27**(6), 779–788 (2015). doi:10.1080/0952813X.2015.1020624
15. Garg, H., Sharma, S.P.: Multi-objective reliability-redundancy allocation problem using particle swarm optimization. J. Comput. Ind. Eng. **64**(1), 247–255 (2013)

An Adaptive Firefly Algorithm for Load Balancing in Cloud Computing

Gundipika Kaur[(⊠)] and Kiranbir Kaur

Department of Computer Engineering and Technology,
Guru Nanak Dev University, Amritsar, India
gunbhatia@yahoo.com

Abstract. Over the past few years, cloud computing has become a popular paradigm that provides computing over the internet. There are umpteen factors that a cloud ecosystem need such as reliability, flexibility, dynamic load balancing etc. With the internet facility, resources are provided dynamically to the end users in an on-demand fashion. Users could be billions in number accessing the cloud. Their need for services have been increasing at an alarming rate. To enhance the performance of the system, resources should be used efficiently. Cloud computing needs to identify different issues and challenges. One of the main issues in cloud computing is Load balancing, in which workload is distributed dynamically to all the nodes. Load balancing not only optimize the resource use, maximize throughput, minimize processing time of datacenters and response time of user base, but also helps in evading the overloading of any single resource. This paper proposes an Adaptive firefly algorithm (ADF) for solving the load balancing problem in cloud computing by performing virtual machine scheduling over datacenters. The results have been compared with Ant Colony Optimization (ACO) algorithm used for load balancing.

Keywords: Cloud computing · Scheduling of virtual machines · Adaptive Firefly Algorithm · Ant Colony Optimization · Cloud analyst

1 Introduction

Cloud computing [15] is a faddish paradigm which has been proliferating both in academia and industry. It starts affecting swarm of industries such as government, finance, telecommunications, and education. Cloud computing is a prominent model in which shared pool of resources such as servers, application, services, storage etc. are accessed in an on-demand fashion. These computing resources are provisioned and released rapidly with the minimal effort of management [16]. Although cloud computing has been gaining popularity in the industry, the research on cloud computing is still at an early stage. Umpteen existing issues are there which have not been fully addressed. One of the important research issues which need to be focused for efficient utilization of resources of datacentres are resource scheduling [13] as well as scheduling the jobs which are being assigned to virtual machines. Here, this paper proposes a load balancing algorithm by scheduling the virtual machines. Scheduling is an important aspect of cloud computing. When resources are allocated optimally

© Springer Nature Singapore Pte Ltd. 2017
K. Deep et al. (eds.), *Proceedings of Sixth International Conference on Soft Computing for Problem Solving*, Advances in Intelligent Systems and Computing 546, DOI 10.1007/978-981-10-3322-3_7

among tasks in a finite time to achieve high quality of service, is called scheduling. It aims in optimizing one or more objectives. Scheduling belongs to the category of NP-hard problems because of large solution space and it is difficult to find an optimal solution. The mapping of tasks [12] on large pool of resources is not an easy job to get an optimal solution. There are no algorithms which provide optimal solution by satisfying polynomial time to solve these problems. Therefore, In cloud environment, it is preferable to find suboptimal solution, but in short period of time. As cloud computing is growing exponentially, it has been widely adopted by the industry and thus making a rapid expansion in availability of resources in the internet. As the cloud size increasing at an alarming rate, cloud computing service providers requires handling of massive requests. Thus in spite of glorious future of cloud computing, many critical problems still need to be explored for its perfect realization. One of these issues is Load balancing. As the incoming requests changes dynamically [9] due to heterogeneity of resources, dynamic resource allocation [16] is required in cloud computing. This inherent dynamism in cloud computing requires efficient load balancing mechanism. Load is nothing but the amount of work that a system performs. Load can be classified as CPU load, memory size and network load.

Load balancing [8] over a cloud is a process of distributing workloads among the multiple resources available in the cloud. With efficient load balancing of resources or virtual machines over the cloud, user can achieve better service with reduced cost. This is the most important factor which keeps a binding between users and cloud service provider. So, to enhance the services of cloud, various new meta-heuristic techniques have been applied to schedule virtual machines for load balancing which gave promising results till now. Some of the meta-heuristic technique [1] which have been applied are Genetic Algorithm [7], Particle Swarm Optimization [17], Ant Colony Optimization [11] and many more. With the emergence of newer nature inspired algorithm in the field of optimization, the field of virtual machine scheduling [19] over cloud computing for load balancing has gained popularity.

The rest of the paper is organised as follows: Sect. 2 shows the related work in this field using Ant Colony Optimization (ACO). Section 3 introduces the proposed methodology of Adaptive Firefly Algorithm (ADF) for virtual machine scheduling in cloud. Section 4 describes experimental setup required for evaluating proposed algorithm. Section 5 analyse the results and finally, Sect. 6 concludes the paper.

2 Related Work

In this section of paper we are discussing Ant Colony Optimization for virtual machine scheduling with which we will compare the result of our proposed technique. Ant Colony Optimization (ACO) scheduling algorithm [14] is inspired by the behaviour of real ants finding the shortest path between their colonies and a source of food. While walking from their colony to the food source, ants leave pheromones on the ways they move. The pheromone intensity on the passages increases with the number of ants passing through and drops with the evaporation of pheromone. As the time goes on, smaller paths draw more pheromone and thus, pheromone intensity helps ants to recognize smaller paths to the food source. ACO methods are useful for solving discrete optimization problems that need to find paths to goals. The basic equation which was proposed in [3] for pheromone representation in terms of Virtual machine parameters are

$$\tau(t = 0) = f\left(MIPS_j, L, BW_j\right) \tag{1}$$

Where $\tau_{ij}(t = 0)$ is the pheromone value in between two nodes i and j at turn $t = 0$, $MIPS_j$ is million instruction per second of virtual machine VM_j and BW_j is the communication bandwidth availability of VM_j.

For updating of pheromone at $(t + 1)$ turn following equation had been proposed in [3].

$$\tau_j(t + 1) = (1 - \rho) * \tau_j(t) + \Delta\tau_j \tag{2}$$

where ρ pheromone trail decay coefficient.

Result of the mention technique had been compared to various existing optimization algorithm at that time which are First Come First Serve (FCFS), Stochastic Hill Climbing Algorithm (SHC) and Genetic Algorithm (GA). As presented in [3] the result was quite promising as compared to others, the whole algorithm was developed on cloud analyst simulator with various cloud computing configuration.

3 The Proposed Methodology

This section proposes an Adaptive Firefly algorithm (ADF) to schedule virtual machines over cloud for better load balancing. As compared to original firefly algorithm [2], adaptive firefly algorithm (ADF) as presented in [4] has an advantage of variable step size for the movement of the firefly. The Basic firefly algorithm is inspired by the behaviour of insects (firefly) which mostly produces short and rhythmic flash with different intensities. The function of such flashes is to attract or alert the neighbouring firefly. Depending upon the intensity of firefly, other firefly moves toward or away from that firefly. The light intensity of the firefly also depends on the distance of the eyes of beholder. It will be simple to say that more is the distance lesser will be the intensity seen by the distant fireflies, thus intensity become less appealing to the distant fireflies.

There are basically two variable on which firefly algorithm depend, one is light intensity and other is attractiveness. The attractiveness is inversely proportional to the distance of firefly from other firefly. In terms of cloud resources, intensity (represented in Eq. 3) is been taken as the combination of parameter which actually represents the basic utilization of virtual machine.

$$I_j(t) = f\left(Mips_j, BW_j, CPU_j, Mem_j, Size_j\right) \tag{3}$$

Where $Mips_j$ is the million instructions per seconds, BW_j is the bandwidth, CPU_j is the processing elements, Mem_j is the memory used and $size_j$ is the available storage of the virtual machine VM_j and I_j is the intensity of the jth virtual machine.

The attractiveness factor (represented in Eq. 4) of the firefly is mapped to virtual machines as following

$$\beta = \beta_{t-1} * \left(e^{-\gamma * r^2}\right) + \alpha * \beta_{t-1} \tag{4}$$

Where β, is the attractiveness factor [18] of firefly, γ is the function of free Processing elements and α is the function for the variability of the movement of firefly toward the other firefly. So the value of β decides the movement of firefly toward the other firefly which actually depends on the newly added parameter α & γ which gives the variability to the attractiveness and depends on the number of processing elements free (represented in Eqs. 5 and 6) at the time of allocation of virtual machine to the opted job which may be different for different virtual machines.

$$\alpha, \gamma = f(PE_k) \tag{5}$$

$$r = f(f_{best}, f_k) \tag{6}$$

Where k is the kth virtual machine, f_{best} is the current best solution having attractiveness of virtual machine and f_k is the current attractiveness of virtual machine. Pseudocode for proposed algorithm has been shown in Algorithm 1.

1. Begin procedure	16. end if
2. Initialize algorithm parameters	17. $\alpha \leftarrow f(PE_j)$
3. MaxGen: the maximal number of generations	18. $\gamma \leftarrow f(PE_i)$
4. α & γ: function of free processing elements	19. $r \leftarrow f(f_{best}, f_j)$
5. r: funtion of distance between two fireflies	20. Attractiveness varies with distance r via $e^{[-\gamma r^2]}$ and α ;
6. Define the objective function of f(V), where V is the (Mips, BW, CPU, Mem, Size) are features of VMs	21. $\beta_i \leftarrow \beta_j * e^{-\gamma r^2} + \alpha * \beta_j$
7. Generate the initial population of fireflies(VMs) or ' 1,2,.. n)	22. Evaluate new solutions and update light intensity;
8. Determine the light intensity of I_i at V_i via f(V$_i$)	23. $I_i \leftarrow f(I_j, \beta_j)$
9. for i = 1 to n (all n fireflies (VMs));	24. End for j;
10. $I_i \leftarrow f(V_i)$	25. Rank the fireflies and find the current best
11. $f_{best} \leftarrow$ Null	26. if $(f_{best} > f_i)$
12. for j = 1 to n (n fireflies(VMs))	27. $f_{best} \leftarrow f_i$
13. $I_j \leftarrow f(V_j)$	28. end if
14. If $(Ij > Ii)$,	29. end for i;
15. choose firefly $(VM_i)i$ over $(VM_j)j$	30. Post process results and visualization;
	31. End procedure

Algorithm 1: Pseudocode of adaptive firefly

4 Experimental Setup

To carry out the simulation of the proposed algorithm of Adaptive Firefly in load balancing over cloud computing, we have used Cloud-Analyst developed by cloudlabs using basic cloudsim toolkit which provide an interface for developing and integrating your proposed approach [5, 6]. In this simulator to get appropriate result we have to set the configuration of the Virtual machine at the end of cloud server over the datacenter part of cloud. Different result has been evaluated using different environment for the same algorithm. In Cloud-Analyst, we have to set the number of data center over the world and the configuration of virtual machine which has been placed over every datacentre. Along with that we need to create client i.e. user base, which actually request datacentres depending upon the policy for choosing datacentre, here we are using closed datacentre policy to select the datacentre to execute user's request over the cloud.

Fig. 1. Cloud analyst configuration C_2 of 15 DC and 25 UB

Three different Configurations (C_1, C_2, C_3) as shown in Table 4 has been set to get the desired output of the simulator. The actual parameter of the configurations is provided in the Tables 1, 2 and 3 below.

Table 1. Physical hardware detail of datacenter

Memory (Mb)	Storage (Mb)	Available BW	No. of processor	Processor speed	VM policy
204800	100000000	1000000	4	10000	TIME_SHARED

Table 2. Datacenter configuration

Arch	OS	VMM	Cost per VM	Memory cost per VM	Storage cost per VM	Datacentre transfer cost	Physical HW unit
X86	Linux	Xen	0.1	0.05	0.1	0.1	2

Table 3. Userbase configuration

Requests per user per hour	Data size per request	Peak hour start	Peak hour end	Average peak user	Average off peak user
60	100	3	9	1000	100

Table 4. Cloud configuration

Configuration	No. of user base (UB)	No. of datacentre (DC)
C_1	25	10
C_2	25	15
C_3	30	20

5 Performance Analysis

After setting the defined configuration over the cloud in the cloud-analyst, different result has been came out which has been displayed in Fig. 1. The graph represents the Overall response time and Data centre processing time against the total execution time for the simulation in mili-seconds. The graphs shows that over all response time kept on decreasing as number of datacenter increases because of the sharing of the load among datacenter which provides parallelism and saves execution time for the job or request. Since request has been shared among different datacentre, the number of request per datacentre become less and hence reduces the processing time of datacentre for the request (Fig. 2).

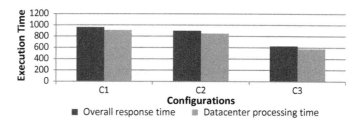

Fig. 2. Graphical representation of results

Figures 3 and 4 represents the response time for the User base 21 for the configuration C_3 for both the algorithm. As it has been cleared from the two figures that response time with ADF is much better than ACO.

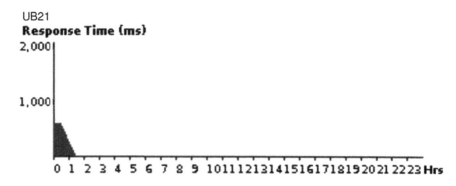

Fig. 3. Response time of UB 21 for AFA

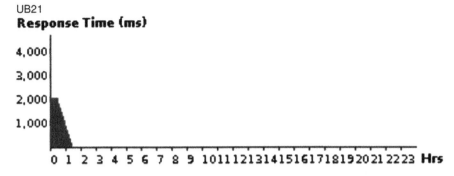

Fig. 4. Response time of UB 21 for ACO

Figures 5 and 6 represent the processing time for datacentre 13 for both the algorithm for same configuration, C_3. The processing time of DC13 with AFA is much better than with ACO.

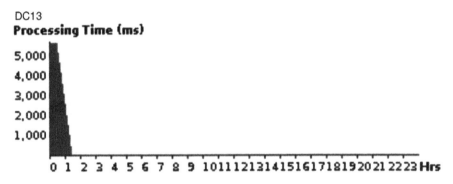

Fig. 5. Processing time of DC13 for ACO

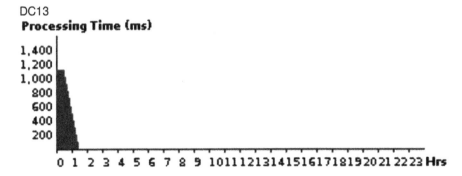

Fig. 6. Processing time of DC13 for AFA

As the number of datacenter increases, load will be balanced in datacenters due to sharing of load. The overall response time of the userbase and data processing time will be decreased as data centers increases and results shows that both parameters(overall response time of userbase and datacenter processing time) decreases when ADF(An adaptive firefly algorithm) outperformed ACO(Ant colony optimization) algorithm as shown in Figs. 7, 8 and 9 below:

	Average (ms)	Minimum (ms)	Maximum (ms)
Overall Response Time:	967.14	37.63	95520.01
Data Center Processing Time:	854.33	0.02	95466.75
	Average (ms)	Minimum (ms)	Maximum (ms)
Overall Response Time:	5389.32	36.63	109589.75
Data Center Processing Time:	5279.12	0.02	109538.25

Fig. 7. Results of ADF and ACO with 10 datacenters and 25 userbase

	Average (ms)	Minimum (ms)	Maximum (ms)
Overall Response Time:	1451.74	37.11	79970.78
Data Center Processing Time:	1402.35	0.01	79917.26
	Average (ms)	Minimum (ms)	Maximum (ms)
Overall Response Time:	4305.23	37.95	135645.76
Data Center Processing Time:	4258.85	0.02	135597.25

Fig. 8. Results of ADF and ACO with 15 datacenters and 25 userbase

	Average (ms)	Minimum (ms)	Maximum (ms)
Overall Response Time:	775.50	35.37	70844.76
Data Center Processing Time:	726.07	0.01	70791.26
	Average (ms)	Minimum (ms)	Maximum (ms)
Overall Response Time:	1778.97	35.64	139245.26
Data Center Processing Time:	1730.36	0.01	139193.26

Fig. 9. Results of ACO and ADF with 20 datacenters and 30 userbase

It has been cleared from the above figures that overall response time and data processing time with ADF is much better than ACO.

6 Conclusion

Scheduling of resources [10] is a major concern in cloud computing so that users get resources in an efficient manner. Need of scheduling is there because of NP-hard problems, which will not give an optimal results. Scheduling can take place in application layer, virtualization layer and deployment layer. Our major concern is with virtualization layer where tasks are mapped to virtual resources in an efficient manner to solve the solution of scheduling. Different scheduling algorithms have been implemented to get the desired results but it needs an efficient VM load balancer. Scheduling of resources leads to one major problem which is called load balancing in datacenters. In our proposed work, we have implemented an algorithm which will load the balance in datacenters by scheduling of virtual machines. There can be many techniques which have not used yet in the field of load balancing. In our work, we have proposed an algorithm i.e. An adaptive firefly algorithm (ADF). This algorithm is compared with Ant colony optimization (ACO) algorithm by taking different parameters. ADF algorithm worked really well as compared to ACO by minimizing the response time of userbase and datacenter processing time. Since various optimization algorithms are being explored in near future which shows that there won't be any end until researcher keep finding the relation between the various nature's algorithm. So there won't be an end to keep finding the better virtual machine scheduling over the cloud computing to balance the load over the cloud.

References

1. Kalra, M., Singh, S.: A review of metaheuristic scheduling techniques in cloud computing. Egypt. Inf. J. **16**, 1–20 (2015)
2. Florence, A.P., Shanthi, V.: A load balancing model using firefly algorithm in cloud computing. J. Comput. Sci. **10**(7), 1156 (2014)
3. Dam, S., Mandal, G., Dasgupta, K., Dutta, P.: An ant colony based load balancing strategy in cloud computing. In: Kumar Kundu, M., Mohapatra, D.P., Konar, A., Chakraborty, A. (eds.) Advanced Computing, Networking and Informatics- Volume 2. SIST, vol. 28, pp. 403–413. Springer, Heidelberg (2014). doi:10.1007/978-3-319-07350-7_45
4. Mohammadi, S., et al.: An adaptive modified firefly optimisation algorithm based on Hong's point estimate method to optimal operation management in a microgrid with consideration of uncertainties. Energy **51**, 339–348 (2013)
5. Ahmed, T., Singh, Y.: Analytic study of load balancing techniques using tool cloud analyst. Int. J. Eng. Res. Appl. **2**, 1027–1030 (2012)
6. Wickremasinghe, B.: CloudAnalyst: A CloudSim-based tool for modelling and analysis of large scale cloud computing environments. MEDC Proj. Rep. **22**(6), 433–659 (2009)
7. Dasgupta, K., et al.: A genetic algorithm (GA) based load balancing strategy for cloud computing. Procedia Technol. **10**, 340–347 (2013)
8. Mesbahi, M., Rahmani, A.M.: Load balancing in cloud computing: a state of the art survey. Int. J. Mod. Educ. Comput. Sci. **8**(3), 64 (2016)
9. Gao, R., Juebo, W.: Dynamic load balancing strategy for cloud computing with ant colony optimization. Future Int. **7**(4), 465–483 (2015)

10. Tan, G., Zheng, W., Du, Y., Xin, D.: A cloud resource scheduling strategy based on ant colony optimization algorithm. In: Control, Mechatronics and Automation Technology: Proceedings of the International Conference on Control, Mechatronics and Automation Technology (ICCMAT 2014), Beijing, China, 24–25 July 2014, vol. 6, p. 189. CRC Press (2015)
11. Wen, W.-T., Wang, C.-D., Wu, D.-S., Xie, Y.-Y.: An ACO-based scheduling strategy on load balancing in cloud computing environment. In: Ninth International Conference on Frontier of Computer Science and Technology (FCST) (2015)
12. Wu, X., et al.: A task scheduling algorithm based on QoS-driven in cloud computing. Procedia Comput. Sci. **17**, 1162–1169 (2013)
13. Singh, S., Chana, I.: QRSF: QoS-aware resource scheduling framework in cloud computing. J. Supercomputing **71**(1), 241–292 (2015)
14. Zuo, L., Shu, L., Dong, S., Zhu, C., Hara, T.: A multi-objective optimization scheduling method based on the ant colony algorithm in cloud computing. IEEE Access **3**, 2687–2699 (2015)
15. Zhan, Z.-H., et al.: Cloud computing resource scheduling and a survey of its evolutionary approaches. ACM Computing Surveys (CSUR) **47**(4), 63 (2015)
16. Manvi, S.S., Shyam, G.K.: Resource management for Infrastructure as a Service (IaaS) in cloud computing: a survey. J. Netw. Comput. Appl. **41**, 424–440 (2014)
17. Liu, Z., Wang, X.: A PSO-based algorithm for load balancing in virtual machines of cloud computing environment. In: Tan, Y., Shi, Y., Ji, Z. (eds.) ICSI 2012. LNCS, vol. 7331, pp. 142–147. Springer, Heidelberg (2012). doi:10.1007/978-3-642-30976-2_17
18. Fister, I., Yang, X.-S., Brest, J.: A comprehensive review of firefly algorithms. Swarm Evol. Comput. **13**, 34–46 (2013)
19. Cho, K.-M., et al.: A hybrid meta-heuristic algorithm for VM scheduling with load balancing in cloud computing. Neural Comput. Appl. **26**(6), 1297–1309 (2015)

Review on Inertia Weight Strategies for Particle Swarm Optimization

Ankush Rathore$^{(\boxtimes)}$ and Harish Sharma

Rajasthan Technical University, Kota, India
ankushrathore777@gmail.com

Abstract. In the category of swarm intelligence based algorithms, Particle Swarm Optimization (PSO) is an effective population-based meta-heuristic used to solve complex optimization problems. In PSO, global optima is searched with the help of individuals. For the efficient search process, individuals have to explore whole search space as well as have to exploit the identified search area. Researchers are continuously working to balance these two contradictory properties i.e. exploration and exploitation and have been modified the PSO in many different ways to improve its solution search capability in the search space. In this regard, incorporation of inertia weight strategy in PSO is a significant modification and after that many researchers have been developed different inertia weight strategies to improve the solution search capability of PSO. This paper presents an analysis of the developed inertia weight strategies in respect to problem-solving capability and their effect in the solution search process of PSO. The effect of 30 recent inertia weight strategies on PSO is measured while comparing over ten well known test functions of having different degree of complexity and modularity.

Keywords: Soft computing · Optimisation · Inertia weight · Swarm intelligence · Nature inspired algorithms

1 Introduction

Particle Swarm Optimization (PSO) algorithm was developed by Eberhart and Kennedy in 1995 [1]. It is inspired by the intelligent behaviour of bird in search of food. The PSO algorithm is used to solve the different complex optimization problems including economics, engineering, complex real-world problems, biology and industry [2]. PSO can be applied to non-linear, non-differentiable, huge search space problems and gives better results with good accuracy [3].

For n- dimensional search space, the velocity and position of the ith particle represents as: $V_i = (v_{i1}, v_{i2}, ..., v_{id})^T$ and $X_i = (x_{i1}, x_{i2}, ..., x_{id})^T$ respectively. Where, v_{id} and x_{id} is the velocity and position of ith particle in d-dimension respectively. The velocity of the swarm (particle) is defined as follows:

$$v_{id}(new) = v_{id}(old) + c_1 r_1 (p_{id} - x_{id}) + c_2 r_2 (p_{gd} - x_{id}) \tag{1}$$

© Springer Nature Singapore Pte Ltd. 2017
K. Deep et al. (eds.), *Proceedings of Sixth International Conference on Soft Computing for Problem Solving*, Advances in Intelligent Systems and Computing 546, DOI 10.1007/978-981-10-3322-3_8

$$x_{id}(new) = x_{id}(old) + v_{id}(new) \qquad (2)$$

where, d = 1, 2, ..., n presents the dimension and i = 1, 2, ..., N represents the particle index, N is the size of the swarm, c_1 and c_2 are called social scaling and cognitive parameters respectively that determines the magnitude of the random force in the direction of particle's previously best visited position (p_{id}) and best particle (p_{gd}) and r_1, r_2 are the uniform random variable between $[0,1]$. The maximum velocity (V_{max}) assists as a constraint to control the position of the swarms within the solution search space.

Further, Shi and Eberhart [4] was developed the concept of an inertia weight (IW) in 1998 to ensure an optimal tradeoff between exploration and exploitation mechanisms of the swarm population. This inertia weight strategy was to be able to eliminate the need of maximum velocity (V_{max}). Inertia weight controls the particles movement by maintaining its previous memory. The velocity update equation is considered as follows:

$$v_{id}(new) = w * v_{id}(old) + c_1 r_1(p_{id} - x_{id}) + c_2 r_2(p_{gd} - x_{id}) \qquad (3)$$

This paper discusses the 30 different inertia weight strategies on 10 benchmark functions for PSO algorithm. A comprehensive review on 30 inertia weight strategies have been presented in next section.

2   A Review on Different Inertia Weight Strategies for PSO

Inertia weight plays an important role in the process of providing a trade-off between diversification and intensification skills of PSO algorithm. When the inertia weight strategy is implemented to PSO algorithm, the particles move around while adjusting their velocities and positions according to Eqs. (1) and (2) in the search space.

In 1998, first time Shi and Eberhart [4] proposed the concept of constant inertia weight. A small inertia weight helps in explore the search space while a large inertia weight facilitates in exploit the search space. Eberhart and Shi [5] proposed a random inertia weight strategy and enhances the performance and efficiency of PSO algorithm.

The linearly decreasing strategy [6] increases the convergence speed of PSO algorithm in early iterations of the search space. The inertia weight starts with some large value and then linearly decreases to some smaller value. The inertia weight provides the excellent results from 0.9 to 0.4. In global-local best inertia weight [7], the inertia weight is based on the global best and local best of the swarms in each generation. It increases the capabilities of PSO algorithm and neither takes a linearly decreasing time-varying value nor a constant value.

Fayek et al. [8] introduced a particle swarm simulated annealing technique (PSOSA). This inertia weight strategy is optimized by using simulated annealing and improves its searching capability.

Chen et al. [9] present two natural exponent inertia weight strategies as e1-PSO and e2-PSO, which are based on the exponentially decreasing the inertia weight. Experimentally, these strategies become a victim of premature convergence, despite its quick convergence speed towards the optimal positions at the early stage of the search process.

Using the merits of chaotic optimization, chaotic inertia weight has been proposed by Feng et al. [10] and PSO algorithm becomes better global search ability, convergence precision and quickly convergence velocity.

Malik et al. [11] presented a sigmoid increasing inertia weight (SIIW) and sigmoid decreasing inertia weight (SDIW). These strategies provide better performance with quick convergence ability and aggressive movement narrowing towards the solution region.

Oscillating Inertia Weight [12] provides a balance between diversification and intensification waves and concludes that this strategy looks to be competitive and, in some cases, better performs in terms of consistency.

Gao et al. [13] proposed a logarithmic decreasing inertia weight with chaos mutation operator. The chaos mutation operator can enhance the ability to jump out the premature convergence and improve its convergence speed and accuracy.

To overcome the stagnation and premature convergence of the PSO algorithm, Gao et al. [14] proposed an exponent decreasing inertia weight (EDIW) with stochastic mutation (SM). The stochastic mutations (SM) is used to enhance the diversity of the swarm while EDIW is used to improve the convergence speed of the individuals (Table 1).

Linearly decreasing inertia weight have been proposed by Shi and Eberhart [4] and greatly improved the accuracy and convergence speed. A large inertia weight facilitates at the inceptive phase of search space while later linearly decreases to a small inertia weight.

Adewumi et al. [25] proposed the swarm success rate random inertia weight (SSRRIW) and swarm success rate descending inertia weight (SSRDIW). These strategies use swarm success rates as a feedback parameter. Further, it enhances the effectiveness of the algorithm regarding convergence speed and global search ability.

Shen et al. [18] proposed the dynamic adaptive inertia weight and used to solve the complex and multi-dimensional function optimization problems. This strategy can timely adjust the particle speed, jump out of a locally optimal solution and improve the convergence speed.

Ting et al. [24] proposed the exponent inertia weight. There exist two important parameters as a local attractor (a) and global attractor (b). This method controls the population diversity by adaptive adjustment of local attractor (a) and global attractor (b).

Chatterjee and Siarry [22] proposed nonlinear decreasing inertia weight strategy with nonlinear modulation index. This strategy is quite effective as well as avoid premature issues. Lei et al. [17] proposed adaptive inertia weight. It furnishes with automatically harmonize global and local search ability and obtained the global optima.

Table 1. Inertia weight strategies

S.No.	Name of inertia weight	Formula of inertia weight
1	Logarithm Decreasing Inertia Weight [13]	$w = w_{max} + (w_{min} - w_{max}).log_{10}(a + \frac{10t}{T})$
2	Exponent Decreasing Inertia Weight [14]	$w = (w_{max} - w_{min} - d_1).exp(\frac{1}{1 + \frac{d_2 t}{T}})$
3	Natural Exponent Inertia Weight Strategy(e2 -PSO) [9]	$w = w_{min} + (w_{max} - w_{min}).e^{-[\frac{t}{(\frac{T}{4})}]^2}$
4	Natural Exponent Inertia Weight Strategy(e1 -PSO) [9]	$w = w_{end} + (w_{start} - w_{end}).e^{[\frac{-t}{(\frac{T}{10})}]}$
5	Global-Local Best Inertia Weight [7]	$w = [1.1 - \frac{gbest_i}{pbest_i}]$
6	Simulated Annealing Inertia Weight [8]	$w = w_{min} + (w_{max} - w_{min}).\lambda^{k-1}$
7	Oscillating Inertia Weight [12]	$w = (\frac{w_{min} + w_{max}}{2} + \frac{w_{max} - w_{min}}{2} cos(\frac{2\Pi t}{T}))$, where $T = \frac{2S_1}{3 + 2k}$
8	Chaotic Random Inertia Weight [10]	$z = 4 * z * (1 - z)$, $w = 0.5 * rand + 0.5 * z$
9	The Chaotic Inertia Weight [10]	$w = (w_{max} - w_{min}) * (\frac{T-t}{T}) + w_{min} * z$, where, $z = 4 * z * (1 - z)$
10	Linear Decreasing Inertia Weight [6]	$w = w_{max} - (w_{max} - w_{min})(\frac{t}{T})$
11	Sigmoid Decreasing Inertia Weight [11]	$w = \frac{(w_{max} - w_{min})}{(1 + e^{-u(k - n*gen)})} + w_{min}$, $u = 10^{log((gen) - 2)}$
12	Sigmoid Increasing Inertia Weight [11]	$w = \frac{(w_{max} - w_{min})}{(1 + e^{u(k - n*gen)})} + w_{min}$, $u = 10^{log((gen) - 2)}$
13	Random Inertia Weight [5]	$w = 0.5 + 0.5 * rand$
14	Constant Inertia Weight [4]	$w = c$, where $c = 0.2$(considered for experiments)
15	Chaotic Adaptive Inertia Weights (CAIWS-D) [15]	$w = [(w_{max} - w_{min})(\frac{T-t}{T}) + w_{min}] * z$, where, $z = 4*SR*(1-SR)$
16	Chaotic Adaptive Inertia Weights (CAIWS-R) [15]	$w = (0.5*SR + 0.5)*z$, where $z = 4*SR*(1-SR)$
17	Decreasing Exponential Function Inertia Weight (DEFIW) [15]	$w = t^{-(\sqrt[t]{t})}$
18	Fixed inertia weight (FIW) [16]	$w = \frac{1}{2ln(2)}$
19	Adaptive Inertia Weight Strategy [17]	$w = [\frac{1 - (\frac{t}{T})}{(1 + S\frac{t}{T})}]$
20	Dynamic Adaptive Inertia Weight [18]	$w = w_{min} + (w_{max} - w_{min})F(t)\Psi(t)$, $\Psi(t) = exp(-\frac{t^2}{(2\sigma^2)})$ and $\sigma = T/3$
21	Decreasing Inertia Weight (DIW) [19]	$w = w_{init} * u^{-t}$
22	Inertia Weight Strategy [20]	$w = \frac{(w_{init} - 0.4)(g_{size} - i)}{(g_{size} + 0.4)}$
23	Double Exponential Dynamic Inertia Weight [2]	$w = exp(-exp(-R))$, where $R = \frac{(T-t)}{T}$
24	Tangent Decreasing Inertia Weight (TDIW) [21]	$w = (w_{max} - w_{min}) * tan(\frac{7}{8}(1 - \frac{t}{T})^k)$
25	Nonlinear Decreasing Inertia Weight (NDIW) [22]	$w = (w_{max} - w_{min})(\frac{T-t}{T})^n + w_{min}$
26	Linear or Non-Linear Decreasing Inertia Weight [23]	$w = (\frac{2}{t})^{0.3}$
27	Exponent Inertia Weight [24]	$w = w_0 e^{-a(\frac{t}{T})}$
28	Swarm Success Rate Random Inertia Weight (SSRRIW) [25]	$w = 0.5 * rand + 0.5 * ssr_{t-1}$
29	Swarm Success Rate Descending Inertia Weight (SSRDIW) [25]	$w = (w_{max} - w_{min})(\frac{T-t}{T}) + w_{min} * ssr_{t-1}$
30	Descending Inertia Weight [25]	$w = w_{min} + (w_{max} - w_{min})(\frac{T-t}{T})$

J. asoc. [23] proposed the linear or non-linear decreasing inertia weight. This strategy has global search ability and also helpful to find a better optimal solution. It overcomes the weakness of premature convergence and converges faster at the early stage of the search process. Jiao et al. [19] proposed the decreasing inertia weight (DIW). This strategy provides the algorithm with dynamic adaptability and controls the population diversity by adaptive adjustment of inertia weight.

Li, L. et al. [21] proposed the tangent decreasing inertia weight (TDIW) based on tangent function (TF). This strategy is to increase the diversity of swarm for more exploration of the search space at initial iterations while later exploit the search area. So that this approach provides better results with accuracy.

Chauhan et al. [2] proposed the double exponential dynamic inertia weight (DEDIW). The inertia weight is calculated for whole swarm iteratively by using gompertz function, and it is capable of providing a stagnation free environment with better accuracy. Peram et al. [20] proposed a new inertia weight that provides the less susceptible to premature convergence and less likely to be stuck in local optima. Sheng-Ta Hsieh et al. [16] introduced fixed inertia weight (FIW). It provides better convergence speed and less computational efforts.

The decreasing exponential function inertia weight (DEFIW) [15] decreases the value of inertia weight iteratively as the algorithm approaches equilibrium state and furnishes the superiority to the competitors in fitness quality.

Arasomwan et al. [15] Proposed chaotic adaptive inertia weights as CAIWS-D and CAIWS-R. These strategies simply combine chaotic mapping with the swarm success rate as a feedback parameter to harness together chaotic and adaptivity characteristics. These approaches provide more refine accuracy, faster convergence speed as well as global search ability.

3   Experimental Results

To evaluate the performance of the inertia weight strategy, it is tested over 10 different benchmark functions (F_1 to F_{10}) as given in Table 2.

3.1   Parameter Settings

Following experimental settings are adopted:

- $G_0 = 100$ and $\alpha = 20$ [26],
- Number of runs = 30,
- Number of populations = 50,
- Maximum number of iterations (T) = 1000,
- Value of c_1 and c_2 are 2.0 [25].

3.2   Results and Discussion

In this section, 30 different inertia weight strategies are analyzed on 10 benchmark problems in terms of average number of function evaluations (AFE's), mean error (ME) and standard deviation (SD). The AFE's, ME and SD are presented in Tables 3, 4 and 5 respectively. Boxplot of AFE's, ME and SD are shown in Figs. 1, 2 and 3 respectively.

Table 2. Test problems, D: Dimensions, AE: Acceptable Error

Test problem	Objective function	Search range	Optimum value	D	AE
Sphere	$f_1(x) = \sum_{i=1}^{D} x_i^2$	[-5.12 5.12]	$f(0) = 0$	30	$1.0E-05$
De Jong f4	$f_2(x) = \sum_{i=1}^{D} i.(x_i)^4$	[-5.12 5.12]	$f(0) = 0$	30	$1.0E-05$
Ackley	$f_3(x) = -20 + e + exp(-\frac{0.2}{D}\sqrt{\sum_{i=1}^{D} x_i^3})$	[-30, 30]	$f(0) = 0$	30	$1.0E-05$
Alpine	$f_4(x) = \sum_{i=1}^{D} |x_i \sin x_i + 0.1x_i|$	[-10, 10]	$f(0) = 0$	30	$1.0E-05$
Michalewicz	$f_5(x) = -\sum_{i=1}^{D} \sin x_i (\sin(\frac{i.x_i^2}{\pi})^{20})$	[0, π]	f_{min}=-9.66015	10	$1.0E-05$
Cosine Mixture	$f_6(x) = \sum_{i=1}^{D} x_i^2 - 0.1(\sum_{i=1}^{D} \cos 5\pi x_i) + 0.1D$	[-1, 1]	$f(0) = -D \times 0.1$	30	$1.0E-05$
Exponential	$f_7(x) = -(exp(-0.5\sum_{i=1}^{D} x_i^2)) + 1$	[-1, 1]	$f(0) = -1$	30	$1.0E-05$
brown3	$f_8(x) = \sum_{i=1}^{D-1}(x_i^{2(x_{i+1})^2+1} + x_{i+1}^{2x_i^2+1})$	[-1 4]	$f(0) = 0$	30	$1.0E-05$
Beale	$f_9(x) = [1.5 - x_1(1-x_2)]^2 + [2.25 - x_1(1-x_2^2)]^2 + [2.625 - x_1(1-x_2^3)]^2$	[-4.5,4.5]	$f(3,0.5) = 0$	2	$1.0E-05$
Colville	$f_{10}(x) = 100[x_2 - x_1^2]^2 + (1-x_1)^2 + 90(x_4 - x_3^2)^2 + (1-x_3)^2 + 10.1[(x_2-1)^2 + (x_4-1)^2] + 19.8(x_2-1)(x_4-1)$	[-10,10]	$f(1) = 0$	4	$1.0E-05$

It is clear from the reported results that most of the Inertia weight strategies produce poor results in case of michalewicz function (F_5). It clear from Fig. 1 that constant inertia weight and linearly decreasing inertia weight (LDIW) is best and worst strategy respectively in terms of AFE's. It is observed from Fig. 2 that the mean error taken by chaotic random inertia weight strategy and global local best inertia weight strategy are minimum and maximum in terms of mean error respectively compared to the other inertia weight strategies.

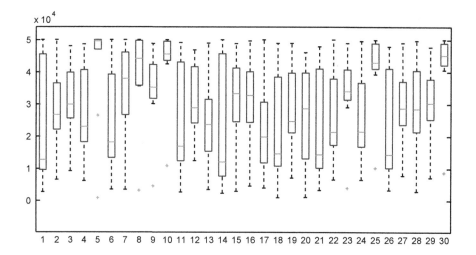

Fig. 1. Boxplots for average number of function evaluations of 30 different Inertia Weight strategies on 10 benchmark functions as per Table 3

Table 3. Average number of function evaluations of different inertia weight strategies for different benchmark functions

Inertia Weight	Sphere	De Jong f4	Ackley	Alpine	Michalewicz	Cosine Mixture	Exponential	brown3	Beale	Colville
1	09851.67	08540.00	15218.33	45575.00	49731.67	50100.00	9611.67	10270.00	2771.667	26260.00
2	22313.33	19670.00	32858.33	30181.67	50100.00	42166.67	22125.00	23351.67	6585.00	36578.33
3	25801.66	24028.33	35703.33	33166.67	48105.00	41483.33	25651.67	26728.33	09263.33	39870.00
4	18846.67	16835.00	29221.67	26516.67	48733.33	40691.67	18248.33	19665.00	06175.00	40780.00
5	50100.00	49196.67	50100.00	26536.67	48840.00	50100.00	50100.00	50100.00	00885.00	47108.33
6	13663.33	11920.00	23591.67	21495.00	50100.00	44928.33	13378.33	14915.00	3565.00	39290.00
7	29278.33	23780.00	50100.00	46190.00	47988.33	45698.33	26725.00	32083.333	3526.67	43920.00
8	39623.33	35648.33	50100.00	50100.00	47638.33	49888.33	35870.00	41895.00	3230.00	46575.00
9	32738.33	30196.67	42308.33	39525.00	48946.67	42490.00	31791.67	33640.00	04591.67	36876.67
10	44853.33	42535.00	50100.00	50096.67	48795.00	49615.00	43596.67	45206.67	10890.00	46073.33
11	13218.33	11451.67	23440.00	19735.00	49196.67	46841.67	12383.33	14135.00	02711.67	43118.33
12	25103.33	23176.67	34655.00	32036.67	46930.00	46618.33	24230.00	25821.67	12451.67	41675.00
13	15840.00	13016.67	29701.67	29538.33	49053.33	48191.67	15398.33	17851.67	03503.33	31575.00
14	8433.33	7330.00	18216.67	15083.33	49228.33	50100.00	07688.33	08975.00	02385.00	45735.00
15	26308.33	23300.00	47571.67	38541.67	48951.67	39000.00	24606.67	28170.00	2968.33	41273.33
16	26348.33	22995	47791.67	37485.00	49750.00	39408.33	24390.00	28135.00	4603.33	40158.33
17	11821.67	14613.33	28691.67	25045.00	50100.00	30740.00	11070.00	12343.33	4011.67	49971.67
18	11116.67	10255.00	20148.33	17286.67	49011.67	47786.67	10798.33	12016.67	01056.67	38490.00
19	21593.33	20201.67	28485.00	27230.00	49003.33	42706.67	21200.00	22236.67	07161.67	39731.67
20	13108.33	11238.33	44071.67	20541.67	46168.33	39801.67	36930.00	14003.33	1121.67	38113.33
21	10585.00	09818.33	17485.00	17256.67	47970.00	47661.67	10198.33	11496.67	03293.33	41001.67
22	18141.67	15730.00	26410.00	23940.00	50100.00	37963.33	17363.33	18841.67	06580.00	45256.67
23	32145	29100	44275.00	40775.00	49000.00	37770.00	31460.00	33501.67	03953.33	34638.33
24	17571.67	15713.33	26353.33	24056.67	49713.33	36716.67	16858.33	19021.67	06556.67	41691.67
25	42213.33	39358.33	50041.67	48980.00	49168.33	47785.00	41083.33	42838.33	10196.67	43158.33
26	10585.00	9818.33	17485.00	17256.67	47970.00	47661.67	10198.33	11496.67	03293.33	41131.67
27	24753.33	22180.00	35250.00	31966.67	49148.33	38595.00	23735.00	25720.00	07821.67	37138.33
28	23265.00	19488.33	40460.00	33003.33	49821.67	35823.33	21410.00	24258.33	02921.67	45111.67
29	26200.00	24011.67	36238.33	33400.00	47765.00	37798.33	25305.00	27360.00	07193.33	38271.67
30	45070.00	42221.67	50100.00	50085.00	49010.00	49023.33	43833.33	45480.00	08773.33	40638.33

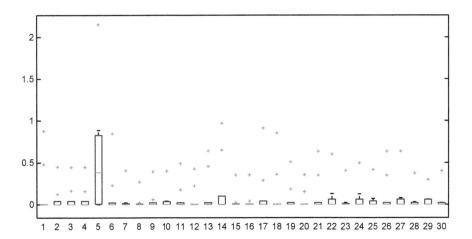

Fig. 2. Mean error value of 30 different Inertia Weight strategies on 10 benchmark functions as per Table 4

Table 4. Mean error value of different inertia weight strategies for different benchmark functions

Inertia Weight	Sphere	De Jong f4	Ackley	Alpine	Michalewicz	Cosine Mixture	Exponential	brown3	Beale	Colville
1	8.90E-06	8.29E-06	5.03E-04	9.14E-04	8.79E-01	4.81E-01	8.40E-06	8.66E-06	4.09E-06	1.50E-03
2	9.15E-06	8.91E-06	9.59E-06	9.57E-06	4.52E-01	1.23E-01	9.23E-06	9.16E-06	3.99E-02	1.57E-03
3	9.32E-06	8.97E-06	9.55E-06	9.64E-06	4.47E-01	1.63E-01	9.17E-06	9.13E-06	3.99E-02	1.34E-03
4	9.28E-06	8.74E-06	9.50E-06	9.54E-06	4.48E-01	1.58E-01	9.28E-06	9.18E-06	3.99E-02	1.22E-03
5	4.54E-01	3.14E-01	5.21E-01	1.15E-04	8.86E-01	2.15E+00	2.19E-01	8.26E-01	5.55E-06	4.84E-03
6	9.08E-06	9.13E-06	9.52E-06	9.25E-06	8.45E-01	2.27E-01	9.15E-06	9.24E-06	2.00E-02	1.18E-03
7	9.04E-06	8.78E-06	3.96E-05	1.37E-05	4.07E-01	1.08E-02	8.77E-06	9.36E-06	2.00E-02	6.33E-03
8	9.44E-06	8.87E-06	4.51E-04	5.51E-05	2.67E-01	5.41E-03	9.33E-06	9.51E-06	2.00E-02	5.13E-03
9	9.25E-06	8.97E-06	9.61E-06	9.13E-06	3.92E-01	5.91E-02	9.28E-06	9.19E-06	2.00E-02	1.84E-03
10	9.26E-06	8.96E-06	1.78E-04	6.26E-05	4.00E-01	3.45E-02	9.30E-06	9.43E-06	3.99E-02	4.95E-03
11	9.15E-06	8.91E-06	9.67E-06	9.56E-06	4.90E-01	1.76E-01	9.20E-06	9.15E-06	2.00E-02	3.64E-03
12	9.22E-06	8.61E-06	9.58E-06	9.40E-06	4.26E-01	2.22E-01	9.11E-06	9.30E-06	4.57E-06	1.54E-03
13	9.17E-06	9.26E-06	9.58E-06	8.69E-06	6.41E-01	4.58E-01	9.13E-06	9.05E-06	2.00E-02	9.58E-04
14	8.96E-06	8.77E-06	1.00E-01	8.07E-06	9.71E-01	6.50E-01	9.23E-06	8.98E-06	2.00E-02	2.55E-03
15	9.13E-06	8.74E-06	1.27E-05	9.31E-06	3.52E-01	4.94E-03	9.14E-06	9.54E-06	2.00E-02	3.33E-03
16	9.43E-06	9.05E-06	1.39E-05	9.62E-06	3.55E-01	9.43E-06	9.26E-06	9.44E-06	3.99E-02	2.86E-03
17	8.52E-06	8.88E-06	8.69E-06	5.52E-04	9.13E-01	4.93E-03	8.08E-06	8.46E-06	3.99E-02	2.88E-01
18	9.06E-06	8.69E-06	9.59E-06	9.31E-06	8.54E-01	3.60E-01	8.94E-06	9.28E-06	5.66E-06	1.32E-03
19	9.12E-06	8.56E-06	9.46E-06	9.36E-06	5.09E-01	1.82E-01	8.96E-06	8.90E-06	2.00E-02	2.06E-03
20	9.32E-06	9.22E-06	9.61E-06	9.38E-06	3.54E-01	1.53E-01	9.18E-06	9.23E-06	4.88E-06	1.09E-03
21	8.84E-06	8.84E-06	9.54E-06	7.78E-06	6.34E-01	3.50E-01	8.64E-06	9.01E-06	2.00E-02	4.69E-03
22	8.98E-06	9.09E-06	9.62E-06	9.38E-06	6.00E-01	1.28E-01	9.12E-06	9.13E-06	5.99E-02	2.13E-03
23	9.32E-06	9.19E-06	9.68E-06	9.52E-06	4.05E-01	9.86E-03	9.26E-06	9.35E-06	2.00E-02	2.02E-03
24	9.24E-06	9.12E-06	9.61E-06	9.63E-06	4.94E-01	1.23E-01	9.19E-06	9.01E-06	5.99E-02	1.94E-03
25	9.11E-06	8.62E-06	3.85E-05	1.29E-05	4.16E-01	6.90E-02	9.33E-06	9.35E-06	2.00E-02	2.93E-03
26	8.84E-06	8.84E-06	9.54E-06	7.78E-06	6.34E-01	3.50E-01	8.64E-06	9.01E-06	2.00E-02	2.01E-03
27	9.17E-06	9.04E-06	9.54E-06	9.50E-06	6.36E-01	7.88E-02	9.49E-06	9.05E-06	5.99E-02	1.18E-03
28	9.26E-06	9.23E-06	9.61E-06	9.67E-06	3.73E-01	2.96E-02	9.43E-06	9.21E-06	2.00E-02	3.46E-03
29	9.27E-06	8.97E-06	9.49E-06	9.65E-06	2.95E-01	5.91E-02	9.27E-06	9.29E-06	5.99E-02	1.54E-03
30	9.31E-06	9.07E-06	1.93E-04	4.63E-05	4.02E-01	1.97E-02	9.52E-06	8.96E-06	2.00E-02	3.62E-03

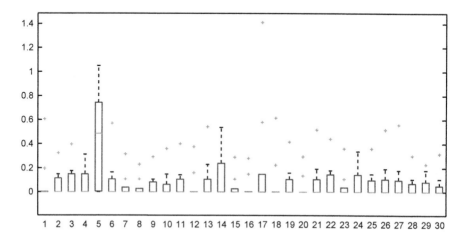

Fig. 3. Standard Deviation value of 30 different Inertia Weight strategies on 10 benchmark functions as per Table 5

Table 5. Standard deviation value of different inertia weight strategies for different benchmark functions

Inertia Weight	Sphere	De Jong f4	Ackley	Alpine	Michalewicz	Cosine Mixture	Exponential	brown3	Beale	Colville
1	1.54E-06	9.56E-07	2.66E-03	2.05E-03	6.06E-01	1.96E-01	1.86E-06	1.38E-06	2.38E-06	1.53E-03
2	8.26E-07	1.08E-06	3.58E-07	3.59E-07	3.24E-01	1.15E-01	6.84E-07	8.14E-07	1.49E-01	1.10E-03
3	6.19E-07	9.40E-07	3.49E-07	2.86E-07	3.96E-01	1.76E-01	7.40E-07	6.89E-07	1.49E-01	8.07E-04
4	6.45E-07	9.15E-07	4.10E-07	3.84E-07	3.13E-01	1.62E-01	5.22E-07	7.74E-07	1.49E-01	3.59E-04
5	4.34E-01	1.05E+00	5.36E-01	4.54E-04	6.88E-01	1.00E+00	1.64E-01	7.43E-01	3.02E-06	4.00E-03
6	8.16E-07	1.19E-06	3.28E-07	7.43E-07	5.70E-01	1.65E-01	5.50E-07	7.21E-07	1.07E-01	7.74E-04
7	7.26E-07	1.17E-06	2.10E-05	9.87E-06	3.15E-01	3.70E-02	1.09E-06	4.36E-07	1.07E-01	6.96E-03
8	5.53E-07	1.19E-06	2.74E-04	3.25E-01	2.29E-01	2.65E-02	6.58E-07	6.79E-07	1.07E-01	4.58E-03
9	6.61E-07	7.83E-07	3.54E-07	1.10E-06	2.92E-01	8.18E-02	4.59E-07	6.97E-07	1.07E-01	1.23E-03
10	6.61E-07	7.48E-07	1.23E-04	8.33E-05	3.60E-01	6.25E-02	6.82E-07	3.49E-07	1.49E-01	3.99E-03
11	7.83E-07	1.07E-06	4.02E-07	3.87E-07	4.01E-01	1.43E-01	5.79E-07	8.31E-07	1.07E-01	4.99E-03
12	5.26E-07	1.20E-06	4.27E-07	5.60E-07	3.77E-01	1.61E-01	6.72E-07	6.50E-07	3.30E-06	8.53E-04
13	8.54E-07	6.95E-07	4.37E-07	1.86E-06	5.43E-01	2.30E-01	8.55E-07	7.03E-07	1.07E-01	7.45E-05
14	1.02E-06	1.37E-06	3.75E-01	2.89E-06	5.38E-01	2.40E-01	4.98E-07	1.17E-06	1.07E-01	2.14E-03
15	8.31E-07	9.18E-07	5.80E-06	5.50E-07	2.90E-01	2.65E-02	5.92E-07	5.28E-07	1.07E-01	3.27E-03
16	6.17E-07	9.55E-07	1.22E-05	3.57E-07	2.81E-01	6.24E-07	6.58E-07	5.21E-07	1.07E-01	2.54E-03
17	1.74E-06	1.50E-06	1.63E-06	2.31E-03	5.80E-01	2.65E-02	2.04E-06	1.77E-06	1.49E-01	1.41E+00
18	9.20E-07	1.10E-06	5.99E-07	6.01E-07	6.15E-01	2.25E-01	8.30E-07	5.78E-07	2.85E-06	6.36E-04
19	7.09E-07	1.23E-06	5.04E-07	6.34E-07	4.16E-01	1.60E-01	9.73E-07	1.15E-06	1.07E-01	1.85E-03
20	6.24E-07	9.46E-07	3.55E-07	5.39E-07	2.94E-01	1.35E-01	5.31E-07	5.89E-07	2.98E-06	3.50E-04
21	8.07E-07	8.84E-07	3.55E-07	3.06E-06	5.18E-01	1.93E-01	1.36E-06	8.22E-07	1.07E-01	4.41E-03
22	7.53E-07	7.83E-07	4.40E-07	5.57E-07	4.41E-01	1.46E-01	8.90E-07	7.62E-07	1.80E-01	1.56E-03
23	6.08E-07	9.61E-07	3.55E-07	4.06E-07	3.60E-01	3.69E-02	6.05E-07	5.29E-07	1.07E-01	2.07E-03
24	6.71E-07	1.13E-06	3.58E-07	3.92E-07	3.38E-01	1.43E-01	6.37E-07	9.25E-07	1.80E-01	1.19E-03
25	9.37E-07	1.17E-06	2.93E-05	7.08E-06	3.60E-01	9.90E-02	4.63E-07	5.80E-07	1.49E-01	2.84E-03
26	8.07E-07	8.84E-07	3.55E-07	3.06E-06	5.18E-01	1.93E-01	1.36E-06	8.22E-07	1.07E-01	1.25E-03
27	6.81E-07	7.45E-07	4.11E-07	4.33E-07	5.58E-01	9.90E-02	4.22E-07	9.90E-02	1.80E-01	4.49E-04
28	6.41E-07	7.43E-07	3.61E-07	3.08E-07	3.01E-01	7.04E-02	5.76E-07	6.86E-07	1.07E-01	2.49E-03
29	6.65E-07	8.10E-07	3.12E-07	3.45E-07	2.30E-01	8.18E-02	5.80E-07	7.32E-07	1.80E-01	1.03E-03
30	5.71E-07	8.76E-07	9.32E-05	3.83E-05	3.18E-01	5.02E-02	3.12E-07	7.68E-07	1.07E-01	4.22E-03

If the comparison is made through standard divisions (SD's) the chaotic random inertia weight produces near optimal solutions in comparison to other inertia weight strategies as shown in Fig. 3. The summary results of inertia weight strategies are shown in Table 6.

Table 6. Summary Results for Inertia Weight

Criterion	Best inertia weight strategy	Worst inertia weight strategy
Average Function Evaluation	Constant Inertia Weight	Linear Decreasing Inertia Weight
Mean Error	Chaotic Random Inertia Weight	Global-Local Best Inertia Weight
Standard Deviation	Chaotic Random Inertia Weight	Global-Local Best Inertia Weight

4 Conclusion

This paper presents the significance of inertia weight strategies in the solution search process of particle swarm optimization (PSO). Here, total 30 inertia weight strategies in PSO are analyzed in terms of efficiency, reliability and robustness while testing over 10 complex test functions. Through boxplots and success rate, it is found that the chaotic random inertia weight is better in terms of accuracy while constant inertia weight performs better in terms of efficiency of PSO among the considered inertia weight strategies.

References

1. Eberhart, R.C., Kennedy, J. et al.: A new optimizer using particle swarm theory. In: Proceedings of the sixth international symposium on micro machine and human science, vol. 1, pp. 39–43. New York (1995)
2. Chauhan, P., Deep, K., Pant, M.: Novel inertia weight strategies for particle swarm optimization. Memetic Comput. 5(3), 229–251 (2013)
3. Bansal, J.C., Singh, P.K., Saraswat, M., Verma, A., Jadon, S.S., Abraham, A.: Inertia weight strategies in particle swarm optimization. In: 2011 Third World Congress on Nature and Biologically Inspired Computing (NaBIC), pp. 633–640. IEEE (2011)
4. Shi, Y., Eberhart, R.: A modified particle swarm optimizer. In: The 1998 IEEE International Conference on Evolutionary Computation Proceedings, IEEE World Congress on Computational Intelligence, pp. 69–73. IEEE (1998)
5. Eberhart, R.C., Shi, Y.: Tracking and optimizing dynamic systems with particle swarms. In: Proceedings of the 2001 Congress on Evolutionary Computation, vol. 1, pp. 94–100. IEEE (2001)
6. Xin, J., Chen, G., Hai, Y.: A particle swarm optimizer with multi-stage linearly-decreasing inertia weight. In: International Joint Conference on Computational Sciences and Optimization, CSO 2009, vol. 1, pp. 505–508. IEEE (2009)
7. Arumugam, M.S., Rao, M.V.C.: On the performance of the particle swarm optimization algorithm with various inertia weight variants for computing optimal control of a class of hybrid systems. Discrete Dyn. Nat. Soc. 2006, 15 (2006)
8. Al-Hassan, W., Fayek, M.B., Shaheen, S.I.: PSOSA: an optimized particle swarm technique for solving the urban planning problem. In: The 2006 International Conference on Computer Engineering and Systems, pp. 401–405. IEEE (2006)
9. Chen, G., Huang, X., Jia, J., Min, Z.: Natural exponential inertia weight strategy in particle swarm optimization. In: The Sixth World Congress on Intelligent Control and Automation, WCICA 2006, vol. 1, pp. 3672–3675. IEEE (2006)
10. Feng, Y., Teng, G-F., Wang, A-X., Yao, Y-M.: Chaotic inertia weight in particle swarm optimization. In: Second International Conference on Innovative Computing, Information and Control, ICICIC 2007, p. 475. IEEE (2007)
11. Malik, R.F., Rahman, T.A., Hashim, S.Z.M., Ngah, R.: New particle swarm optimizer with sigmoid increasing inertia weight. Int. J. Comput. Sci. Secur. 1(2), 35–44 (2007)
12. Kentzoglanakis, K., Poole, M.: Particle swarm optimization with an oscillating inertia weight. In: Proceedings of the 11th Annual conference on Genetic and evolutionary computation, pp. 1749–1750. ACM (2009)
13. Gao, Y-l., An, X-h., Liu, J-m.: A particle swarm optimization algorithm with logarithm decreasing inertia weight and chaos mutation. In: International Conference on Computational Intelligence and Security, CIS 2008, vol. 1, pp. 61–65. IEEE (2008)
14. Li, H-R., Gao, Y-L.: Particle swarm optimization algorithm with exponent decreasing inertia weight and stochastic mutation. In: Second International Conference on Information and Computing Science, ICIC 2009, vol. 1, pp. 66–69. IEEE (2009)

15. Arasomwan, M.A., Adewumi, A.O.: On adaptive chaotic inertia weights in particle swarm optimization. In: 2013 IEEE Symposium on Swarm Intelligence (SIS), pp. 72–79. IEEE (2013)

16. Hsieh, S.-T., Sun, T.-Y., Liu, C.-C., Tsai, S.-J.: Efficient population utilization strategy for particle swarm optimizer. IEEE Trans. Syst. Man Cybern. Part B Cybern. **39**(2), 444–456 (2009)

17. Lei, K., Qiu, Y., He, Y.: A new adaptive well-chosen inertia weight strategy to automatically harmonize global and local search ability in particle swarm optimization. In: 1st International Symposium on Systems and Control in Aerospace and Astronautics, ISSCAA 2006, p. 4. IEEE (2006)

18. Shen, X., Chi, Z., Yang, J., Chen, C.: Particle swarm optimization with dynamic adaptive inertia weight. In: 2010 International Conference on Challenges in Environmental Science and Computer Engineering (CESCE), vol. 1, pp. 287–290. IEEE (2010)

19. Jiao, B., Lian, Z., Xingsheng, G.: A dynamic inertia weight particle swarm optimization algorithm. Chaos Solitons Fractals **37**(3), 698–705 (2008)

20. Peram, T., Veeramachaneni, K., Mohan, C.K.: Fitness-distance-ratio based particle swarm optimization. In: Proceedings of the 2003 IEEE on Swarm Intelligence Symposium, SIS 2003, pp. 174–181. IEEE (2003)

21. Li, L., Xue, B., Niu, B., Tan, L., Wang, J.: A novel particle swarm optimization with non-linear inertia weight based on tangent function. In: Huang, D.-S., Jo, K.-H., Lee, H.-H., Kang, H.-J., Bevilacqua, V. (eds.) ICIC 2009. LNCS (LNAI), vol. 5755, pp. 785–793. Springer, Heidelberg (2009). doi:10.1007/978-3-642-04020-7_84

22. Chatterjee, A., Siarry, P.: Nonlinear inertia weight variation for dynamic adaptation in particle swarm optimization. Comput. Oper. Res. **33**(3), 859–871 (2006)

23. Fan, S.-K.S., Chiu, Y.-Y.: A decreasing inertia weight particle swarm optimizer. Eng. Optim. **39**(2), 203–228 (2007)

24. Ting, T,O., Shi, Yuhui, Cheng, Shi, Lee, Sanghyuk: Exponential inertia weight for particle swarm optimization. In: Tan, Ying, Shi, Yuhui, Ji, Zhen (eds.) ICSI 2012. LNCS, vol. 7331, pp. 83–90. Springer, Heidelberg (2012). doi:10.1007/978-3-642-30976-2_10

25. Adewumi, A.O., Arasomwan, A.M.: An improved particle swarm optimiser based on swarm success rate for global optimisation problems. J. Exp. Theor. Artif. Intell. **28**, 441–483 (2016)

26. Rashedi, E., Nezamabadi-Pour, H., Saryazdi, S.: GSA: a gravitational search algorithm. Inf. Sci. **179**(13), 2232–2248 (2009)

Hybridized Gravitational Search Algorithms with Real Coded Genetic Algorithms for Integer and Mixed Integer Optimization Problems

Amarjeet Singh[(✉)] and Kusum Deep

Department of Mathematics,
Indian Institute of Technology Roorkee,
Roorkee 247667, Uttarakhand, India
amarjeetiitr@gmail.com,
kusumdeep@gmail.com

Abstract. In this paper, the Gravitational Search Algorithm (GSA) is hybridized with real coded Genetic Algorithm to solve Integer and Mixed Integer programming problems. The idea is based on two earlier papers of the authors. In the first paper, the authors proposed a methodology in which the Laplace Crossover and Power Mutation were embedded in Gravitational Search Algorithm and in the second paper, these algorithms were extended for the case of constrained optimization problems. In order to deal with integer variables, a special method is adopted. For dealing with the constraints the Deb's technique is implemented. The original GSA and three new variants are tested on a set of benchmark problems available in literature. Based on the extensive numerical and graphical analysis of results it is concluded that one of the proposed variants outperform the original GSA and the other proposed variants.

Keywords: Gravitational search algorithm · Constrained optimization problems · Integer and mixed integer programming problems · Laplace crossover · Power mutation

1 Introduction

In many real life optimization problems, decision variables are desired to take integral values. Mathematical models of these problems are known as integer programming problems (IPPs) or mixed integer programming problems (MIPPs) depending upon whether all or only some of the decision variables are to have integral values. Such problems frequently arise in various applications such as finance, plant operation, process flow sheet, chemical engineering, automobile engineering, VLSI manufacturing, optimal design of gas and water distribution networks, scheduling [1–3] etc.

© Springer Nature Singapore Pte Ltd. 2017
K. Deep et al. (eds.), *Proceedings of Sixth International Conference on Soft Computing for Problem Solving*, Advances in Intelligent Systems and Computing 546, DOI 10.1007/978-981-10-3322-3_9

A mixed integer optimization problem is usually written as:

Minimize $f(x, y)$,
Subject to

$$g_j(x, y) \leq 0, \; j = 1, \ldots, J$$
$$h_k(x, y) = 0, \; k = 1, \ldots, K$$
$$x_l^i \leq x^i \leq x_u^i, \; i = 1, 2, \ldots, m_1.$$
$$y_l^i \leq y^i \leq y_u^i : \text{ integer}, \; i = 1, 2, \ldots, m_2.$$
$$x = (x^1, x^2, \ldots, x^{m_1})$$
$$y = (y^1, y^2, \ldots, y^{m_2}).$$

(1)

where x represents a vector of continuous variables, y represents a vector of integer variables, (x_i^l, x_i^u) are lower and upper bounds of x_i and (y_i^l, y_i^u) are lower and upper bounds of y_i.

In literature several classical techniques are available to handle such problems (such as branch and bound techniques, cutting plane technique relaxation technique etc.), but these techniques are applicable to a restricted class of problems.

Since the last few decades, many heuristic algorithms are developed and used to solve mixed integer optimization problems. Mohan and Nguyen [4] extended the concept of controlled random search technique algorithm and proposed RST2ANU algorithm for solving integer and mixed integer global optimization problems. Differential Evolution [5], Line-up competition algorithm [6], Particle Swarm Optimization [7] are also extended and used for integer and mixed integer programing problems. Deep and Thakur [8, 9] proposed LXPM algorithm for global optimization. Later on [1] Deep et al. modified and extended it and proposed MI-LXPM algorithm for integer and mixed integer programing problems. Costa and Oliveria [10] presented a work in which a comparison is made between an algorithm based on simulated annealing (M-SIMPSA) and two evolutionary algorithms: Genetic Algorithms and Evolutionary Strategies using a test bed of seven mixed integer non-linear programing problems. Hong and Zhang [11] developed a discrete hybrid Differential Evolution (DHDE) algorithm to solve global numerical optimization problems with discrete variables. Zhu and Fan [12] proposed an algorithm to find a global minimizer of the box constrained nonlinear integer programming problem. It minimizes the auxiliary function from random initial points and this auxiliary function has the same discrete global minimizers as the problem. Zhua and Ali [13] proposed an algorithm based on minimizing the auxiliary function for constrained nonlinear integer programming problems and auxiliary function is constructed based on a penalty function. Misra and Sharma [14] proposed Misra Integer Programing (MIP) to solve integer programming problems arising in the system reliability design. Tan et al. [15] combined a novel chaotic local search in genetic algorithm and proposed MSCLSGA. Zhou et al. [16] used Quesada Grossmann (QG) and Tabu search (TS) algorithm to solve mixed integer nonlinear programming (MINLP) in a heterogeneous parallel structure simultaneously. In [17], Ten et al. hybridized particle swarm optimization with a novel chaotic search and proposed CLSPSO to solve integer and mixed integer programming problems. Newby and Ali [18] developed a trust-region-based derivative free algorithm for solving bound constrained mixed integer nonlinear programs. Fahim and Hedar [19] proposed a hybrid scatter search method for solving unconstrained or constrained integer programming problems. Jun et al. [20] proposed an improved Differential Evolution algorithm for constrained mixed integer programming problems. Gao et al. [21] proposed a modified Differential Evolution algorithm for constrained nonlinear mixed integer programing problems in which the positions of variation particles are self-adaptively adjusted. Lin et al. [22] developed a hybrid Differential Evolution (MIHDE) deal with the mixed-integer optimization problems in which individuals clustering is avoided by the migration operation. Yokota et al. [23] proposed a method for solving non-linear mixed integer programming (NMIP) problems using genetic algorithm (GAs). Schlueter and Munetomo [24] discussed two different parallelization strategies of evolutionary algorithms for mixed integer nonlinear programming (MINLP). In first strategy, some internal parts of the evolutionary algorithm is parallelized and in second strategy the MINLP function calls outside and independently of the evolutionary algorithm is parallelized. Omran and Engelbrecht [25] investigated the performance of two variants of Differential Evolution (DE), Self-adaptive DE and DE on integer programing problems.

Gravitational Search Algorithm (GSA) [26] is a population based stochastic search algorithm which mimic the Newton law of gravity. The concept of GSA was introduced by Rashedi. Initially, it is used to solve unconstrained and constrained continuous global optimization problems. Singh and Deep [27, 28] hybridized GSA with real coded genetic algorithm operators namely Laplace Crossover and Power mutation and proposed LXGSA, PMGSA, LXPMGSA variants for unconstrained continuous optimization problems and in [29] same variants are extended for constrained optimization problems. The major contribution of this research paper is two-folds. Firstly, it proposes a MI-GSA which is extension of the GSA for Integer and Mixed Integer Programming Problems. Secondly, the extension of LXGSA, PMGSA and LXPMGSA proposed by the authors earlier are further extended for the case of Integer and Mixed Integers Optimization Problems. A comparison analysis of all these four versions is performed on a set of test benchmarks problems available in literature.

This paper is organized as follows: In Sect. 2, Gravitational Search Algorithm is described. Constrained handling mechanism is described in Sect. 3. In Sect. 4, modified Laplace Crossover, modified Power Mutation and the three proposed variants of GSA for Integer and Mixed Integer variables are described. In Sect. 5, numerical results are analyzed and finally in Sect. 6, the conclusion are drawn.

2 Gravitational Search Algorithm

Gravitational Search Algorithm (GSA) [26, 30, 31] is a newly developed Nature Inspired Algorithm based on the metaphor of gravity and mass interactions. In this algorithm, the solution of the problem is represented by the position of the particle at the specified dimension and quality of the solution is represented by the mass of the particle, higher the mass better solution. Each iteration of GSA passes through three steps: (a) Initialization (b) Force calculation (c) Motion. Consider a system of N particle in which the position of ith particle is represented by

$$X_i = \left(x_i^1, x_i^2, \ldots, x_i^d, \ldots, x_i^m\right) \quad \text{for} \quad i = 1, 2, \ldots, N \tag{2}$$

where x_i^d is the position of ith particle in d^{th} direction.

In Initialization, a population of N particles is generated randomly in the search space. The velocities of each particle are initialized to zero (it could be non-zero if desired). Fitness value is evaluated using the objective function $fit(X_i)$. The position of the best particle at step t is denoted by X_{best}.

In Force calculation, first mass of each particle is calculated using a function of fitness of particle i.e. $M_i = g(f(X_i))$. Where $M_i \in (0, 1]$ and $g(.)$ is bounded and monotonically decreasing. The function g is defined in such a way that the best particle has the largest value (normalized) and worst particle has smallest value. Thus, after evaluating the current population fitness, the gravitational mass and inertia mass of each particle are calculated as follows:

$$M_{ai} = M_{pi} = M_{ii} = M_i \tag{3}$$

$$m_i = \frac{fit_i(t) - worst(t)}{best(t) - worst(t)}, \quad i = 1, 2, \ldots, N. \tag{4}$$

$$M_i(t) = \frac{m_i(t)}{\sum_{j=1}^{N} m_j(t)} \tag{5}$$

where M_{ai} is the active gravitational mass, M_{pi} is the passive gravitational mass, M_{ii} is the inertia mass of particle i, $fit_i(t)$ is the fitness value of the ith particle at time t. Also, $best(t)$ and $worst(t)$ are the best and worst particle with regard to their fitness value.

For minimization problem

$$best(t) = \min_{j \in \{1,2,\ldots,N\}} fit_j(t) \quad \text{and} \quad worst(t) = \max_{j \in \{1,2,\ldots,N\}} fit_j(t) \tag{6}$$

For maximization problem

$$best(t) = \max_{j \in \{1,2,\ldots,N\}} fit_j(t) \quad \text{and} \quad worst(t) = \min_{j \in \{1,2,\ldots,N\}} fit_j(t) \tag{7}$$

Then the force acting on mass 'i' from 'j' is evaluated by

$$F_{ij}^d(t) = G(t) \frac{M_{pi}(t) \times M_{aj}(t)}{R_{ij}(t) + \varepsilon} \left(x_j^d(t) - x_i^d(t)\right) \tag{8}$$

where $M_{aj}(t)$ is the active gravitational mass related to particle j, $M_{pi}(t)$ is the passive gravitational mass related to particle i. ε is a small value. $G(t)$ is the gravitational constant and it is calculated by

$$G(t) = G_0 \exp(-\alpha t/\max\_iter) \qquad (9)$$

$R_{ij}(t)$ is the Euclidean distance between i and j particles and it is defined as follows:

$$R_{ij}(t) = \left\| X_i(t), X_j(t) \right\|_2 \qquad (10)$$

The total force acting on i^{th} particle in dimension d is calculated by

$$F_i^d(t) = \sum_{j \in Kbest, j \neq i} rand_j F_{ij}^d(t) \qquad (11)$$

where $rand_j$ is randomly distributed random number in interval $(0, 1]$, Kbest is the set of first k particles with the best fitness value and k is a decreasing function with time. Initially k is set to the number of particles in the system and it decreases linearly in such a way that at the last iteration $k = 1$.

In Motion, first acceleration of each particle is calculated by

$$a_i^d(t) = \frac{F_i^d(t)}{M_i(t)} \qquad (12)$$

where $a_i^d(t)$ is the acceleration of particle i in the dimension d at time t. Then the velocities and next position of particles i in the dth dimension are updated by

$$v_i^d(t+1) = rand_i \times v_i^d(t) + a_i^d(t) \qquad (13)$$

$$x_i^d(t+1) = x_i^d(t) + v_i^d(t+1) \qquad (14)$$

where $rand_i$ is randomly distributed random number in the interval $(0, 1]$.

3 Constraint Handling Mechanism

To overcome the shortcoming of general penalty function methods, Deb [32] proposed an efficient penalty parameter free constraint handling technique. In this method, a penalty term is added to the objective function to penalize infeasible solutions. Therefore the fitness function of (1) is defined as

$$fit_i = \begin{cases} f(X_i), & \text{if } X_i \text{ is feasible} \\ f_{worst} + \sum_{j=1}^{m} |\phi_j(X_i)|, & \text{otherwise} \end{cases} \qquad (15)$$

where f_{worst} is the objective function value of the worst feasible solution available in the current population and $\phi_j(X_i)$ refers to the amount of constraint violation of ith particle in j^{th} inequality constraint. In this method the fitness of feasible solution is fixed to its objective function value and the fitness of infeasible solution depends on amount of constraint violation and worst feasible solution of the current population. If all solution are infeasible in the population then f_{worst} is set to zero. Hence two solutions are compared by the following rules:

1. A feasible solution is always preferred over an infeasible solution.
2. Between two feasible solutions, the one having a better objective function value is preferred.
3. Between two infeasible solutions, the one having the smaller constraint violation is preferred.

Hence infeasible solutions are pushed towards the feasible region.
Equality constraints are transformed into inequalities of the form

$$|h_k(x,y)| - \varepsilon \leq 0, \;\; for \; k = 1, \ldots, K \tag{16}$$

A solution (x, y) is regarded as feasible if $g_j(x, y) \leq 0$ for $j = 1, \ldots, J$ and $|h_k(x,y)| - \varepsilon \leq 0$, $for \; k = 1, \ldots, K$. Here, ε is set to 0.0001.

4 Proposed Algorithms

In Singh and Deep [27], GSA is hybridized with Laplace Crossover [8] which is a real coded genetic algorithm crossover operator and with Power Mutation [9] which is a real coded genetic algorithm mutation operator. The resulting three hybridized version namely LXGSA, PMGSA and LXPMGSA are compared with the original GSA, with an objective to increase the efficiency and reliability of the original GSA. In the present study, GSA, LXGSA, PMGSA and LXPMGSA are extended to solve integer and mixed integer constrained optimization problems and four variants are proposed namely MI-GSA, MI-LXGSA, MI-PMGSA and MI-LXPMGSA. Before proposing the algorithm it is important to note how the Laplace Crossover and Power Mutation have to be modified to deal with integer values.

4.1 Modified Laplace Crossover

Originally, Laplace Crossover (LX) is defied by Deep and Thakur [8] based on Laplace distribution but to handle integer and mixed integer problems Deep et al. [1] modified it. The working of the modified LX is described below. A pair of offspring $y_1 = (y_1^1, y_1^2, \ldots, y_1^m)$ and $y_2 = (y_2^1, y_2^2, \ldots, y_2^m)$ is generated from a pair of parents $x_1 = (x_1^1, x_1^2, \ldots, x_1^m)$ and $x_2 = (x_2^1, x_2^2, \ldots, x_2^m)$ in the following way. First, two uniformly distributed random numbers $r_i, s_i \in [0, 1]$ are generated and a random number β_i, following Laplace distribution, is generated as:

$$\beta_i = \begin{cases} a - b \log_e(r_i), & s_i \leq 0.5 \\ a + b \log_e(r_i), & s_i > 0.5 \end{cases} \tag{17}$$

where a is location parameter and $b > 0$ is scaling parameter. b is different for integer and real decision variables. It is $b = b_{int}$ for integer decision variable otherwise $b = b_{real}$.

Then offspring is created by the equations:

$$\begin{aligned} y_1^i &= x_1^i + \beta_i |x_1^i - x_2^i|, \\ y_2^i &= x_2^i + \beta_i |x_1^i - x_2^i|, \end{aligned} \tag{18}$$

Let x_{lower}^i and x_{upper}^i to be the lower and upper bounds of the unknown variables x^i. If $x^i < x_{lower}^i$ or $x^i > x_{upper}^i$ for some i, then x^i is assigned a random value in the interval $[x_{lower}^i, x_{upper}^i]$.

4.2   Modified Power Mutation

Power Mutation (PM) operator introduced by Deep and Thakur [9] is based on power distribution. Later on it is modified by Deep et al. [1] to handle integer and mixed integer problems. PM operator creates a solution $y = (y^1, y^2, \ldots, y^m)$ in the vicinity of a parent solution $\bar{x} = (\bar{x}^1, \bar{x}^2, \ldots, \bar{x}^m)$ as follows. First, a uniformly distributed random number $r \in [0, 1]$ is generated. Then a random number w^i following power distribution, is generated by $w^i = r^p$, where p is the index of distribution. $p = p_{real}$ or $p = p_{int}$ depending on integer or real restriction on the decision variable. Offspring y is created by the formula:

$$y^i = \begin{cases} \bar{x}^i - w^i(\bar{x}^i - x_{lower}), & if\ t < v^i \\ \bar{x}^i + w^i(x_{upper} - \bar{x}^i), & if\ t \geq v^i \end{cases}, \ for\ i = 1, 2, \ldots, m \tag{19}$$

where $v^i \in [0, 1]$ is a uniformly distributed random number, $t = \frac{\bar{x}^i - x_{lower}}{x_{upper} - x_{lower}}$ and x_{lower} and x_{upper} are lower and upper bounds of decision variables.

If a decision variable x^i has integer restrictions, then it uses $[x^i]$ or $[x^i] + 1$.

The working procedure of MI-GSA, MI-LXGSA, MI-PMGSA and MI-LXPMGSA are as follows:

4.3   The Proposed MI-GSA

MI-GSA is a modified version of original GSA for integer and mixed integer constrained optimization problems. In the working procedure of MI-GSA, the fitness of particles is evaluated using Deb's rule during the initial generation of random population as well as during each iteration and Lbest is updated if (i) Lbest is infeasible and best(t) is feasible (ii) both are feasible or infeasible but best(t) is better in term of fitness (Fig. 3).

4.4   Proposed MI-LXGSA

Each iteration of MI-LXGSA is carried out as follows. Firstly, the steps of MI-GSA are performed. Then Laplace Crossover is applied on Lbest and a randomly selected particle as shown in Fig. 1 and Lbest is updated as per Fig. 3 Then the iteration is incremented.

Select parents x_1 and x_2 where x_1 is Lbest and x_2 is a randomly selected particle
Obtain offspring be y_1 and y_2 by Laplace
Crossover using eq. (18)
Apply integer retraction if applicable on y_1^i and y_2^i, for i=1,2,...,m
Determine worst particle of the population
If fit(y_1) < fit(worst)
        {   worst $\rightarrow y_1$
                update worst
                update Lbest
        }

If fit(y_2) < fit(worst)
        {   worst $\rightarrow y_2$
                update Lbest
        }

Fig. 1. Implementation of Laplace Crossover

4.5   Proposed MI-PMGSA

Each iteration of MI-PMGSA is carried out as follows. Firstly, the steps of MI-GSA are performed. Then Power Mutation is applied on Lbest as shown in Fig. 2. Lbest is updated as per Fig. 3. Then the iteration is incremented.

Select parent x where x is Lbest
Obtain offspring be y by Power Mutation using eq. (19)
Apply integer retraction if applicable on y^i, for i=1,2,...,m
Determine worst particle of the population
If fit(y) <fit(worst)
        {   worst $\rightarrow$ y
                update Lbest
        }

Fig. 2. Implementation of Power Mutation

```
If  Lbest and best(t) is infeasible and best(t)< Lbest
       {      Lbest→ best(t)
       }
elseif Lbest is infesible and best(t) is feasible
       {      Lbest→ best(t)
       }
elseif Lbest is fesible and best(t) is feasible and best(t)< Lbest
       {      Lbest→ best(t)
       }
```

Fig. 3. Lbest Updating Procedure

4.6 Proposed MI-LXPMGSA

Each iteration of MI-LXPMGSA is carried out as follows. Firstly, the steps of MI-GSA are performed. Then Laplace Crossover is applied on Lbest and a randomly selected particle as shown in Fig. 1. Lbest is updated as per Fig. 3. Then, Power Mutation is applied on Lbest as shown in Fig. 2. Lbest is updated as per Fig. 3. Then the iteration is incremented. The pseudo of the algorithm is given Fig. 4.

5 Experimental Results and Benchmark Functions

The performance of the proposed algorithms is tested on 18 problems, taken from literature, arising in chemical engineering, reliability, etc. [1, 4, 33] and are listed in Appendix A. Problem set consists of linear and nonlinear constraints integer and mixed integer problems and the number of unknown decision variables varies from 2 to 100. Problems F_1 to F_{13} are minimization problem and F_{14} to F_{18} are maximization problems. The environment for running computer programs of the experiments is processor: Intel(R) Xeon(R) CPU E5645 @ 2.40 GHz 2.39 GHz RAM: 12 GB, Operating System Window, Integrated Development Environment: MATLAB 2008. The parameters of the algorithms are $G_0 = 100$, $\alpha = 20$, $a = 0, b_{real} = 0.15, b_{int} = 0.35, p_{real} = 10, p_{int} = 4$,(as in literature) and population size = 50.

In the present study, two types of experiments are performed. Since the global optima, reported in Appendix A, is based on literature wherein these problems are solved by heuristic techniques. Hence, it is hoped that optima may improve by proposed algorithms. Therefore, the first experiment is carried out with stopping criteria: maximum iterations = 5000. Second experiment is carried out to observe that out of the four algorithms which is more reliable, robust and efficient. In this experiment, the stopping criteria is either algorithm reaches 5000 iterations or absolute error is less than 0.01, if the optimal value of function is zero or relative error is less than 0.01, if the optimal value of function is non-zero. Each problem is executed for 100 runs by all four algorithms (MI-GSA, MI-LXGSA, MI-PMGSA and MI-LXPMGSA), making sure that the first randomly generated population is used for first run of all algorithms, second randomly generated population is used for second run of all algorithms, etc. A run is considered a success if achieved value of the objective function is within 1% of the

---

Set number of particle N, problem dimension m,
parameter value G_0, α, maximum iteration = max_iter

Deploy N particles randomly in the search space

Let $x_i(t) = \left(x_i^1(t), \ldots, x_i^d(t), \ldots, x_i^m(t) \right)$ be the position of

particle i at iteration t

Evaluate fitness fit_i of all particles using eq. (16)

$best(t) = \min\limits_{j \in \{1,\ldots,N\}} fit(x_j(t)), worst(t) = \max\limits_{j \in \{1,\ldots,N\}} fit(x_j(t)),$

Set Lbest= best(t);

t=0

while (t $\leq$ max_iter) **do:**

  { $G(t) = G_0 exp(-\alpha t / max\_iter)$

    $best(t) = \min\limits_{j \in \{1,\ldots,N\}} fit(x_j(t)), worst(t) = \max\limits_{j \in \{1,\ldots,N\}} fit(x_j(t)),$ msum=0;

  for i =1 to N

    { $m_i(t) = \dfrac{fit(x_i(t)) - worst(t)}{best(t) - worst(t)}$; msum=msum + $m_i(t)$; }

  for i =1 to N

    { $M_i(t) = m_i(t)/\text{msum}$; }

  for each particle i = 1 to N **do:**

    { **for** d = 1 to m **do:**

      { $F_i^d(t) = \sum\limits_{j \in kbest, j \neq i} rand_j\, G(t) \dfrac{M_{pi}(t) \times M_{qi}(t)}{R_{ij}(t) + \varepsilon} \left(x_j^d(t) - x_i^d(t) \right)$

      $a_i^d(t) = F_i^d(t)/M_{ii}(t)$

      $v_i^d(t+1) = rand_i\, v_i^d(t) + a_i^d(t)$

      $x_i^d(t+1) = x_i^d(t) + v_i^d(t+1)$

Apply integer restriction if applicable on $x_i^d(t+1)$

      }

    }

Evaluate fitness fit_i of all particles using eq. (16)

$best(t) = \min\limits_{j \in \{1,\ldots,N\}} fit(x_j(t)), worst(t) = \max\limits_{j \in \{1,\ldots,N\}} fit(x_j(t)),$

Update Lbest as per Fig 3

Apply Laplace Crossover as per Fig 1 ⎫ //for MI-LXGSA ⎫
Update Lbest as per Fig 3   ⎬   ⎬ //MI-LXPMGSA
Apply Power Mutation as per Fig 2 ⎫ //for MI-PMGSA ⎭
Update Lbest as per Fig 3   ⎭

  }

t=t+1

}

Fig. 4. Pseudo code of algorithms

known optimal value (in case the optimal value of the objective is zero, a run is considered success if the achieved absolute value of objective function is less than 0.01).

5.1   Analysis of Experiment I

In order to compare the quality of the solutions found by proposed algorithms Best, Worst, Average, Median and standard deviation (STD) of objective function values are calculated for experiment I and shown in Table 1 and the best entries are shown using boldface. From the Table 1, it is concluded that the performance of MI-GSA, MI-PMGSA, MI-LXGSA and MI-LXPMGSA are approximately same on F4, F5, F6, F14 and F15. All four algorithms get optimal solution of these functions. MI-PMGSA, MI-LXGSA and MI-LXPMGSA solve the functions F1, F2 and F11 with 100%, but MI-GSA could not solve them with 100% success. All algorithms get equal Best and equal worst on function F3, F7. As per average value of F3 the performance order of algorithms is MI-LXGSA > MI-GSA > MI-LXPMGSA > MI-PMGSA but based on the median, performance order is MI-LXPMGSA = MI-LXGSA > MI-PMGSA > MI-GSA and based on average value of F7 the performance order is MI-LXPMGSA = LXGSA > MI-GSA > MI-PMGSA but MI-GSA has worst median in comparison to others. On F8, the best solution is found by MI-GSA and MI-LXGSA and best median and best average are found by MI-LXGSA.

Algorithms have equal Best and equal Median on F9, F10 and F13. As per average value, performance order on F9 is MI-LXGSA > MI-LXPMGSA > MI-PMGSA > MI-GSA, on F10 is MI-LXPMGSA = MI-PMGSA > MI-LXGSA > MI-GSA and on F13 is MI-LXPMGSA > MI-GSA > MI-LXGSA > MI-PMGSA. The performance order of algorithms on F12 is MI-LXGSA > MI-PMGSA > MI-LXPMGSA > MI-GSA. On F16, performance of MI-LXPMGSA and MI-PMGSA are same and it is better than MI-LXGSA and MI-GSA. On F17, performance order is MI-LXPMGSA > MI-PMGSA > MI-LXGSA > MI-GSA. On F18, best solution and Median, obtained by MI-LXPMGSA and MI-LXGSA is same but MI-PMGSA and MI-GSA get a slightly worse solution in comparison to others and based on average value performance order is MI-LXPMGSA > MI-LXGSA > MI-PMGSA > MI-GSA. From the Table 1, it is concluded that the overall performance of MI-LXPMGSA is better than other proposed algorithms and on functions F16 and F17, the best optima found is better than the literature [1, 4, 33].

To study the convergence behaviour towards optimal solution of MI-GSA, MI-LXGSA, MI-PMGSA and MI-LXPMGSA, iteration wise convergence graphs are plotted and shown in Figs. 5 and 6 for F1 to F13 (minimization problems) and Fig. 7 for F15 to F18 (maximization problems). On the horizontal axis the iterations are shown, whereas on the vertical axis the average best-so-far is shown. Average best-so-far is the average value of objective function in each iteration over 100 runs. From these convergence plots it is concluded that MI-LXPMGSA has faster convergence rate on all problems except F3 and F9. On F3 and F9, MI-LXGSA has better convergence rate.

5.2   Analysis of Experiment II

For the experiment II percentage of success, average number of function evaluation of successful run and average time in seconds of successful run are shown in Tables 2, 3 and 4, respectively and best entries are shown by boldface.

Table 1. Results of experiment I showing Best, Worst, Average, Median and STD of objective function values

Pro.	Algorithm	Best	Worst	Median	Average	STD
F1	MI-GSA	2.00000774	2.19600954	2.00229751	2.00714121	0.02305478
	MI-LXGSA	2	2	2	2	2.2407E-12
	MI-PMGSA	2	2	2	2	**2.3617E-16**
	MI-LXPMGSA	2	2	2	2	3.0761E-16
F2	MI-GSA	**2.12446758**	2.21534922	2.12663433	2.13120049	0.01207187
	MI-LXGSA	**2.12446758**	**2.12446758**	**2.12446758**	**2.12446758**	**9.9763E-13**
	MI-PMGSA	**2.12446758**	**2.12446758**	**2.12446758**	**2.12446758**	1.7042E-11
	MI-LXPMGSA	**2.12446758**	**2.12446758**	**2.12446758**	**2.12446758**	3.3709E-12
F3	MI-GSA	**1.07654308**	**1.25**	1.08893722	1.10468264	0.04533669
	MI-LXGSA	**1.07654308**	**1.25**	**1.07654308**	**1.08521593**	**0.03799451**
	MI-PMGSA	**1.07654308**	**1.25**	1.07654309	1.12858016	0.07988839
	MI-LXPMGSA	**1.07654308**	**1.25**	**1.07654308**	1.11817274	0.07445373
F4	MI-GSA	**−6961.81388**	**−6961.81386**	**−6961.81387**	**−6961.81387**	3.7835E-06
	MI-LXGSA	**−6961.81388**	**−6961.81386**	**−6961.81387**	**−6961.81387**	2.7931E-06
	MI-PMGSA	**−6961.81388**	**−6961.81386**	**−6961.81387**	**−6961.81387**	3.1469E-06
	MI-LXPMGSA	**−6961.81388**	**−6961.81386**	**−6961.81387**	**−6961.81387**	**2.4882E-06**
F5	MI-GSA	**−68**	**−68**	**−68**	**−68**	**0**
	MI-LXGSA	**−68**	**−68**	**−68**	**−68**	**0**
	MI-PMGSA	**−68**	**−68**	**−68**	**−68**	**0**
	MI-LXPMGSA	**−68**	**−68**	**−68**	**−68**	**0**
F6	MI-GSA	**−6**	**−6**	**−6**	**−6**	**0**
	MI-LXGSA	**−6**	**−6**	**−6**	**−6**	**0**
	MI-PMGSA	**−6**	**−6**	**−6**	**−6**	**0**
	MI-LXPMGSA	**−6**	**−6**	**−6**	**−6**	**0**
F7	MI-GSA	**99.23963505**	**103.0236383**	99.23974429	99.91642038	1.44059308
	MI-LXGSA	**99.23963505**	**103.0236383**	**99.23963505**	**99.61803538**	**1.14091991**
	MI-PMGSA	**99.23963505**	**103.0236383**	**99.23963505**	99.92075564	1.46109039
	MI-LXPMGSA	**99.23963505**	**103.0236383**	**99.23963505**	**99.61803538**	**1.14091991**
F8	MI-GSA	**3.557461258**	3.896223807	3.557461664	3.564684354	0.04770295
	MI-LXGSA	**3.557461258**	**3.557468743**	**3.557461258**	**3.557461694**	**1.4281E-06**
	MI-PMGSA	3.557485108	3.89626245	3.557819894	3.562148961	0.03401905
	MI-LXPMGSA	3.557461673	3.579585973	3.557686563	3.559477061	0.00596375
F9	MI-GSA	**8**	**18**	**14**	13.17	**3.07171850**
	MI-LXGSA	**8**	**18**	**14**	**11.68**	3.47248928
	MI-PMGSA	**8**	**18**	**14**	12.6	3.30900211
	MI-LXPMGSA	**8**	**18**	**14**	12.39	3.94634468
F10	MI-GSA	14	20	14	14.87	1.64319841
	MI-LXGSA	14	20	14	14.49	1.48047904
	MI-PMGSA	14	20	14	**14.34**	1.20788654
	MI-LXPMGSA	14	18	14	**14.34**	**1.10298035**
F11	MI-GSA	**−42.63212056**	−38.86466472	**−42.63212056**	−42.48142232	0.74198681
	MI-LXGSA	**−42.63212056**	**−42.63212056**	**−42.63212056**	**−42.63212056**	**8.5695E−14**
	MI-PMGSA	**−42.63212056**	**−42.63212056**	**−42.63212056**	**−42.63212056**	**8.5695E−14**
	MI-LXPMGSA	**−42.63212056**	**−42.63212056**	**−42.63212056**	**−42.63212056**	**8.5695E−14**

(continued)

Table 1. (*continued*)

Pro.	Algorithm	Best	Worst	Median	Average	STD
F12	MI-GSA	8.67552E-07	0.002136916	0.000374392	0.000432375	0.00037064
	MI-LXGSA	**1.44308E-29**	**0.000334796**	9.14037E-07	**1.15206E-05**	**4.237E-05**
	MI-PMGSA	3.43878E-29	0.001167474	**2.79266E-08**	1.36187E-05	0.00011734
	MI-LXPMGSA	9.09685E-29	0.001167474	6.09026E-08	1.47824E-05	0.00011746
F13	MI-GSA	**807**	**1778**	**892**	883.16	**104.033864**
	MI-LXGSA	**807**	2388	**892**	895.17	220.530859
	MI-PMGSA	**807**	2388	**892**	899.71	242.73997
	MI-LXPMGSA	**807**	2162	**892**	**873.75**	143.251166
F14	MI-GSA	**32217.42778**	**32217.42778**	**32217.42778**	**32217.42778**	**1.8282E-11**
	MI-LXGSA	**32217.42778**	**32217.42778**	**32217.42778**	**32217.42778**	**1.8282E-11**
	MI-PMGSA	**32217.42778**	**32217.42778**	**32217.42778**	**32217.42778**	**1.8282E-11**
	MI-LXPMGSA	**32217.42778**	**32217.42778**	**32217.42778**	**32217.42778**	**1.8282E-11**
F15	MI-GSA	**0.9434705**	**0.9434705**	**0.9434705**	**0.9434705**	**0**
	MI-LXGSA	**0.9434705**	**0.9434705**	**0.9434705**	**0.9434705**	**0**
	MI-PMGSA	**0.9434705**	**0.9434705**	**0.9434705**	**0.9434705**	**0**
	MI-LXPMGSA	**0.9434705**	**0.9434705**	**0.9434705**	**0.9434705**	**0**
F16	MI-GSA	1352439	1350118	1352439	1352193.49	439.943167
	MI-LXGSA	1352439	1351252	1352439	1352300.63	304.498716
	MI-PMGSA	**1352439**	**1352439**	**1352439**	**1352439**	**0**
	MI-LXPMGSA	**1352439**	**1352439**	**1352439**	**1352439**	**0**
F17	MI-GSA	304152552	304078794	304145249.5	304143673	7690.61132
	MI-LXGSA	304153193	304140258	304149757.5	304149509.5	**2171.65575**
	MI-PMGSA	304159895	304132587	304156147.5	304154380.2	4437.1472
	MI-LXPMGSA	**304160077**	**304148020**	**304157276.5**	**304156854**	2183.32289
F18	MI-GSA	0.99995467	0.99655553	0.99978734	0.99953678	0.00053881
	MI-LXGSA	**0.99995468**	0.99966102	**0.99994069**	0.99992311	5.044E-05
	MI-PMGSA	0.99995465	0.99962064	0.99993714	0.99991699	5.0978E-05
	MI-LXPMGSA	**0.99995468**	**0.99980700**	**0.99994069**	**0.99992920**	**3.0466E-05**

In these tables best entries are shown using boldface. From the Table 2 it is concluded that out of 18 problems considered, MI-GSA solves 10 problems with 100% success, MI-LXGSA solves 13 problems with 100% success, MI-PMGSA solves 12 problems with 100% success and MI-LXPMGSA solves 13 problems with 100% success.

Further, for a comparative study of these algorithms, Performance Index is plotted as follows:

A specified weighted importance is given to the success rate, number of function evaluations and computational time corresponding to each of the algorithms considered.

The value of Performance Index PI_j for j^{th} algorithm is evaluated by:

$$PI_j = \frac{1}{N} \sum_{i=1}^{N} \left(w_1 \alpha_1^i + w_2 \alpha_2^i + w_3 \alpha_3^i \right) \qquad (20)$$

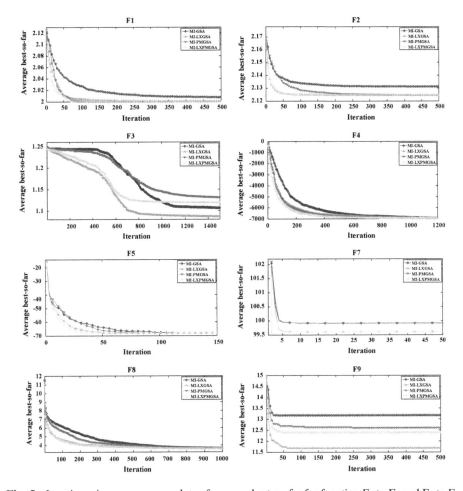

Fig. 5. Iteration wise convergence plots of average best-so-far for function F_1 to F_5 and F_7 to F_9

$$\text{where} \quad \alpha_1^i = \frac{Sr^i}{Tr^i} \tag{21}$$

$$\alpha_2^i = \begin{cases} \dfrac{Mf^i}{Af^i} & \text{if } Sr^i > 0 \\ 0 & \text{if } Sr^i = 0 \end{cases} \tag{22}$$

$$\alpha_3^i = \begin{cases} \dfrac{Mt^i}{At^i} & \text{if } Sr^i > 0 \\ 0 & \text{if } Sr^i = 0 \end{cases} \tag{23}$$

here, N is the total number of considered problems and $i = 1, \ldots, N$. Tr^i represents the total number of times the problem i is solved and Sr^i is the number of times problem i is solved successfully. Af^i is the average number of function evaluations used by

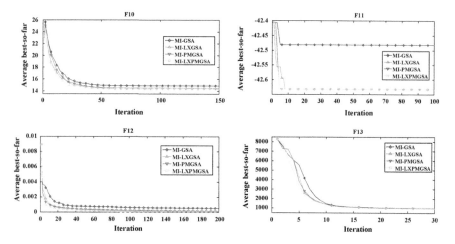

Fig. 6. Iteration wise convergence plots of average best-so-far for function F_{10} to F_{13}

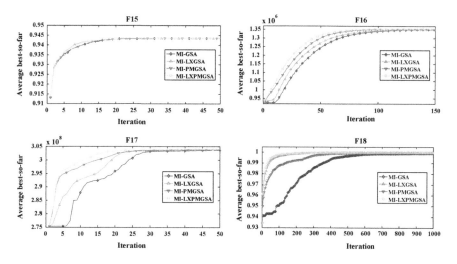

Fig. 7. Iteration wise convergence plots of average best-so-far for function F_{15} to F_{18}

algorithm j to obtain the optimal solution of problem i in case of successful runs, and Mf^i is the minimum of the average number of function evaluation of successful run. Similarly, At^i is average time required by algorithm j to obtain the optimal solution of problem i in case of successful runs, and Mt^i is minimum of the average time by all the algorithms under comparison to obtain the optimal solution of problem i. Further w_1, w_2 and w_3 are nonnegative assigned weight to the percentage of success, average number of function evaluations used in successful run and the average execution time of successful runs respectively with $w_1 + w_2 + w_3 = 1$. Algorithm having largest PI is the winner, amongst the considered algorithms. In order to analyze the relative performance of MI-GSA, MI-LXGSA, MI-PMGSA and MI-LXPMGSA, equal weights

Table 2. Percentage of success obtained for all variants considered

Problem	MI-GSA	MI-LXGSA	MI-PMGSA	MI-LXPMGSA
F1	92	**100**	**100**	**100**
F2	93	**100**	**100**	**100**
F3	52	**89**	79	72
F4	**100**	**100**	**100**	**100**
F5	**100**	**100**	**100**	**100**
F6	**100**	**100**	**100**	**100**
F7	81	**90**	75	82
F8	99	**100**	99	**100**
F9	23	**51**	38	47
F10	63	**87**	82	85
F11	**100**	**100**	**100**	**100**
F12	**100**	**100**	**100**	**100**
F13	26	33	36	**48**
F14	**100**	**100**	**100**	**100**
F15	**100**	**100**	**100**	**100**
F16	**100**	**100**	**100**	**100**
F17	**100**	**100**	**100**	**100**
F18	**100**	**100**	**100**	**100**

Table 3. Average function evaluations of successful runs by all variants considered

Problem	MI-GSA	MI-LXGSA	MI-PMGSA	MI-LXPMGSA
F1	4711	1442	1444	**899**
F2	875	354	1042	**301**
F3	57135	36857	60599	**35084**
F4	46797	42632	35980	**31075**
F5	1882	**899**	1253	931
F6	**52**	**52**	**52**	**52**
F7	**143**	144	157	145
F8	57521	43842	50371	**42435**
F9	**50**	486	819	498
F10	1014	**842**	1005	880
F11	2550	**51**	53	**51**
F12	101	82	84	**70**
F13	14892	13225	15383	**13037**
F14	**100**	101	**100**	102
F15	**164**	170	181	188
F16	4924	4607	3577	**3407**
F17	1151	926	561	**511**
F18	14579	1756	4881	**1447**

Table 4. Average execution time (sec.) of successful runs by all variants considered

Problem	MI-GSA	MI-LXGSA	MI-PMGSA	MI-LXPMGSA
F1	0.6186	0.1849	0.1877	**0.1131**
F2	0.1132	0.0417	0.1382	**0.0349**
F3	7.3553	4.8375	7.7205	**4.6344**
F4	6.1147	5.5524	4.7740	**4.1169**
F5	0.2693	**0.1232**	0.1758	0.1276
F6	0.0018	0.0018	**0.0017**	**0.0017**
F7	**0.0173**	**0.0173**	0.0196	0.0175
F8	8.2834	6.3904	7.3320	**6.1998**
F9	**0.0018**	0.0670	0.1177	0.0689
F10	0.1672	**0.1365**	0.1651	0.1426
F11	0.2775	**0.0017**	0.0021	**0.0017**
F12	0.0122	0.0087	0.0091	**0.0066**
F13	2.2749	1.9994	2.3330	**1.9693**
F14	**0.0112**	0.0114	0.0113	0.0115
F15	**0.0212**	0.0220	0.0241	0.0250
F16	0.8417	0.8140	0.6334	**0.6240**
F17	0.3932	0.3171	0.1914	**0.1741**
F18	2.4496	0.2920	0.8232	**0.2413**

are assigned to two terms (w_1, w_2 and w_3) at a time. Therefore PI_j becomes a function of single variable. Following three cases are possible

$$
\begin{aligned}
(i)\ & w_1 = w,\ w_2 = w_3 = (1 - w)/2;\ 0 \leq w \leq 1 \\
(ii)\ & w_2 = w,\ w_1 = w_3 = (1 - w)/2;\ 0 \leq w \leq 1 \\
(ii)\ & w_3 = w,\ w_1 = w_2 = (1 - w)/2;\ 0 \leq w \leq 1
\end{aligned}
\qquad (24)
$$

Figures 8, 9, 10 shows the Performance Index graphs corresponding to each of these three cases on 18 integer and mixed integer problems. Figure 8 corresponds to the weight assigned for success rate w is varied. Figure 9 corresponds to weight assigned

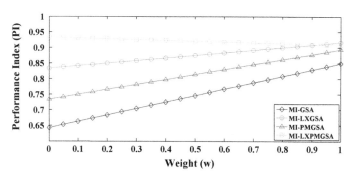

Fig. 8. Performance Index of MI-GSA, MI-LXGSA, MI-PMGSA and MI-LXPMGSA when percentage of success is varied

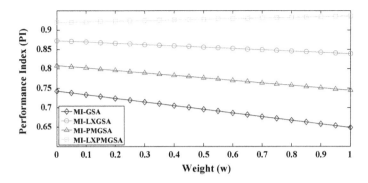

Fig. 9. Performance Index of MI-GSA, MI-LXGSA, MI-PMGSA and MI-LXPMGSA when average function evaluations of successful runs is varied

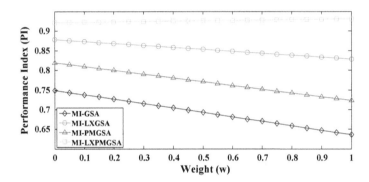

Fig. 10. Performance Index of MI-GSA, MI-LXGSA, MI-PMGSA and MI-LXPMGSA when average computational time of successful run is varied

for average function evaluations w is varied and Fig. 10 corresponds to the weight assigned for average time of the successful runs w is varied. It is clear from the figures that MI-LXPMGSA is best among them.

6 Conclusion

This paper presents, three hybridized variants of Gravitational Search Algorithm with Real Coded Genetic Algorithm operators for solving Integer and Mixed Integer Programming Problems. The idea has been extended from Singh and Deep [27] which proposes three hybridized variants for the unconstrained continuous optimization problems and Singh and Deep [29] which propose three hybridized variants for constrained optimization case. As earlier the three new variants are compared with the similar version of GSA based on a set of benchmarks problems taken from literature. A number of numerical and graphical analysis is performed and it is concluded that the best performer is the variant in which the Laplace Crossover and Power Mutation are incorporated into Gravitational Search Algorithm.

Acknowledgment. The first author would like to thank Council for Scientific and Industrial Research (CSIR), New Delhi, India, for providing him the financial support vide grant number 09/143(0824)/2012-EMR-I.

Appendix A: Benchmark Functions

Problem 1:

$$\text{Min } f(x, y) = 2x + y$$

$$\text{Subject to} : 1.25 - x^2 - y \leq 0$$
$$x + y \leq 1.6$$
$$0 \leq x \leq 1.6$$
$$y \in \{0, 1\}$$

The global optima is $(x, y; f) = (0.5, 1; 2)$.

Problem 2:

$$\text{Min } f(x, y) = -y + 2x - \ln(x/2)$$

$$\text{Subject to} : -x - \ln(x/2) + y \leq 0,$$
$$0.5 \leq x \leq 1.5,$$
$$y \in \{0, 1\}.$$

The global optima is $(x, y; f) = (1.375, 1; 2.124)$.

Problem 3:

$$\text{Min } f(x, y) = -0.7y + 5(x_1 - 0.5)^2 + 0.8$$

$$\text{Subject to} : -\exp(x_1 - 0.2) - x_2 \leq 0,$$
$$x_2 + 1.1y \leq -1.0,$$
$$x_1 - 1.2y \leq 0.2,$$
$$0.2 \leq x_1 \leq 1.0,$$
$$-2.22554 \leq x_2 \leq -1.0,$$
$$y \in \{0, 1\}.$$

The global optima is $(x_1, x_2, y; f) = (0.94194, -2.1, 1; 1.07654)$.

Problem 4:

$$\text{Min } f(x) = (x_1 - 10)^3 + (x_2 - 20)^3$$

$$\text{Subject to} : (x_1 - 5)^2 + (x_2 - 5)^2 - 100 \geq 0.0,$$
$$- (x_1 - 6)^2 - (x_2 - 5)^2 + 82.81 \geq 0.0,$$
$$13 \leq x_1 \leq 100,$$
$$0 \leq x_2 \leq 100.$$

The global optima is $(x_1, x_2; f) = (14.095, 0.84296; -6961.81381)$.

Problem 5:

$$\text{Min } f(x) = x_1^2 + x_1 x_2 + 2x_2^2 - 6x_1 - 2x_2 - 12x_3,$$

$$\text{Subject to} : 2x_1^2 + x_2^2 \leq 15.0,$$
$$- x_1 + 2x_2 + x_3 \leq 3.0,$$
$$0 \leq x_i \leq 10, \text{ integer } i = 1, \ldots, 3.$$

The global optima is $(x_1, x_2, x_3; f) = (2, 0, 5; -68)$.

Problem 6:

$$\text{Min } f(x) = (x_1 + 2x_2 + 3x_3 - x_4)(2x_1 + 5x_2 + 3x_3 - 6x_4),$$

$$\text{Subject to} : x_1 + 2x_2 + x_3 + 3x_4 \geq 4.0,$$
$$x \in \{0, 1\}^4$$

The global optima is $(x_1, x_2, x_3, x_4; f) = (0, 0, 1, 1; -6)$.

Problem 7:

$$\text{Min } f(y_1, v_1, v_2) = 7.5y_1 + 5.5(1 - y_1) + 7v_1 + 6v_2$$
$$+ 50 \frac{y_1}{0.9[1 - \exp(-0.5v_1)]}$$
$$+ 50 \frac{1 - y_1}{0.8[1 - \exp(-0.4v_2)]}$$

$$\text{Subject to} : 0.9[1 - \exp(-0.5v_1)] - 2y_1 \leq 0,$$
$$0.8[1 - \exp(-0.4v_2)] - 2(1 - y_1) \leq 0,$$
$$v_1 \leq 10y_1,$$
$$v_2 \leq 10(1 - y_1),$$
$$v_1, v_2 \geq 0,$$
$$y_1 \in \{0, 1\}.$$

The objective is to select one between two candidate reactors in order to minimize the production cost. The global optima is $(y_1, v_1, v_2; f) = (1, 3.514237, 0; 99.239635)$

Problem 8:

$$\text{Min } f(x, y) = (y_1 - 1)^2 + (y_2 - 1)^2 + (y_3 - 1)^2 - \ln(y_4 + 1)$$
$$+ (x_1 - 1)^2 + (x_2 - 2)^2 + (x_3 - 3)^2$$

Subject to:

$$y_1 + y_2 + y_3 + x_1 + x_2 + x_3 \le 5.0,$$
$$y_3^2 + x_1^2 + x_2^2 + x_3^2 \le 5.5,$$
$$y_1 + x_1 \le 1.2,$$
$$y_2 + x_2 \le 1.8,$$
$$y_3 + x_3 \le 2.5,$$
$$y_4 + x_1 \le 1.2,$$
$$y_2^2 + x_2^2 \le 1.64,$$
$$y_3^2 + x_3^2 \le 4.25,$$
$$y_2^2 + x_3^2 \le 4.64,$$
$$x_1, x_2, x_3 \ge 0,$$
$$y_1, y_2, y_3, y_4 \in \{0, 1\}.$$

The global optima is $(x_1, x_2, x_3, y_1, y_2, y_3, y_4; f) = (0.2, 1.280624, 1.954483, 1, 0, 0, 1; 3.557463)$. Our algorithm achieves solution $(x_1, x_2, x_3, y_1, y_2, y_3, y_4; f) = (0.084607, 0.798719, 2.116424, 1, 1, 0, 1; 3.3685783)$.

Problem 9:

$$\text{Min } f(x) = x_1^2 + x_2^2 + x_3^2 + x_4^2 + x_5^2,$$

Subject to : $x_1 + 2x_2 + x_4 \ge 4.0,$
$$x_2 + 2x_3 \ge 3.0,$$
$$x_1 + 2x_5 \ge 5.0,$$
$$x_1 + 2x_2 + 2x_3 \le 6.0,$$
$$2x_1 + x_3 \le 4.0,$$
$$x_1 + 4x_5 \le 13.0,$$
$$0 \le x_i \le 3, \ i = 1, 2, \ldots, 5; \text{ integer.}$$

The global optimal solution is $(x_1, x_2, x_3, x_4, x_5; f) = (1, 1, 1, 1, 2; 8)$.

Problem 10:

$$\text{Min } f(x) = x_1x_7 + 3x_2x_6 + x_3x_5 + 7x_4,$$

Subject to:

$$x_1 + x_2 + x_3 \geq 6.0,$$
$$x_4 + x_5 + 6x_6 \geq 8.0,$$
$$x_1x_6 + x_2 + 3x_5 \geq 7.0,$$
$$4x_2x_7 + 3x_4x_5 \geq 25.0,$$
$$3x_1 + 2x_3 + x_5 \geq 7.0,$$
$$3x_1x_3 + 6x_4 + 4x_5 \leq 20.0,$$
$$4x_1 + 2x_3 + x_6x_7 \leq 15.0,$$
$$0 \leq x_1, x_2, x_3 \leq 4,$$
$$0 \leq x_4, x_5, x_6 \leq 2,$$
$$0 \leq x_7 \leq 6,$$
$$x_i; i = 1, 2, \ldots, 7; \text{ integers.}$$

The global optimal solution is $(x_1, x_2, x_3, x_4, x_5, x_6, x_7; f) = (0, 2, 4, 0, 2, 1, 4; 14)$.

Problem 11:

$$\text{Min } f(x) = \exp(-x_1) + x_1^2 - x_1x_2 - 3x_2^2 - 6x_2 + 4x_1,$$

$$\text{Subject to} : 2x_1 + x_2 \leq 8.0,$$
$$- x_1 + x_2 \leq 2.0,$$
$$0 \leq x_1, x_2 \leq 3,$$
$$x_1, x_2 \text{ integers.}$$

The global optimal solution is $(x_1, x_2; f) = (1, 3; -42.632)$.

Problem 12:

$$\text{Min } f(x) = \sum_{i=1}^{9} \left[\exp\left(-\frac{(u_i - x_2)^{x_3}}{x_1} \right) - 0.01i \right]^2,$$

$$where, u_i = 25 + (-50\ln(0.01i))^{2/3},$$

$$\text{Subject to} : 0.1 \leq x_1 \leq 100.0,$$
$$0.0 \leq x_2 \leq 25.6,$$
$$0.0 \leq x_3 \leq 5.0,$$
$$x_1, x_2 \text{ integers.}$$

The global optimal solution is $(x_1, x_2, x_3; f) = (50, 25, 1.5; 0.0)$.

Problem 13:

$$\text{Min } f(x) = x_1^2 + x_2^2 + 3x_3^2 + 4x_4^2 + 2x_5^2 - 8x_1 - 2x_2$$
$$- 3x_3 - x_4 - 2x_5,$$

$$\text{Subject to} : x_1 + x_2 + x_3 + x_4 + x_5 \leq 400,$$
$$x_1 + 2x_2 + 2x_3 + x_4 + 6x_5 \leq 800,$$
$$2x_1 + x_2 + 6x_3 \leq 200,$$
$$x_3 + x_4 + 5x_5 \leq 200,$$
$$x_1 + x_2 + x_3 + x_4 + x_5 \geq 55,$$
$$x_1 + x_2 + x_3 + x_4 \geq 48,$$
$$x_2 + x_4 + x_5 \geq 34,$$
$$6x_1 + 7x_5 \geq 104,$$
$$0 \leq x_i \leq 99; \text{ integers } i = 1, 2, \ldots, 5.$$

the global optimal solution is $(x_1, x_2, x_3, x_4, x_5; f) = (16, 22, 5, 5, 7; 807)$.

Problem 14:

$$\text{Max } f(x, y) = -5.357854x_1^2 - 0.835689y_1x_3$$
$$- 37.29329y_1 + 40792.141,$$

$$\text{Subject to} : a_1 + a_2y_2x_3 + a_3y_1x_2 - a_4x_1x_3 \leq 92.0,$$
$$a_5 + a_6y_2x_3 + a_7y_1y_2 + a_5x_1^2 \leq 110.0,$$
$$a_9 + a_{10}x_1x_3 + a_{11}y_1x_1 + a_{12}x_1x_2 \leq 25.0,$$
$$27 \leq x_1, x_2, x_3 \leq 45,$$
$$y_1 \in \{78, 79, \ldots, 102\},$$
$$y_2 \in \{33, 34, \ldots, 45\}.$$

The global optima is $(x_1, x_3, y_1, :f) = (27, 27, 78; 32217.4)$ and it is obtained with various different feasible combination of (x_2, y_2).

Problem 15:

$$\text{Max } f(y) = r_1r_2r_3,$$
$$r_1 = 1 - 0.1^{y_1}0.2^{y_2}0.15^{y_3},$$
$$r_2 = 1 - 0.05^{y_4}0.2^{y_5}0.15^{y_6},$$
$$r_3 = 1 - 0.02^{y_7}0.06^{y_8},$$

Subject to:

$$y_1 + y_2 + y_3 \geq 1,$$
$$y_4 + y_5 + y_6 \geq 1,$$
$$y_7 + y_8 \geq 1,$$
$$3y_1 + y_2 + 2y_3 + 3y_4 + 2y_5 + y_6 + 3y_7 + 2y_8 \leq 10,$$
$$y \in \{0, 1\}^8.$$

The global optimal solution is $(y; f) = (0, 1, 1, 1, 0, 1, 1, 0; 0.94347)$.

Problem 16:

$$\begin{aligned}
\text{Max } f(x) = {} & 215x_1 + 116x_2 + 670x_3 + 924x_4 + 510x_5 \\
& + 600x_6 + 424x_7 + 942x_8 + 43x_9 + 369x_{10} \\
& + 408x_{11} + 52x_{12} + 319x_{13} + 214x_{14} + 851x_{15} \\
& + 394x_{16} + 88x_{17} + 124x_{18} + 17x_{19} + 779x_{20} \\
& + 278x_{21} + 258x_{22} + 271x_{23} + 281x_{24} + 326x_{25} \\
& + 819x_{26} + 485x_{27} + 454x_{28} + 297x_{29} + 53x_{30} \\
& + 136x_{31} + 796x_{32} + 114x_{33} + 43x_{34} + 80x_{35} \\
& + 268x_{36} + 179x_{37} + 78x_{38} + 105x_{39} + 281x_{40}
\end{aligned}$$

Subject to:

$$\begin{aligned}
& 8x_1 + 11x_2 + 6x_3 + x_4 + 7x_5 + 9x_6 + 10x_7 + 3x_8 + 11x_9 \\
& + 11x_{10} + 2x_{11} + x_{12} + 16x_{13} + 18x_{14} + 2x_{15} + x_{16} + x_{17} \\
& + 2x_{18} + 3x_{19} + 4x_{20} + 7x_{21} + 6x_{22} + 2x_{23} + 2x_{24} + x_{25} \\
& + 2x_{26} + x_{27} + 8x_{28} + 10x_{29} + 2x_{30} + x_{31} + 9x_{32} + x_{33} \\
& + 9x_{34} + 2x_{35} + 4x_{36} + 10x_{37} + 8x_{38} + 6x_{39} \\
& + x_{40} \leq 25,000, \\
& 5x_1 + 3x_2 + 2x_3 + 7x_4 + 7x_5 + 3x_6 + 6x_7 + 2x_8 + 15x_9 \\
& + 8x_{10} + 16x_{11} + x_{12} + 2x_{13} + 2x_{14} + 7x_{15} + 7x_{16} + 2x_{17} \\
& + 2x_{18} + 4x_{19} + 3x_{20} + 2x_{21} + 13x_{22} + 8x_{23} + 2x_{24} + 3x_{25} \\
& + 4x_{26} + 3x_{27} + 2x_{28} + x_{29} + 10x_{30} + 6x_{31} + 3x_{32} + 4x_{33} \\
& + x_{34} + 8x_{35} + 6x_{36} + 3x_{37} + 4x_{38} + 6x_{39} + 2x_{40} \leq 25,000 \\
& 3x_1 + 4x_2 + 6x_3 + 2x_4 + 2x_5 + 3x_6 + 7x_7 + 10x_8 + 3x_9 \\
& + 7x_{10} + 2x_{11} + 16x_{12} + 3x_{13} + 3x_{14} + 9x_{15} + 8x_{16} + 9x_{17} \\
& + 7x_{18} + 6x_{19} + 16x_{20} + 12x_{21} + x_{22} + 3x_{23} + 14x_{24} + 7x_{25} \\
& + 13x_{26} + 6x_{27} + 16x_{28} + 3x_{29} + 2x_{30} + x_{31} + 2x_{32} + 8x_{33} \\
& + 2x_{34} + 2x_{35} + 7x_{36} + x_{37} + 2x_{38} + 6x_{39} + 5x_{40} \leq 25,000 \\
& 10 \leq x_i \leq 99; \ i = 1, 2, \ldots, 20, \\
& 20 \leq x_i \leq 99; \ i = 21, 22, \ldots, 40.
\end{aligned}$$

Initially, this problem is solved by Monte Carlo technique on a random sample of 2000 points [33] and best solution is obtained at

$$\begin{pmatrix} 48 & 73 & 16 & 86 & 49 & 99 & 94 & 79 & 98 & 86 \\ 94 & 33 & 95 & 80 & 53 & 86 & 87 & 50 & 39 & 78 \\ 47 & 72 & 97 & 98 & 73 & 86 & 99 & 81 & 77 & 95 \\ 28 & 95 & 58 & 23 & 55 & 70 & 35 & 82 & 32 & 94 \end{pmatrix}$$

with $f_{max} = 1030361$. But proposed algorithm found the optimal solution at

$$\begin{pmatrix} 99 & 99 & 99 & 99 & 99 & 99 & 99 & 99 & 99 & 99 \\ 99 & 99 & 99 & 99 & 99 & 99 & 99 & 99 & 99 & 99 \\ 99 & 99 & 99 & 99 & 99 & 99 & 99 & 99 & 99 & 99 \\ 99 & 99 & 99 & 99 & 99 & 99 & 99 & 99 & 99 & 99 \end{pmatrix}$$

with $f_{max} = 1352439$.

Problem 17:

$$\begin{aligned}
\text{Max } f(x) = {} & 50x_1 + 150x_2 + 100x_3 + 92x_4 + 55x_5 + 12x_6 \\
& + 11x_7 + 10x_8 + 8x_9 + 3x_{10} + 114x_{11} + 90x_{12} \\
& + 87x_{13} + 91x_{14} + 58x_{15} + 16x_{16} + 19x_{17} + 22x_{18} \\
& + 21x_{19} + 32x_{20} + 53x_{21} + 56x_{22} + 118x_{23} + 192x_{24} \\
& + 52x_{25} + 204x_{26} + 250x_{27} + 295x_{28} + 82x_{29} \\
& + 30x_{30} + 29x_{31}^2 - 2x_{32}^2 + 9x_{33}^2 + 94x_{34} + 15x_{35}^3 \\
& + 17x_{36}^2 - 15x_{37} - 2x_{38} + x_{39}^3 + 3x_{40}^4 + 52x_{41} + 57x_{42}^2 \\
& - x_{43}^3 + 12x_{44} + 21x_{45} + 6x_{46} + 7x_{47} - x_{48} + x_{49} + x_{50} \\
& + 119x_{51} + 82x_{52} + 75x_{53} + 18x_{54} + 16x_{55} + 12x_{56} \\
& + 6x_{57} + 7x_{58} + 3x_{59} + 6x_{60} + 12x_{61} + 13x_{62} + 18x_{63} \\
& + 7x_{64} + 3x_{65} + 19x_{66} + 22x_{67} + 3x_{68} + 12x_{69} + 9x_{70} \\
& + 18x_{71} + 19x_{72} + 12x_{73} + 8x_{74} + 5x_{75} + 2x_{76} + 16x_{77} \\
& + 17x_{78} + 11x_{79} + 12x_{80} + 9x_{81} + 12x_{82} + 11x_{83} \\
& + 14x_{84} + 16x_{85} + 3x_{86} + 9x_{87} + 10x_{88} + 3x_{89} + x_{90} \\
& + 12x_{91} + 3x_{92} + 12x_{93} - 2x_{94}^2 - x_{95} + 6x_{96} + 7x_{97} \\
& + 4x_{98} + x_{99} + 2x_{100}
\end{aligned}$$

$$\text{Subject to : } \sum_{i=1}^{100} x_i \le 7500,$$

$$\sum_{i=1}^{50} 10x_i + \sum_{i=51}^{100} x_i \le 42,000,$$

$$0 \le x_i \le 99; \ i = 1, 2, \ldots, 100.$$

This is a nonlinear optimization problem with one hundred decision variables. Initially, it is solved by Monte Carlo technique on a random sample of 10000 points [33] and the global optimal solution of this problem is achieved at

$$\begin{pmatrix} 51 & 10 & 90 & 85 & 35 & 36 & 75 & 98 & 99 & 30 \\ 56 & 23 & 10 & 56 & 98 & 94 & 63 & 8 & 27 & 92 \\ 10 & 66 & 69 & 10 & 39 & 38 & 49 & 8 & 95 & 96 \\ 86 & 14 & 1 & 55 & 98 & 64 & 8 & 1 & 18 & 99 \\ 84 & 78 & 4 & 19 & 85 & 33 & 59 & 95 & 57 & 48 \\ 37 & 95 & 62 & 82 & 62 & 62 & 87 & 38 & 95 & 14 \\ 91 & 21 & 72 & 85 & 68 & 69 & 30 & 30 & 85 & 93 \\ 73 & 19 & 26 & 62 & 94 & 59 & 53 & 11 & 0 & 1 \\ 2 & 26 & 43 & 50 & 42 & 93 & 27 & 71 & 61 & 93 \\ 44 & 94 & 15 & 92 & 8 & 18 & 42 & 27 & 66 & 49 \end{pmatrix}$$

with $f_{max} = 303062435$.

The global optima of problem 17 is improved by MI-LXPMGSA and it is found at

$$\begin{pmatrix} 99 & 99 & 99 & 99 & 99 & 99 & 71 & 99 & 0 & 0 \\ 99 & 99 & 99 & 99 & 99 & 99 & 99 & 99 & 99 & 99 \\ 99 & 99 & 99 & 99 & 99 & 99 & 99 & 99 & 99 & 99 \\ 99 & 0 & 99 & 99 & 99 & 99 & 0 & 0 & 99 & 99 \\ 99 & 99 & 0 & 99 & 99 & 0 & 0 & 0 & 0 & 0 \\ 99 & 99 & 99 & 99 & 99 & 99 & 99 & 99 & 4 & 99 \\ 99 & 99 & 99 & 99 & 0 & 99 & 99 & 99 & 99 & 99 \\ 99 & 99 & 99 & 99 & 0 & 0 & 99 & 99 & 99 & 99 \\ 99 & 99 & 99 & 99 & 99 & 0 & 99 & 99 & 0 & 0 \\ 99 & 0 & 99 & 0 & 0 & 99 & 99 & 0 & 0 & 0 \end{pmatrix}$$

with $f_{max} = 304160077$.

Problem 18:

$$\text{Max } R(m, r) = \prod_{j=1}^{4} \left\{ 1 - (1 - r_j)^{m_j} \right\},$$

Subject to:

$$g_1(m) = \sum_{j=1}^{4} v_j . m_j^2 \leq v_Q,$$

$$g_2(m, r) = \sum_{j=1}^{4} C(r_j) . \left(m_j + \exp\left(m_j / 4 \right) \right) \leq C_Q,$$

$$g_3(m) = \sum_{j=1}^{4} w_j \cdot \left(m_j \cdot \exp\left(m_j/4\right)\right) \leq w_Q,$$

$$1 \leq m_j \leq 10 : \text{ intger}; j = 1, 2, \ldots, 4,$$

$$0.5 \leq r_j \leq 1 - 10^{-6}; j = 1, 2, \ldots, 4,$$

where, v_j is the product of weight and volume per element at stage j, w_j is the weight of each component at stage j, and $C(r_j)$ is the cost of each component with reliability r_j at stage j as follows:

$$C(r_j) = \alpha_j \cdot \left(\frac{-T}{\ln(r_j)} \right)^{\beta_j}$$

where α_j and β_j are constants representing the physical characteristic of each component at stage j and T is the operating time during which the component must not fail. The known optimal solution is $R(m, r) = 0.999955, m = [5, 5, 4, 6]$ and $r = [0.899845, 0.887909, 0.948990, 0.851017]$. the design data is given below. $C_Q = 400.0, w_Q = 500.0, v_Q = 250.0, T = 1000\,h.$

Subsys.	$10^5 \cdot \alpha_j$	β_j	v_j	w_j
1	1.0	1.5	1	6
2	2.3	1.5	2	6
3	0.3	1.5	3	8
4	2.3	1.5	2	7

References

1. Deep, K., Singh, K. P., Kansal, M.L., Mohan, C.: A real coded genetic algorithm for solving integer and mixed integer optimization problems. Appl. Math. Comput. **212**(2), 505–518 (2009)
2. Floudas, C.A., Lin, X.: Mixed integer linear programming in process scheduling: Modelling, algorithms, and applications. Ann. Oper. Res. **139**(1), 131–162 (2005)
3. Grossmann, I.E.: Mixed-integer programming approach for the synthesis of integrated process flowsheets. Comput. Chem. Eng. **9**(5), 463–482 (1985)
4. Mohan, C., Nguyen, H.T.: A controlled random search technique incorporating the simulated annealing concept for solving integer and mixed integer global optimization problems. Comput. Optim. Appl. **14**(1), 103–132 (1999)
5. Babu, B.V., Angira, R.: A differential evolution approach for global optimization of MINLP problems. In: Proceedings of 4th Asia-Pacific Conference on Simulated Evolution and Learning, vol. 2, pp. 880–884 (2002)
6. Yan, L., Shen, K., Hu, S.: Solving mixed integer nonlinear programming problems with line-up competition algorithm. Comput. Chem. Eng. **28**(12), 2647–2657 (2004)

7. Yiqing, L., Xigang Y., Yongjian, L.: An improved PSO algorithm for solving non-convex NLP/MINLP problems with equality constraints. Comput. Chem. Eng. **31**(3), 153–162 (2007)
8. Deep, K., Thakur, M.: A new crossover operator for real coded genetic algorithms. Appl. Math. Comput. **188**(1), 895–911 (2007)
9. Deep, K., Thakur, M.: A new mutation operator for real coded genetic algorithms. Appl. Math. Comput. **193**(1), 211–230 (2007)
10. Costa, L., Oliveira, P.: Evolutionary algorithms approach to the solution of mixed integer non-linear programming problems. Comput. Chem. Eng. **25**(2), 257–266 (2001)
11. Li, H., Zhang, L.: A discrete hybrid differential evolution algorithm for solving integer programming problems. Eng. Optim. **46**(9), 1238–1268 (2014)
12. Zhu, W., Fan, H.: A discrete dynamic convexized method for nonlinear integer programming. J. Comput. Appl. Math. **223**(1), 356–373 (2009)
13. Zhu, W., Ali, M. M.: Discrete dynamic convexized method for nonlinearly constrained nonlinear integer programming. Comput. Oper. Res. **36**(10), 2723–2728 (2009)
14. Misra, K.B., Sharma, U.: An efficient algorithm to solve integer-programming problems arising in system-reliability design. IEEE Trans. Reliab. **40**(1), 81–91 (1991)
15. Tan, Y., Tan, G., Tan, Q., Yang B., Ye, Y.: Hybrid genetic algorithm with chaotic local search for mixed integer programming problems. J. Comput. inf. Syst. **8**(24), 10223–10230 (2012)
16. Zhou, K., Chen, X., Shao, Z., Wan, W., Biegler, L.T.: Heterogeneous parallel method for mixed integer nonlinear programming. Comput. Chem. Eng. **66**, 290–300 (2014)
17. Tan, Y., Tan, G. Z., Deng, S.G.: Hybrid particle swarm optimization with chaotic search for solving integer and mixed integer programming problems. J. Central South Univ. **21**, 2731–2742 (2014)
18. Newby, E., Ali, M.M.: A trust-region-based derivative free algorithm for mixed integer programming. Comput. Optim. Appl. **60**(1), 199–229 (2015)
19. Fahim, A., Hedar, A.R.: Hybrid scatter search for integer programming problems. In: 9th International Conference on Informatics and Systems (INFOS2014), Operations Research and Decision Support Track, pp. 61–69. IEEE (2014)
20. Jun, W., Yuelin, G., Lina, Y.: An improved differential evolution algorithm for mixed integer programming problems. In: 9th International Conference on Computational Intelligence and Security (CIS2013), pp. 31–35. IEEE (2013)
21. Gao, Y., Ren, Z., Gao, Y.: Modified differential evolution algorithm of constrained nonlinear mixed integer programming problems. Inf. Technol. J. **10**(11), 2068–2075 (2011)
22. Lin, Y.C., Hwang, K.S., Wang, F.S.: A mixed-coding scheme of evolutionary algorithms to solve mixed-integer nonlinear programming problems. Comput. Math. Appl. **47**(8), 1295–1307 (2004)
23. Yokota, T., Gen, M., Li, Y.X.: Genetic algorithm for nonlinear mixed integer programming problems and its applications. Comput. Ind. Eng. **30**(4), 905–917 (1996)
24. Schlueter, M., Munetomo, M.: Parallelization strategies for evolutionary algorithms for MINLP. In: 2013 IEEE Congress on Evolutionary Computation (CEC2013), pp. 635–641. IEEE (2013)
25. Omran, M.G., Engelbrecht, A.P.: Differential evolution for integer programming problems. In: 2007 Congress on Evolutionary Computation, (CEC 2007), pp. 2237–2242. IEEE (2007)
26. Rashedi, E., Nezamabadi-Pour, H., Saryazdi, S.: GSA: a gravitational search algorithm. Inf. Sci. **179**(13), 2232–2248 (2009)
27. Singh, A., Deep, K.: Real coded genetic algorithm operators embedded in gravitational search algorithm for continuous optimization. Intl. J. Intell. Syst. Appl. **7**(12), 1–22 (2015)

28. Singh, A., Deep, K., Deep, A.: Curve fitting using gravitational search algorithm and its hybridized variants. In: Proceedings of Fifth International Conference on Soft Computing for Problem Solving, pp. 823–837. Springer, Singapore (2016)
29. Singh, A., Deep, K.: Novel hybridized variants of gravitational search algorithm for constraint optimization. Intl. J. Swarm Intell. (in press)
30. Singh, A., Deep, K., Nagar, A.: A new improved gravitational search algorithm for function optimization using a novel "best-so-far" update mechanism. In: Second International Conference on Soft Computing and Machine Intelligence (ISCMI), pp. 35–39. IEEE (2015)
31. Jiang, S., Wang, Y., Ji, Z.: Convergence analysis and performance of an improved gravitational search algorithm. Appl. Soft Comput. **24**, 363–384 (2014)
32. Deb, K.: An efficient constraint handling method for genetic algorithms. Comput. Methods Appl. Mech. Eng. **186**(2), 311–338 (2000)
33. Conley, W.: Computer Optimization Techniques. Petrocelli Books, Princeton (1980)

Spider Monkey Optimization Algorithm Based on Metropolis Principle

Garima Hazrati[1], Harish Sharma[1(✉)], Nirmala Sharma[1], Vani Agarwal[2], and D.C. Tiwari[2]

[1] Rajasthan Technical University, Kota, India
harish.sharma0107@gmail.com
[2] Jiwaji University, Gwalior, India

Abstract. Spider Monkey Optimization (SMO) Algorithm being a stretcher to the domain of meta-heuristics is performing well but has a flaw of converging early. For eradicating this flaw and improving exploration capability, a new modification is intended which is named as SMO based on metropolis principle (SMOM). The Metropolis principle is taken from simulated annealing in expectation to improve exploration capability of SMO. In this intended modification, non-prominent solutions also get a chance to upgrade themselves and reach global optima. This amendment enhances the global search capability of global leader phase which helps in sustaining exploration and exploitation of algorithm while maintaining the convergence speed. The intended algorithm is analyzed with SMO, one of its recent variant namely, self-adaptive spider monkey optimization (SaSMO) and another rooted algorithm i.e. particle swarm optimization (PSO) over 12 benchmark functions and recorded outcomes depicts that SMOM is a noted variant among them.

Keywords: Swarm intelligence · Nature inspired algorithms · Simulated annealing · Metropolis principle

1 Introduction

Nature always being a good teacher and by taking inspiration from it, humans evolve unique algorithms commonly known as nature-inspired algorithms [13]. Population-based meta-heuristics are one of its dominant class. A bevy of particles when reform their location based on their intellectual and unified behavior fall under this category [8]. Artificial bee colony (ABC) [2], Particle swarm optimization (PSO) [7], Gravitational search algorithm (GSA) [10] are few population-based meta-heuristics. Spider Monkey Optimization (SMO) being a new addition to this arena lessens the flaws of older meta-heuristics like stagnation in its basic design and an efficient algorithm. SMO is inspired by the intelligent food foraging behavior of spider monkeys portraying the concept of fission-fusion society and is developed by J. C. Bansal et. al [3]. Besides, being an efficient algorithm it has some flaws too like being stuck in local optima [9].

© Springer Nature Singapore Pte Ltd. 2017
K. Deep et al. (eds.), *Proceedings of Sixth International Conference on Soft Computing for Problem Solving*, Advances in Intelligent Systems and Computing 546, DOI 10.1007/978-981-10-3322-3_10

This paper exhibits a newly developed variant of SMO namely Spider Monkey Optimization Algorithm based on Metropolis principle. This variant is evolved to strengthen the exploration capability of all spider monkeys which guide them to reach the optimal solution while maintaining their convergence speed.

Further, the structure of the paper is as: Sect. 2 presents an overview of the SMO algorithm proceeding to Sect. 3 representing SMO based on metropolis principle. Section 4 depicts the performance evaluation followed by conclusion in Sect. 5.

2   Spider Monkey Optimization

Stimulated from the intellectual conduct of spider monkeys, researchers evolved spider monkey optimization algorithm that portrays a perfect fission-fusion structure. It has six phases except initialization that is interpreted below, and its brief can be read in [3].

2.1   Local Leader Phase(LLP)

This phase presents the location amendment process of all spider monkeys (SM) which depends on SM's persistence and social influence. Its social influence is based on their local leader and local group members experience. Location amendment depends on greedy approach by which prominent SM is chosen. Location amendment process is given in Eq. 1:

$$SMnew_{ij} = SM_{ij} + r_1 \times (LL_{kj} - SM_{ij}) + r_2 \times (SM_{rj} - SM_{ij}) \qquad (1)$$

where, SM_{ij} is the persistence of ith SM in j^{th} dimension, LL_{kj} represents the local leader of kth group and SM_{rj} is r^{th} randomly selected SM. r_1 is random number between (0,1) and r_2 varies in the range of (−1,1).

2.2   Global Leader Phase (GLP)

Taking inspiration from global leader and get influenced from neighbour, SM update its position. In this phase, location is updated on the basis of fitness as SM having high fitness get more chance to update itself as compared to less fit SM. The location amendment process is given in Eq. 2:

$$SMnew_{ij} = SM_{ij} + r_1 \times (GL_j - SM_{ij}) + r_2 \times (SM_{rj} - SM_{ij}) \qquad (2)$$

Here, GL_j is the location of global leader of the bevy. After amendment, greedy selection is prescribed on the selection of individuals, i.e. if the fitness of monkey is high, then its new location is selected else old one.

$$\text{if } fit_{new} \text{ (SM)} > fit_{old}\text{(SM)}$$
$$\text{select } new$$
$$\textbf{else}$$
$$\text{select } old$$

2.3 Global Leader Learning Phase (GLLP)

This phase is about learning of global leader in whole bevy and monkey with highest fitness get elected as global leader. If global leader doesn't amend its location then global limit count is set to 0 else incremented by 1.

2.4 Local Leader Learning Phase (LLLP)

Every sub-group has its leader that is elected in this phase. If a local leader doesn't amend itself then, a counter named local limit count is increased by 1.

2.5 Local Leader Decision Phase (LLDP)

If local leader of any bevy doesnt get relocated to a distinct edge known as Local Leader Limit (LLL), then all the monkeys of that group amend their locations either by random initialization or by using global leader wisdom through pr i.e. perturbation rate given in Eq. 3:

$$SMnew_{ij} = SM_{ij} + r_1 \times (GL_j - SM_{ij}) + r_1 \times (SM_{ij} - LL_{kj}) \qquad (3)$$

2.6 Global Leader Decision Phase (GLDP)

If the global leader doesnt get relocated to a distinct edge known as Global Leader Limit (GLL), then the global leader splits the bevy into smaller subgroups or fuse subgroups into one unit group.

3 Spider Monkey Optimization Algorithm Based on Metropolis Principle

SMO has a major flaw of being stuck in local optima due to which it bounces the global optima. In global leader phase, there are possibility that global leader get stuck or do not explore the search space properly. For enhancing the exploration characteristics of the algorithm, spider monkey optimization algorithm based on metropolis principle is depicted.

3.1 Modified Global Leader Phase

In global leader phase, the location amendment process of SM is given in Eq. 2. From this equation, a new location of SM is evaluated, and then we have one old position represented by SM_{ij} and new location by $SMnew_{ij}$. In the basic version of SMO greedy approach is used. The greedy approach has a dominant flaw that if the strength of the new solution is greater, then it is elected. Whereas, it may be possible that a non-prominent solution also covers the possibility of reaching to global optima.

For eradicating this flaw, new modification is intended i.e. based on metropolis principle taken from simulated annealing [5,6]. It simulates the cooling behavior of the material. By this principle, non-prominent solutions are also accepted with a probability in the intended modification which is evaluated as shown in Eq. 5.

$$\Delta C = C_{new} - C_{old} \tag{4}$$
$$P(\Delta C) = \exp(-\Delta C/T) \tag{5}$$

In Eq. 4, C_{new} and C_{old} are cost of new and old solutions respectively and their difference is saved in ΔC. T represents the temperature i.e. used to evaluate probability exponentially. From Eqs. 4 and 5, it is possible that if the cost of a newly generated solution is less then also it is confirmed. This principle is the backbone of simulated annealing [4] as by this chance of stagnation is less because it has the power of accepting non-prominent solutions with probability.

Now, greedy selection of amended global leader phase is shown as:

if $(fit_{new}$ (SM)$> fit_{old})$(SM) $\|$ r $> \exp(-(fitness_{new} - fitness_{old})/T)$
select *updated solution*
else
select *previous solution*

Here, r is a random number between (0,1), and T is the temperature which is at 20. In above selection, there are two conditions of election reflecting the newly selected location of a SM. Firstly, if an SM attains high strength at an altered location, then it is elected. Secondly, in the case of non-prominent solutions if the random number is greater than probability then also the non-prominent solution is considered but with some probability. Through this modification, SM that are not coming in the range of strength are also elected that is solution giving high strength are chosen, but non-prominent are elected too which overcome the flaw of global leader phase. It results in bettering global search capability of global leader phase. By this modification, exploration ability is upgraded because of which premature convergence speed is maintained, and there are better possibilities to reach global optima.

4   Experimental Outcomes

4.1   Considered Test Problems

The proposed algorithm SMOM is tested over 12 benchmark functions to examine its indulgence among other rooted algorithms. These 12 benchmark functions are taken from reference papers of taken algorithms for testing and are depicted in Table 1.

Table 1. Test problems

Test Problem	Objective function	Search Range	Optimum Value	D	Acceptable Error
Alpine	$f_1 = \sum_{i=1}^{D} |x_i \sin x_i + 0.1 x_i|$	[-10, 10]	$f(0) = 0$	30	$1.0E-05$
Michalewicz	$f_2 = -\sum_{i=1}^{D} \sin x_i (\sin(\frac{i \cdot x_i^2}{\pi})^{20})$	$[0, \pi]$	$f_{min} = -9.66015$	10	$1.0E-05$
Salomon Problem	$f_3 = 1 - \cos(2\pi\sqrt{\sum_{i=1}^{D} x_i^2}) + 0.1(\sqrt{\sum_{i=1}^{D} x_i^2})$	[-100, 100]	$f(0) = 0$	30	$1.0E-01$
Step function	$f_4 = \sum_{i=1}^{D} (\lfloor x_i + 0.5 \rfloor)^2$	[-100, 100]	$f(-0.5 \leq x \leq 0.5) = 0$	30	$1.0E-05$
Inverted cosine wave	$f_5 = -\sum_{i=1}^{D-1}\left(exp\left(\frac{-(x_i^2 + x_{i+1}^2 + 0.5 x_i x_{i+1})}{8}\right) \times I\right)$ where, $I = \cos\left(4\sqrt{x_i^2 + x_{i+1}^2 + 0.5 x_i x_{i+1}}\right)$	[-5, 5]	$f(0) = -D + 1$	10	$1.0E-05$
Levy montalvo 1	$f_6 = \frac{\Pi}{D}(10\sin^2(\Pi y_1)) + \sum_{i=1}^{D-1}(y_i-1)^2 \times (1 + 10\sin^2(\Pi y_{i+1})) + (y_D - 1)^2)$, where $y_i = 1 + \frac{1}{4}(x_i + 1)$	[-10, 10]	$f(-1) = 0$	30	$1.0E-05$
Shifted Rosenbrock	$f_7 = \sum_{i=1}^{D-1}(100(z_i^2 - z_{i+1})^2 + (z_i - 1)^2) + f_{bias}$, $z = x - o + 1$, $x = [x_1, x_2, ...x_D]$	[-100, 100]	$f(0) = f_{bias} = 390$	10	$1.0E-01$
Shifted Griewank	$f_8 = \sum_{i=1}^{D} \frac{z_i^2}{4000} - \prod_{i=1}^{D} \cos(\frac{z_i}{\sqrt{i}}) + 1 + f_{bias}$, $x = [x_1, x_2, ...x_D]$, $o = [o_1, o_2, ...o_D]$	[-600, 600]	$f(0) = f_{bias} = -180$	10	$1.0E-05$
Shifted Ackley	$f_9 = -20 \exp(-0.2\sqrt{\frac{1}{D}\sum_{i=1}^{D} z_i^2}) - \exp(\frac{1}{D}\sum_{i=1}^{D} \cos(2\pi z_i)) + 20 + e + f_{bias}$, $z = (x - o)$, $x = (x_1, x_2, ...x_D)$, $o = (o_1, o_2, ...o_D)$	[-32, 32]	$f(0) = f_{bias} = -140$	10	$1.0E-05$
Shubert	$f_{10} = -\sum_{i=1}^{5} i \cos((i+1)x_1 + 1)\sum_{i=1}^{5} i\cos((i+1)x_2 + 1)$	[-10, 10]	$f(7.0835, 4.8580) = -186.7309$	2	$1.0E-05$
Sinusoidal	$f_{11} = -[A\prod_{i=1}^{D} \sin(x_i - z) + \prod_{i=1}^{D} \sin(B(x_i - z))]$, $A = 2.5$, $B = 5$, $z = 30$	[0, 180]	$f(90 + z) = -(A+1)$	10	$1.0E-02$
Pressure Vessel	$f_{12} = 0.6224 x_1 x_3 x_4 + 1.7781 x_2 x_3^2 + 3.1661 x_1^2 x_4 + 19.84 x_1^2 x_3$	$x_1 = [1.1, 12.5]$ $x_2 = [0.6, 12.5]$ $x_3 = [0, 240]$ $x_4 = [0, 240]$	0	6	$1.0E-05$

4.2 Experimental setting

To substantiate that SMOM is a competed member in arena of population based meta-heuristics, comparative examination is performed among SMOM, SMO [3], PSO [7] and SaSMO [11]. Following experimental setup is done:

- Population of Spider Monkeys (N) = 50;
- Max number of groups = 5;
- LLL = 1500
- GLL = 50
- Settings of SMO, PSO, and SaSMO are taken from their original papers [3,7,11].

4.3 Results

Table 2 unfolded the attained outcomes of all the taken algorithms SMO, PSO, SaSMO and SMOM based on above parameter settings. All taken algorithms are tested on 100 runs in C++. Results are shown in the form of standard deviation (SD), mean error (ME), average function evaluation (AFE) and success rate (SR).

Results in above Table 2 exhibits that SMOM is a better variant than SMO, PSO, and SaSMO regarding reliability and accuracy at a cost of function evaluations in some benchmarks because it is giving a remarkable success rate. In addition to above results box-plots analysis of compared algorithms in terms of success rate is presented. Box-plots analysis [12] of SMO, PSO, SaSMO, and

Table 2. Comparison of outcomes of test problems

T P	Algorithm	SD	ME	AFE	SR
f_1	SMO	3.46E−06	9.79E−06	79051.30	99
	SaSMO	2.82E−04	4.17E−04	98985.87	0
	PSO	1.55E+00	2.30E−01	90070.00	72
	SMOM	5.16E−07	9.44E−06	78543.94	100
f_2	SMO	4.21E−03	4.95E−04	56524.47	98
	SaSMO	4.88E−04	5.45E−05	54914.81	98
	PSO	4.20E−01	4.20E−01	99402.50	2
	SMOM	3.56E−06	4.84E−06	50766.33	100
f_3	SMO	2.55E−02	1.93E−01	200862.84	7
	SaSMO	1.35E−01	1.56E+00	101746.98	0
	PSO	8.01E−02	3.98E−01	100003.00	1
	SMOM	3.36E−02	1.87E−01	195533.78	13
f_4	SMO	0.00E+00	0.00E+00	16239.24	100
	SaSMO	0.00E+00	0.00E+00	21085.75	100
	PSO	0.00E+00	0.00E+00	36092.50	100
	SMOM	0.00E+00	0.00E+00	13021.23	100
f_5	SMO	5.21E−02	5.25E−03	80817.68	99
	SaSMO	1.56E−01	5.05E−02	91340.72	45
	PSO	6.05E−01	1.58E+00	99659.50	2
	SMOM	1.76E−06	8.09E−06	73350.00	100
f_6	SMO	1.03E−02	1.05E−03	18723.73	99
	SaSMO	1.74E−06	8.12E−06	39102.07	100
	PSO	6.00E−07	9.30E−06	33252.50	100
	SMOM	8.32E−07	8.92E−06	11838.42	100
f_7	SMO	9.67E+00	2.50E+00	172472.86	39
	SaSMO	1.35E+00	9.54E−01	94387.43	23
	PSO	2.51E+01	8.38E+00	98430.50	3
	SMOM	3.20E+00	1.33E+00	162517.47	50
f_8	SMO	4.79E−03	1.79E−03	132298.41	81
	SaSMO	2.83E−03	9.19E−04	59303.68	86
	PSO	5.61E−02	6.59E−02	100050.00	0
	SMOM	3.33E−03	9.95E−04	127305.92	87
f_9	SMO	1.21E−06	8.46E−06	9126.81	100
	SaSMO	1.30E−06	8.46E−06	33002.44	100
	PSO	9.86E−07	8.95E−06	24719.00	100
	SMOM	1.06E−06	8.60E−06	9097.11	100

Table 2. *(Continued)*

T P	Algorithm	SD	ME	AFE	SR
f_{10}	SMO	5.29E−06	4.61E−06	4647.06	100
	SaSMO	5.52E−06	4.82E−06	8944.04	100
	PSO	2.49E−03	7.10E−04	46715.00	67
	SMOM	5.24E−06	4.57E−06	4255.02	100
f_{11}	SMO	6.37E−03	1.23E−02	158030.72	58
	SaSMO	1.31E−02	2.40E−02	100020.13	11
	PSO	3.47E−01	7.13E−01	96757.00	9
	SMOM	1.22E−02	1.26E−02	147513.7	72
f_{12}	SMO	2.15E−04	6.02E−05	114575.91	55
	SaSMO	2.02E+00	1.86E+00	103297.65	0
	PSO	3.28E−05	3.22E−05	59489.50	60
	SMOM	9.29E−04	1.51E−04	97740.30	64

SMOM is shown in Fig. 1 representing the empirical distribution of data graphically. Figure 1 shows that variation, interquartile range and medians of developed SMOM is higher than other three. After this, a comparison is made by using the performance indices (PI) graph [1] based on ME, SR, and AFE. The computed values of PI for SMO, SaSMO, PSO and SMOM are portrayed in Fig. 2.

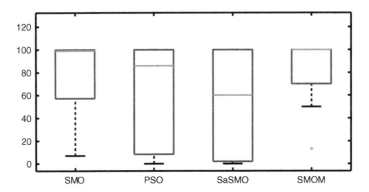

Fig. 1. Boxplot graph for success rate

Figure 2(a), (b) and (c) show the performance index of success rate, an average number of function evaluations and mean error respectively. Figure 2 indicates that PI of SMOM is notable as compared to other variants.

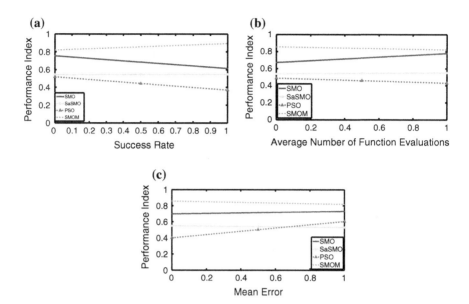

Fig. 2. Performance index for test problems; (a) for case (1) (b) for case (2) and (c) for case (3).

5 Conclusion

Eradicating the pitfalls of SMO, Metropolis step is applied to improve the exploration capability of SMO and avoiding stagnation in the population. This paper presents the modified version of SMO that is more reliable and accurate, namely metropolis operator based spider monkey optimization. This modification helps the global leader to achieve an optimal solution by accepting a non-prominent solution with some probability by using metropolis step. For testing the intensity of SMOM, it is examined over 12 benchmark functions, and results show that it is spell variant. Through statistical analysis, it is demonstrated that the proposed strategy is more reliable (better success rate) at the cost of function evaluations. In future, it can be applied to real-world optimization problems and complex optimization problems of continuous in nature.

References

1. Bansal, J.C., Sharma, H.: Cognitive learning in differential evolution and its application to model order reduction problem for single-input single-output systems. Memetic Comput. **4**(3), 209–229 (2012)
2. Bansal, J.C., Sharma, H., Jadon, S.S.: Artificial bee colony algorithm: a survey. Int. J. Adv. Intell. Paradigms **5**(1), 123–159 (2013)
3. Bansal, J.C., Sharma, H., Jadon, S.S., Clerc, M.: Spider monkey optimization algorithm for numerical optimization. Memetic Comput. **6**(1), 31–47 (2014)

4. Bertsimas, D., Tsitsiklis, J.: Simulated annealing. Stat. Sci. **8**(1), 10–15 (1993)
5. Tong, C.: Gravitational search algorithm based on simulated annealing. J. Convergence Inf. Technol. (JCIT) 9(2) (2014)
6. Hwang, C.-R.: Simulated annealing: theory and applications. Acta Applicandae Math. **12**(1), 108–111 (1988)
7. Kennedy, J.: Particle swarm optimization. In: Sammut, C., Webb, G.I. (eds.) Encyclopedia of Machine Learning, pp. 760–766. Springer, Heidelberg (2010)
8. Kennedy, J., Kennedy, J.F., Eberhart, R.C., Shi, Y.: Swarm Intelligence. Morgan Kaufmann, San Francisco (2001)
9. Lenin, K., Ravindhranath Reddy, B., Surya Kalavathi, M.: Modified monkey optimization algorithm for solving optimal reactive power dispatch problem. Indonesian J. Electr. Eng. Inform. (IJEEI) **3**(2), 55–62 (2015)
10. Rashedi, E., Nezamabadi-Pour, H., Saryazdi, S.: GSA: a gravitational search algorithm. Inf. Sci. **179**(13), 2232–2248 (2009)
11. Kumari, R., Kumar, S., Sharma, V.K.: Self-adaptive spider monkey optimization algorithm for engineering optimization problems. Int. J. Inf. Commun. Comput. Technol. II(II) (2014)
12. Williamson, D.F., Parker, R.A., Kendrick, J.S.: The box plot: a simple visual method to interpret data. Annals Intern. Med. **110**(11), 916–921 (1989)
13. Yang, X.-S.: Nature-inspired Metaheuristic Algorithms. Luniver press, Bristol (2010)

Introducing Biasedness in NSGA-II to Construct Boolean Function Having Best Trade-Off Among Its Properties

Rajni Goyal$^{(\boxtimes)}$ and Anupama Panigrahi

Department of Mathematics, University of Delhi, Delhi 110016, India
goyalrajni2584@gmail.com, anupama.panigrahi@gmail.com

Abstract. To construct Boolean function, many heuristic techniques have been used like NSGA-II, PSO, Ant Colony Method etc., but results are good only for few variables and complexity of these methods are very high. So, to reduced the complexity and to get desired results instead of all solutions, we have introduced a new concept of biasedness in our proposed method. We have used NSGA-II as our heuristic technique with concept of biasedness and got desired Boolean functions for 6 and 7 variables.

Keywords: MOOP · Genetic algorithms · Boolean functions · Biasedness · Nonlinearity

1 Introduction

In literature, there are many methods (heuristic as well as concatenation) to construct optimized Boolean functions. But heuristic techniques are mostly used as complexity of these methods are comparatively less and we can generate Boolean functions on higher variables also. In [1] Aguirre et al. have given a very good approach for multiobjectives. They took two and three objectives, and compared the results with two stage optimization. In [2], Camion et al. mentioned a new approach based on orthogonal arrays and constructed Boolean functions having good correlation immunity. In [3], Clark et al. gave a new two stage method based on simulated annealing, and result listed in this paper was better than previous results. But that method was able to get the optimum Boolean functions only for some limited variables. In [10,13,14], Maitra et al. constructed correlation immune functions keeping their nonlinearity optimal. First time they have constructed 1-resilient Boolean function with maximum nonlinearity for 8 variables and that method was based on concatenation. In [6,8,9], Clark et al. have found some functions with best tradeoff among Boolean function's properties. In [11,12,15], there are some construction methods but these methods are not applicable for multiobjectives and complexity of these methods is not considerable. In [7], we have given a method based on multiobjective optimization (based on genetic algorithms) but were able to get the functions only for 4, 5, 6 and 7 variables and complexity of method was high.

© Springer Nature Singapore Pte Ltd. 2017
K. Deep et al. (eds.), *Proceedings of Sixth International Conference on Soft Computing for Problem Solving*, Advances in Intelligent Systems and Computing 546, DOI 10.1007/978-981-10-3322-3_11

There are many others heuristic and other types of techniques available in literature. But to find good trade-off among the properties, only a heuristic technique is not sufficient. We want a technique that is having less complexity and should work for large number of variables. If we want to optimize many properties simultaneously, technique should be multiobjective also. So, by introducing biasedness concept [4] in heuristic technique, we tried to find good Boolean functions with less complexity. So, in present paper, we have given a new concept (biasedness) and got some optimum results.

2 Some Definition [1, 7]

2.1 Boolean Function

Any function $g : \mathbb{K}_2^n \to \mathbb{K}_2$ is called a Boolean function of n-variables. $\mathbb{K}_2^n$ is vector space (n-dimensional) over $\mathbb{K}_2$ where $\mathbb{K}_2$ represents a field of two elements. $\mathcal{Z}_n$ is called the set of all Boolean functions (n-variables).

2.2 Balancedness

If number of 0's in truth table representation is same as the number of 1's, than function is called balanced and the property is known as the balancedness.

2.3 Walsh Hadamard Transform

Boolean function can be represent in term of Walsh Hadamard Transform (WHT) also. If L_λ is linear function, specified by $\lambda \in \mathbb{K}_2^n$, the we denote WHT by $H_g(\lambda)$ and can be defined as

$$H_g(\lambda) = \sum_{x \in \mathbb{K}_2^n} (-1)^{g(x) \oplus \lambda . x}. \tag{1}$$

2.4 Non-linearity

Nonlinearity of a Boolean function is minimum hamming distance of that function from the set of all affine functions. It can be given by

$$nl(g) = (2^n - \max_{\lambda \in \mathbb{K}_2^n} |H_g(\lambda)|)/2. \tag{2}$$

2.5 Autocorrelation

The derivative of Boolean function g(x), with respect to a vector s, is defined as $g(x) \oplus g(x + s)$, where x and $s \in \mathbb{K}_2^n$. So, in polar form, derivative can be defined as $\hat{g}(x)\hat{g}(x + s)$. The autocorrelation of a function g is denoted by $A_g(s)$ and is defined by

$$A_g(s) = \sum_{x \in \mathbb{K}_2^n} \hat{g}(x)\hat{g}(x + s),$$

where $\hat{g}(x) = (-1)^{g(x)}$.

For a good Boolean function g, value of A_g should small.

2.6   Correlation Immunity

A Boolean function $g \in \mathcal{B}_n$ is said to be correlation immune (order m) if $H_g(\alpha) = 0$ for all $\alpha \in \mathbb{K}_2^n$ such that $1 \leq w_H(\alpha) \leq m$. Moreover, if g is balanced than it is called the m-resilient.

3   Non-dominated Sorting Genetic Algorithm II (NSGA-II) with Biasedness

Deb et al. [5] developed NSGA-II, that is a generational Multiobjective Optimization Evolutionary Algorithm (MOEA). It is based on three modules and we have explained the method in [7]. We have applied the algorithms on our developed method and got some good Boolean function [7]. But only this technique was not sufficient to get desired Boolean functions as complexity of method was comparatively high. Deb [4] discussed a sharing approach which uses a biased distance metric. By introducing biasedness means we give extra weightage to some specific objective function by introducing a constraint (same as objective function) into MOOP. In present paper, we have introduced a new concept of biasedness in NSGA-II to reduce the complexity. In our MOOP, we have formed first objective to optimize nonlinearity and nonlinearity is most important property here to optimize. So, in our MOOP, first objective and first constraint are same.

4   Formulation of MOOP

It consists of (i) Introduction of biasedness into MOOP and (ii) Application of NSGA-II.

(i) **Formulation of MOOP with biasedness:** Our main task is to form objective functions. To get optimum value of Nonlinearity, balancedness, autocorrelation and resiliency is our motive. We have formed first objective to optimize nonlinearity, second to optimize resiliency, and we have optimized autocorrelation by third objective. To get balanced functions, we have introduced two constraints. Nonlinearity is very important property. Hence to give extra weightage to first objective we have introduced concept of biasedness and added another constraints that is same as first objective.

First objective function: Based on the definition of nonlinearity [1, 7]

$$nl = 2^{(n-1)} - 1/2(\max_{\lambda} H_g(\lambda)),$$

We know maximum value of nonlinearity for 6 variables is 48 and for 7 variables is 56. So, to form first objective function we have introduced a new constant say T. Now, we want nl to take the value equal to T. So, first objective can be formed as follows:

$$g^1 = |nl - T|, \tag{3}$$

g^1 is our the first objective function, where T is constant value for a fixed number of variables. (Here we take its value as 48 for 6 variables and as 56 for seven variables.)

Second objective function: Second objective is to optimize autocorrelation. So, we have directly assigned the value of autocorrelation equal to second objective.

To formulate second objective, we have used definition of autocorrelation (Definition 2.5). According to the above definition of autocorrelation, we have formulated

$$A_g(\lambda) = \sum_{x \in \mathbb{K}_2^n} (-1)^{g(x) \oplus g(x+\lambda)},$$

and $A_g(0)$ is maximum,
    So,

$$g^2 = \max_\lambda |A_g(\lambda)| \tag{4}$$

is our second objective function, where $\lambda \in \mathbb{K}_2^n$ and $\lambda \neq zero$
    Now,

$$g^1 = |nl - T|,$$

$$g^1 = |2^{(n-1)} - 1/2 \sum_{x \in \mathbb{K}_2^n} (-1)^{g(x) \oplus \lambda.x} - T|, \tag{5}$$

Similarly, for all $\lambda \in \mathbb{K}_2^n$,

$$g^2 = \max_\lambda \sum_{x \in \mathbb{K}_2^n} (-1)^{g(x) \oplus g(x+\lambda)}. \tag{6}$$

Now

Third objective function: According to the definition 2.6, for a Boolean function to be m resilient, value of Walsh Hadamard Transform should be zero corresponding to all $x \in \mathbb{K}_2^n$ having weight $\leq$ m. So, to form out third objective, we take all WHT corresponding to all such $x \in \mathbb{K}_2^n$. We added all WHT and assigned them to the third objective. Now our purpose is to minimize this third objective (equal to zero). This is because with zero value of third objective, we will get m-resilient functions. So, our third objective is,

$$g^3 = \sum_\lambda |H_f(\lambda)| \tag{7}$$

where $w_H(\lambda) \leq m$ for $\lambda \in \mathbb{F}_2^n$
    So, we design MOOP as:

$$\left.\begin{array}{l} \min F = (g^1, g^2, g^3) \\ subject \quad to \\ \sum_{x \in \mathbb{K}_2^n} g(x) = 2^{n-1}, \\ nl = T. \end{array}\right\} \tag{8}$$

$\sum_{x \in \mathbb{K}_2^n} g(x)$ should be equal to 2^{n-1} for balanced function. To use biasedness sharing technique, the second constraint $nl = T$ is taken to give more weightage to the first objective.

(ii) **Application of above method:** After applying above method (with biasedness concepts) to the MOOP, we get the desired results. Results are given in Sect. 5. The list of parameters are listed in Table 2 (for 6 variables) and 3 (7 variables).

5   Result and Discussion

We got desired results by applying our method (In Sect. 4) on MOOP and got some good Boolean functions from cryptography point of view. These balanced functions have the best trade-off among non-linearity, autocorrelation and resiliency. In Table 1, we have listed those functions for 6 and 7 variables and parameters are given in Table 2 respectively. We have compared our results with literature [1,3] and can conclude that our results are at least as better.

Table 1. Obtained results

No. of variables	Previous results	Our results
6	$nl = 48$, $A_g = 8$, resiliency $= 1$	$nl = 48$, $A_g = 4$, resiliency $= 1$.
7	$nl = 56$, $A_g = 8$, resiliency $= 1$	$nl = 56$, $A_g = 8$, resiliency $= 1$

Table 2. Parameters

Parameters	For 6 variables	For 7 variables
Size of generation	2000	4000
Size of population	500	2000
Probability of crossover	0.8	0.8
Probability of mutation	0.1	0.11
Random seed number	0.9876	0.9976
Number of bits (for binary variables)	1	1
How many objective functions	3	3
How many constraints	2	2

6   Conclusion

In present paper, we have developed a new method to design good Boolean functions from cryptography point of view. We got Boolean functions for 6 and 7 variables that are better or at least comparable with [1,3]. So, we can conclude, our method is at least as better as the methods available in the literature.

References

1. Aguirre, H., Okazaki, H., Fuwa, Y.: An evolutionary multiobjective approach to design highly non-linear boolean functions. In: GECCO 2007, pp. 749–756 (2007)
2. Camion, P., Carlet, C., Charpin, P., Sendrier, N.: On correlation-immune functions. In: Feigenbaum, J. (ed.) CRYPTO 1991. LNCS, vol. 576, pp. 86–100. Springer, Heidelberg (1992). doi:10.1007/3-540-46766-1_6
3. Clark, J.A., Jacob, J.L., Stepney, S., Maitra, S., Millan, W.: Evolving Boolean functions satisfying multiple criteria. In: Menezes, A., Sarkar, P. (eds.) INDOCRYPT 2002. LNCS, vol. 2551, pp. 246–259. Springer, Heidelberg (2002). doi:10.1007/3-540-36231-2_20
4. Deb, K.: Multi-objective evolutionary algorithms: introducing bias among pareto-optimal solutions. In: Ghosh, A., Tsutsui, S. (eds.) Advances in Evolutionary Computing, pp. 263–292. Springer, New York (2003)
5. Deb, K., Pratap, A., Agarwal, S., Meyarivan, T.: A fast and elitist multiobjective genetic algorithm. IEEE Trans. Evol. Comput. $6(2)$, 182–197 (2002)
6. Filiol, E., Fontaine, C.: Highly nonlinear balanced Boolean functions with a good correlation-immunity. In: Nyberg, K. (ed.) EUROCRYPT 1998. LNCS, vol. 1403, pp. 475–488. Springer, Heidelberg (1998). doi:10.1007/BFb0054147
7. Goyal, R., Yadav, S.P.: Design of Boolean functions satisfying multiple criteria by NSGA-II. In: Deep, K., Nagar, A., Pant, M., Bansal, J.C. (eds.) SocProS 2011. AISC, vol. 130, pp. 461–468. Springer, Heidelberg (2011). doi:10.1007/978-81-322-0487-9_45
8. Dobbertin, H.: Construction of bent functions and balanced Boolean functions with high nonlinearity. In: Preneel, B. (ed.) FSE 1994. LNCS, vol. 1008, pp. 61–74. Springer, Heidelberg (1995). doi:10.1007/3-540-60590-8_5
9. Kurosawa, K., Satoh, T.: Design of SAC/PC (l) of order k Boolean functions and three other cryptographic criteria. In: Fumy, W. (ed.) EUROCRYPT 1997. LNCS, vol. 1233, pp. 434–449. Springer, Heidelberg (1997). doi:10.1007/3-540-69053-0_30
10. Maitra, S., Pasalic, E.: Further constructions of resilient boolean functions with very high nonlinearity. IEEE Trans. Inf. Theory $48(7)$, 1825–1834 (2002)
11. Millan, W., Clark, A., Dawson, E.: Heuristic design of cryptographically strong balanced Boolean functions. In: Nyberg, K. (ed.) EUROCRYPT 1998. LNCS, vol. 1403, pp. 489–499. Springer, Heidelberg (1998). doi:10.1007/BFb0054148
12. Chee, S., Lee, S., Lee, D., Sung, S.H.: On the correlation immune functions and their nonlinearity. In: Kim, K., Matsumoto, T. (eds.) ASIACRYPT 1996. LNCS, vol. 1163, pp. 232–243. Springer, Heidelberg (1996). doi:10.1007/BFb0034850
13. Sarkar, P., Maitra, S.: Nonlinearity bounds and constructions of resilient boolean functions. In: Bellare, M. (ed.) CRYPTO 2000. LNCS, vol. 1880, pp. 515–532. Springer, Heidelberg (2000). doi:10.1007/3-540-44598-6_32
14. Su, S., Tang, X., Zeng, X.: A systematic method of constructing Boolean functions with optimal algebraic immunity based on the generator matrix of the ReedMuller code. Des. Codes Crypt. $72(3)$, 653–673 (2014)
15. Wang, Q., Tan, C.H.: A new method to construct Boolean functions with good cryptographic properties. Inf. Process. Lett. $113(14–16)$, 567–571 (2013). Elsevier

Generating Distributed Query Plans Using Modified Cuckoo Search Algorithm

T.V. Vijay Kumar$^{(\boxtimes)}$ and Monika Yadav

School of Computer and Systems Sciences,
Jawaharlal Nehru University, 110067 New Delhi, India
tvvijaykumar@hotmail.com, monika71990@gmail.com

Abstract. In distributed databases, data is replicated and fragmented across multiple disparate sites spread across a computer network. Consequently, there can exist large numbers of possible query plans for a distributed query. This number increases with increase in the number of sites containing the replicated data. For large numbers of sites, computing an efficient query processing plan becomes a computationally expensive task. This necessitates the devising of a distributed query processing strategy capable of generating good quality query plans, from amongst all possible query plans, which minimize the total cost of processing a distributed query. This distributed query plan generation (*DQPG*) problem, being a combinatorial optimization problem, has been addressed in this paper using the modified cuckoo search algorithm. Accordingly, a modified *CSA* (*mCSA*) based *DQPG* algorithm (*DQPG$_{mCSA}$*), which aims to generate good quality Top-*K* query plans for a given distributed query, has been proposed herein. Experimental based comparison of *DQPG$_{mCSA}$* with the existing *GA* based *DQPG* algorithm (*DQPG$_{GA}$*) shows that the former is able to generate comparatively better quality *Top-K* query plans, which, in turn, would result in a reduction in the query response time and thereby enabling efficient decision making.

Keywords: Distributed Query Processing · Distributed Query Plan Generation (*DQPG*) · Swarm intelligence · Cuckoo Search Algorithm (*CSA*)

1 Introduction

In distributed databases, data is stored across multiple disparate sites distributed across a computer network [2, 7]. Queries posed on such data would require transmission of data between these sites. Since data is replicated at multiple sites, there could be several possible semantically equivalent query plans for answering any distributed query. Amongst these, the distributed query processing (*DQP*) strategy aims to compute such query plans that would reduce the total query processing cost. This cost includes the local processing cost as well as the site-to-site communication cost [4, 21]. Local processing cost comprises the *CPU* cost and the *I/O* cost whereas, the site-to-site communication cost depends on the communication network and the amount of data transfer between sites. Since communication cost is the dominant cost, its reduction becomes the prime goal of any *DQP* strategy [7]. Further, since the data is fragmented and replicated at multiple disparate sites, the number of possible query plans increases

© Springer Nature Singapore Pte Ltd. 2017
K. Deep et al. (eds.), *Proceedings of Sixth International Conference on Soft Computing for Problem Solving*, Advances in Intelligent Systems and Computing 546, DOI 10.1007/978-981-10-3322-3_12

exponentially with an increase in the number of sites required for processing a distributed query. For a large number of sites, the computation of an efficient query processing plan becomes a computationally expensive task. One such problem regarding the computation of efficient distributed query plans, from amongst large numbers of distributed query plans, has been given in [15, 16]. This problem, referred to as the Distributed Query Plan Generation (DQPG) problem, has been addressed in this paper. The DQPG problem given in [15, 16] is concerned with the generation of 'close' distributed query plans that minimize the query proximity cost. The close query plans, as defined in [15, 16], are those that involve fewer sites that have higher relation concentrations at the participating sites. These query plans incur a lesser site-to-site communication cost. As an example, consider the distributed SQL query given in Fig. 1. The relation–site matrix, for the six relations accessed by this distributed query, is also shown in Fig. 1. An entry '1' or an entry '0' in the relation-site matrix indicates the presence or absence respectively of a relation at the corresponding site. Some of the possible legal query plans are shown in Fig. 2. For example, the first query plan indicates that relation R1 is in site S6, relation R2 in site S1, relation R3 in site S4, relation R4 in site S1, relation R5 in site S2 and relation R6 at site S1.

Select A1, A2, A3, A4, A5 From R1, R2, R3, R4, R5, R6 Where R1.A1=R2.A1 and R2.A2=R3.A2 and R3.A3=R4.A3 and R4.A4=R5.A4 and R5.A5=R6.A5							
Relation\ Site	**S1**	**S2**	**S3**	**S4**	**S5**	**S6**
R1	0	1	1	0	0	1
R2	1	0	1	1	0	1
R3	1	1	0	1	0	1
R4	1	0	0	1	0	1
R5	0	1	0	0	1	0
R6	0	1	1	1	0	1

Fig. 1. Distributed query and relation-site matrix

No.	Query Plans						$QPC=\sum_{i=1}^{s}\dfrac{S_i}{N}\left(1-\dfrac{S_i}{N}\right)$	QPC
1	6	1	4	1	2	1	$\dfrac{3}{6}\left(1-\dfrac{3}{6}\right)+\dfrac{1}{6}\left(1-\dfrac{1}{6}\right)+\dfrac{1}{6}\left(1-\dfrac{1}{6}\right)+\dfrac{1}{6}\left(1-\dfrac{1}{6}\right)$	0.77
2	5	3	4	4	5	3	$\dfrac{2}{6}\left(1-\dfrac{2}{6}\right)+\dfrac{2}{6}\left(1-\dfrac{2}{6}\right)+\dfrac{2}{6}\left(1-\dfrac{2}{6}\right)$	0.66
3	1	4	2	3	2	6	$\dfrac{2}{6}\left(1-\dfrac{2}{6}\right)+\dfrac{1}{6}\left(1-\dfrac{1}{6}\right)+\dfrac{1}{6}\left(1-\dfrac{1}{6}\right)+\dfrac{1}{6}\left(1-\dfrac{1}{6}\right)+\dfrac{1}{6}\left(1-\dfrac{1}{6}\right)$	0.94
4	6	3	3	3	5	6	$\dfrac{3}{6}\left(1-\dfrac{3}{6}\right)+\dfrac{2}{6}\left(1-\dfrac{2}{6}\right)+\dfrac{1}{6}\left(1-\dfrac{1}{6}\right)$	0.60
5	5	4	4	4	5	5	$\dfrac{3}{6}\left(1-\dfrac{3}{6}\right)+\dfrac{3}{6}\left(1-\dfrac{3}{6}\right)$	0.50
6	1	1	2	1	2	1	$\dfrac{4}{6}\left(1-\dfrac{4}{6}\right)+\dfrac{2}{6}\left(1-\dfrac{2}{6}\right)$	0.44

Fig. 2. Legal query plans and their QPC values

To compute the optimal query, from amongst all query plans, is computationally expensive. For N relations, where each relation is stored at M sites, the total number of possible query plans is M^N. This value increases with an increase in the number of relations and the number of sites and, thus, for higher values of N and M, it becomes computationally infeasible to compute an optimal query plan. In order to address this, the $DQPG$ problem aims to generate a set of good quality query plans that minimize the query proximity cost (QPC) defined below [15, 16]:

$$QPC = \sum_{i=1}^{S} \frac{S_i}{N}\left(1 - \frac{S_i}{N}\right)$$

where, S is the number of sites involved in the query plan, N is the number of relations accessed by the query and S_i is the number of times the ith site is used in the query plan. The value of QPC varies from 0 to $(N-1)/N$. A lower QPC value is desirable, as it involves fewer sites and higher concentrations of relations. The QPC computation of the query plans given in Fig. 2, are given in Fig. 3. The 6th query plan is considered as the most 'close' query plan, as it involves the least number of sites, i.e. 2, and has a higher concentration of relations. On the other extreme, the 3rd query plan is the worst query plan, as it involves the maximum number of sites, i.e. 5.

This $DQPG$ problem, being a combinatorial optimization problem, has already been addressed using PSO in [12], $HBMO$ in [11], ACO in [14] and BCO in [13]. In this paper, the modified Cuckoo Search Algorithm ($mCSA$), given in [17], has been adapted and discretized to address the $DQPG$ problem. Further, the query plans generated using the proposed $mCSA$ based $DQPG$ algorithm ($DQPG_{mCSA}$) shall be compared with those generated using the GA based $DQPG$ algorithm ($DQPG_{GA}$) [15, 16].

The paper is organized as follows: Sect. 2 discusses the $DQPG$ using $mCSA$ followed by an illustrative example based on it in Sect. 3. Section 4 discusses the experimental results. The conclusion is given in Sect. 5.

2 $DQPG$ Using $mCSA$

Cuckoos are captivating birds. They are not only known for their melodious voice, but also exhibit interesting breeding behavior. Cuckoos, instead of creating their own nests, lay their eggs in nest of birds belonging to different species. They have the ability to mimic the colors and patterns of the eggs of the birds, whose nests they use to lay their eggs. The host bird is unable to distinguish the cuckoo's eggs from their own eggs [8]. The cuckoo eggs hatch slightly earlier than those of the host bird's egg whereupon the cuckoo chicks dislodge the host birds eggs out of the nest [19]. Also, these cuckoo chicks are adept in mimicking the sound of the host bird chicks in order to get a greater share of the food brought in by the host bird. Such behavior of cuckoos has been the major inspiration behind the cuckoo search algorithm (CSA) [19]. CSA has been successfully applied to various engineering optimization problems like the travelling salesman problem [5], the Knapsack problem [22], the independent test path generation for software testing [10], the optimizing the web service composition process [3], the job scheduling problem [1], design of a welded beam [20] etc. In this paper,

Inputs: Maximum number of iterations t_{max}, population size P, dimension of cuckoo egg n, Number of worst cuckoo eggs to be selected for abandonment p_a, Number of top cuckoo eggs to be selected for interaction p_β, Lévy flight constant A

Begin

  Initialize $t=1$

  Randomly generate an initial population of P cuckoo eggs, $X=\{x_1, x_2, x_3 ...x_P\}$, of dimension n and compute their fitness $f(x_i)$ where, $i \in P$.

  While $t < t_{max}$

    Select p_a fraction of worst cuckoo eggs to be abandoned.

    For each such cuckoo egg x_i,do:

      Compute the step size for Lévy Flight, $\alpha = A/\sqrt{t}$

      Perform Lévy flight (with step size α) on cuckoo egg x_i to generate new cuckoo egg $x_i{'}$ and evaluate its fitness. Replace x_i with $x_i{'}$.

    End **For**

    Select p_β fraction of top cuckoo eggs.

    For each such cuckoo egg x_i, do:

      Compute the step size for Lévy Flight, $\alpha = A/t^2$

      Perform Lévy flight (with step size α) on cuckoo egg x_i to generate new cuckoo egg $x_i{'}$ and evaluate its fitness.

      Choose a random cuckoo egg x_j from the population.

      Compare $f(x_j)$ and $f(x_i{'})$. Keep the fitter cuckoo egg in the population

    End **For**

    Increment t by 1

    Rank the cuckoo eggs according to their fitness.

  End **While**

  Post-process results and visualize.

  End

Fig. 3. Algorithm $mCSA$

the modified cuckoo search algorithm ($mCSA$), given in [17], has been used to address the $DQPG$ problem. $mCSA$ modifies CSA by considering different step sizes for nest abandonment and nest replacement. Also, $mCSA$ considers the interaction between the top cuckoos before they lay their eggs. Algorithm $mCSA$ [17, 18] is given in Fig. 3.

$mCSA$ commences with the initialization of the iteration count t to 1. An initial random population of cuckoo eggs, of dimension n, is generated in the subsequent step. Next, p_α fraction of the worst cuckoo eggs are selected for abandonment. For every cuckoo egg x_i, amongst the worst cuckoo eggs, the step size α for Lévy Flight is computed using the following equation:

$$\alpha = A/\sqrt{t}$$

where A is a constant, which is generally kept equal to 1, and t is the iteration counter. The Lévy flight [9] is, thereafter, applied on cuckoo egg x_i to generate a new cuckoo egg $x_i{'}$, which replaces x_i. The Lévy flight operator is applied in a similar manner, as applied in [19], using the step size defined above. Next, p_β fraction of top cuckoo eggs are selected for interaction. For each such top cuckoo egg x_i, the step size α for Lévy Flight is computed using the following equation:

$$\alpha = A/t^2$$

where A is a constant, which is generally kept equal to 1, and t is the iteration counter. Next, Lévy flight is applied to generate x_i' from x_i. A random cuckoo egg x_j from the population is selected in the subsequent step. Thereafter, if the condition $f(x_j) > f(x_i')$ is satisfied, x_j replaces x_i'. Next, the cuckoo eggs are ranked according to their fitness values and the above steps are repeated for a pre-specified maximum number of iterations. Thereafter, the best cuckoo eggs are produced as the output. In this paper, $mCSA$ has been adapted and discretized to solve the $DQPG$ problem.

$mCSA$ [17] is a modified version of the original CSA [19], which involves interaction between top cuckoo eggs. So, in the context of the $DQPG$ problem, there would be interaction between the top query plans. A different Lévy flight operator is used in nest replacement and the abandonment of the worst query plans. The Lévy flight operator is discussed below:

The Lévy flight equation to replace the worst cuckoo eggs is given as:

$$x_{i+1} = x_i \oplus \alpha . t^{-1/2}$$

where x_{i+1} represents the new solution generated around x_i, which is the original solution, $\oplus$ denotes the entry wise multiplication and α denotes the step size, which is taken as 1 for most problems. In this adaptation, the change in query plan is represented by b, which is computed using the following equation:

$$b = \alpha . t^{-1/2}, \text{ where } \alpha = N$$

The Lévy flight operator used to replace the worst cuckoo eggs is given below:

$$x_{i+1} = x_i \oplus \alpha . t^{-2}$$

The value of b is computed by using the following equation.

$$b = \alpha . t^{-2}, \text{ where } \alpha = N$$

The value of α is taken as N, i.e. the magnitude of change lies between 1 and N.

$mCSA$, with the above mentioned Lévy distribution, has been adapted and discretized to solve the $DQPG$ problem. Accordingly, the $mCSA$ [17] based $DQPG$ algorithm ($DQPG_{mCSA}$) is proposed and is discussed next.

2.1   DQPG$_{mCSA}$

$DQPG_{mCSA}$ is given in Fig. 4. $DQPG_{mCSA}$ takes the relation-site matrix, the number of relations accessed by the distributed query N, fraction of nests to be abandoned p_α, population size P, maximum number of iterations t_{max}, total query plans selected for applying Lévy flight m and the number of top query plans p_β, as input, and produces the Top-K query plans, as output.

In Step 1, the iteration counter t is initialized to 1. In Step 2, an initial population of P query plans is randomly generated using the relation-site matrix. Size of each query plan is N, where N is the total number of relations accessed by the distributed query. For each query plan, its QPC is computed. In Step 3, p_α fraction from among the worst query plans is selected. For each such query plan, Step 5 to Step 7 is performed as under:

In Step 5, Lévy flight is applied and the value of b is computed by using the following equation:

$$b = \alpha \cdot t^{-1/2}, \text{ where } \alpha = N$$

Next, the new query plan qpi' is generated, by changing b bits randomly in the original query plan qp_i. qpi' replaces qp_i in Step 6. In Step 8, p_β fraction of the top query plans are selected. As per Step 9, for each such query plan, Step 10 and Step 11 are performed.

Inputs: relation-site matrix, set of relations used in the query of size N, maximum number of iteration, t_{max}, size of initial population P, fraction of worst query plans p_α, fraction of best query p_β

Output: *Top-K* query plans

Begin

Step 1: Initialize t=1

Step 2: Randomly generate an initial population of query plans, QP={qp_1, qp_2,..., qp_P}, of size P and compute fitness QPC of each query plan, using the following equation:

$$QPC = \sum_{i=1}^{S} \frac{S_i}{N} \left(1 - \frac{S_i}{N}\right)$$

Where N is the total number of relations and S_i denotes the number of occurrence of a site in the query plan.

Step 3: Select p_α fraction of worst query plans to be abandoned from the population

Step 4: **For** each query plan qp_i, do:

Step 5: Apply Lévy Flight on it to generate a new solution qp_i'.

    Step 5.1: Compute the value of b as follows:

        $b=\alpha.t^{1/2}$, *Where* $\alpha= N$

    Step 5.2: Compute the new solution qp_i' by changing b bits in the original query plan qp_i and evaluate its fitness.

Step 6: Replace qp_i with qp_i'

Step 7: End **For**

Step 8: Select p_β fraction of best query plans.

Step 9: **For** each query plan qp_i, do:

Step 10: Apply Lévy Flight on it to generate a new solution qp_i'.

    Step 10.1. Compute the value of b as follows:

        $b=\alpha.t^{2}$, where $\alpha= N$

    Step 10.2 Compute the new solution qp_i' by changing b bits in the original query plan qp_i and evaluate its QPC.

Step 11: Compare the QPC of qp_i' with the QPC of another randomly selected query plan qp_j, from the population and keep the fitter query plan in the population.

Step 12: End **For**

Step 13: $t=t+1$

Step 14: Rank the query plans and repeat step 3 to 13 for a maximum number of iteration, i.e. t_{max}

Step 15: Return *Top-K* query plans.

End

Fig. 4. Algorithm $DQPG_{mCSA}$

In Step 10, Lévy flight is performed, for which the value of b is computed by using the following equation:

$$b = \alpha.t^{-2}, \text{ Where } \alpha = N$$

After the new query plan qpi' is generated, by changing b bits in the original query plan qp_i, its QPC value is evaluated. In Step 11, another query plan qp_j is randomly selected from the population and its QPC is compared with the QPC of query plan qpi' and the fitter amongst them is retained in the population. In Step 13, the iteration counter t is incremented by 1. In Step 14, the query plans are ranked according to their QPC whereafter Step 3 to Step 13 are repeated for a pre-specified maximum number of iterations t_{max}. Thereafter, in Step 15, the *Top-K* query plans are produced, as output.

Next, an example illustrating the use of $DQPG_{mCSA}$ to generate *Top-K* query plans is given.

3   An Example

Consider the distributed *SQL* query and the relation-site matrix given in Fig. 1. Let the value of P_α and P_β be *0.7* and *0.25* respectively. The generation of *Top-5* query plans using $DQPG_{mCSA}$ is given below:

Initialize the iteration counter $t = 1$. Next, randomly generate an initial population of $P = 10$ query plans, $QP = \{qp_1, qp_2 \ldots qp_{10}\}$ and compute their QPC. The *10* query plans, along with their QPC are given in Fig. 5. Select $p_\alpha = 0.7$ fraction of worst query plans to be abandoned from the population. The 7 worst query plans selected are given in Fig. 6. On each of these worst query plans qp_i, Lévy Flight is applied to generate a new query plan qpi'. The value of b is computed, as given below:

$$b = \alpha.t^{-1/2} = 6 \times 1^{-1/2} = 6$$

A new query plan qpi' is generated by changing b bits in the original query plan qp_i and its QPC is evaluated. The modified query plans, along with their QPC, are given in Fig. 7. Suppose $p_\beta = 0.25$ fraction of the top query plans are to be selected. Then the *Top-2* query plans selected are given in Fig. 8.

Query Plan No.	Query Plan	QPC
qp_1	[6, 6, 5, 6, 5, 6]	0.44
qp_2	[2, 1, 1, 1, 2, 2]	0.50
qp_3	[3, 4, 4, 4, 2, 3]	0.60
qp_4	[2, 4, 1, 1, 2, 4]	0.66
qp_5	[2, 3, 4, 4, 5, 2]	0.74
qp_6	[3, 6, 1 ,6, 5, 6]	0.74
qp_7	[2, 4, 5, 1, 2, 3]	0.77
qp_8	[6, 1, 5, 4, 5, 3]	0.77
qp_9	[6, 1, 2, 4, 5, 3]	0.94
qp_{10}	[2, 6, 1, 4, 5, 3]	0.94

Fig. 5. Randomly generated population

Query Plan No.	Query Plan	QPC
qp_4	[2, 4, 1, 1, 2, 4]	0.66
qp_5	[2, 3, 4, 4, 5, 2]	0.74
qp_6	[3, 6, 1 ,6, 5, 6]	0.74
qp_7	[2, 4, 5, 1, 2, 3]	0.77
qp_8	[6, 1, 5, 4, 5, 3]	0.77
qp_9	[6, 1, 2, 4, 5, 3]	0.94
qp_{10}	[2, 6, 1, 4, 5, 3]	0.94

Fig. 6. Worst selected query plans

Query Plan	Random string	Modified Query Plan	QPC
qp_4 = [2, 4, 1, 1, 2, 4]	[1 1 1 1 1 1]	[3 1 1 6 5 2]	0.77
qp_5 = [2, 3, 4, 4, 5, 2]	[1 1 1 1 1 1]	[6 3 2 4 5 3]	0.77
qp_6 = [3, 6, 1 ,6, 5, 6]	[1 1 1 1 1 1]	[2 1 5 2 5 6]	0.72
qp_7 = [2, 4, 5, 1, 2, 3]	[1 1 1 1 1 1]	[6 1 2 1 2 6]	0.66
qp_8 = [6, 1, 5, 4, 5, 3]	[1 1 1 1 1 1]	[2 3 1 1 2 6]	0.72
qp_9 = [6, 1, 2, 4, 5, 3]	[1 1 1 1 1 1]	[3 3 1 1 2 6]	0.72
qp_{10} = [2, 6, 1, 4, 5, 3]	[1 1 1 1 1 1]	[6 3 2 1 2 6]	0.72

Fig. 7. Modified query plans using lévy flight

Query Plan	QPC
qp_1 = [6 6 5 6 5 6]	0.44
qp_2 = [2 1 1 1 2 2]	0.50

Fig. 8. *Top-2* query plan

On each such query plan qp_i, Lévy flight is applied to generate a new query plan qpi'. The value of b is computed as given below:

$$b = \alpha . t^{-2} = 6 \times 1^{-2} = 6$$

The new query plan qpi' is computed by changing 6 bits in the original query plan qp_i. The modified query plans are given in Fig. 9.

Query Plan	Random string	Modified Query Plan	QPC
qp_1 = [6 6 5 6 5 6]	[1 1 1 1 1 1]	$qp_1{}'$ = [3 1 2 6 2 4]	0.77
qp_2 = [2 1 1 1 2 2]	[1 1 1 1 1 1]	$qp_2{}'$ = [6 6 6 6 2 6]	0.27

Fig. 9. Modified query plans using lévy flight

Next, the fitness of the query plan qpi' is compared with the fitness of another randomly selected query plan, say qp_j, from the population and the fitter query plan, among these two, is retained in the population. Let the randomly selected query plans be qp_7 and qp_2. The selection procedure of query plans to be retained in the population is given in Fig. 10.

Query Plan $qp_i{}'$	QPC	Randomly Selected Query Plan	QPC	Query Plan Selected
$qp_1{}' = [\,3\ 1\ 2\ 6\ 2\ 4\,]$	0.77	$qp_7 = [\,6\ 1\ 2\ 1\ 2\ 6\,]$	0.66	qp_7
$qp_2{}' = [\,6\ 6\ 6\ 6\ 2\ 6\,]$	0.27	$qp_2 = [\,2\ 1\ 1\ 1\ 2\ 2\,]$	0.50	$qp_2{}'$

Fig. 10. Replaced query plans

The iteration counter t is then incremented by 1, i.e. $t = t+1 = 2$. The resulting query plans are ranked, based on their *QPC*, as given in Fig. 11.

Query Plan No.	Query Plan	QPC
qp_2	[6, 6, 6, 6, 2, 6]	0.27
qp_1	[6, 6, 5, 6, 5, 6]	0.44
qp_3	[3, 4, 4, 4, 2, 3]	0.60
qp_7	[6, 1, 2, 1, 2, 6]	0.66
qp_6	[2, 1, 5, 2, 5, 6]	0.66
qp_8	[2, 3, 1, 1, 2, 6]	0.74
qp_9	[3, 3, 1, 1, 2, 6]	0.74
qp_{10}	[6, 3, 2, 1, 2, 6]	0.74
qp_4	[3, 1, 1, 6, 5, 2]	0.77
qp_5	[6, 3, 2, 4, 5, 3]	0.77

Fig. 11. Population after first iteration

The above process is repeated for a pre-specified maximum number of iterations $t_{max} = 1000$. Thereafter, the *Top*-5 query plans are produced, as output.

4   Experimental Results

$DQPG_{mCSA}$ and $DQPG_{GA}$ were implemented in MATLAB 7.12.0 in a Windows 8.1 environment. The two algorithms were compared by conducting experiments on an Intel based 4 GHz PC having 2 GB RAM. The comparisons were carried out on parameters like number of relations, *Average QPC (AQPC)*, *Top-K* query plans and the number of iterations. The population of 20 [20] query plans and a 50 × 50 relation-site matrix are considered for experimentation. First graphs of *AQPC* vs. Iterations, for 10, 20, 30 and 40 relations over 1000 iterations, were plotted to determine the appropriate values of (p_α, p_β) for $DQPG_{mCSA}$. The values obtained are given in Fig. 12 and

$DQPG_{mCSA}$ (p_α, p_β)								
N_R	(0.25, 0.2)	(0.25, 0.25)	(0.25, 0.3)	(0.5, 0.2)	(0.5, 0.25)	(0.5, 0.3)	(0.75, 0.2)	(0.75, 0.25)	(0.75, 0.3)
10	0.5160	0.3792	0.4296	0.4728	0.5000	0.4280	0.5136	0.4504	**0.3696**
20	0.5856	0.6128	0.6360	0.6418	0.6266	0.5878	0.5908	**0.5774**	0.6294
30	0.7422	0.6961	0.7262	0.7142	0.7474	0.7241	**0.6720**	0.7219	0.7096
40	0.7175	0.7233	0.7410	0.7411	0.7091	0.7612	0.7298	**0.6990**	0.7373

Fig. 12. Observed parameter values for $DQPG_{mCSA}$

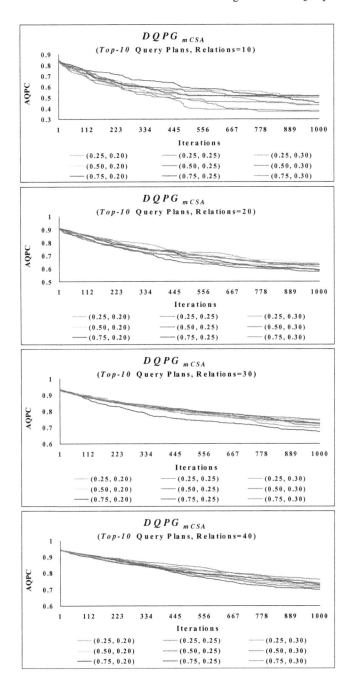

Fig. 13. $DQPG_{mCSA}$ – $AQPC$ Vs. Iterations for different p_α and p_β (10, 20, 30 and 40 relations over 1000 iterations)

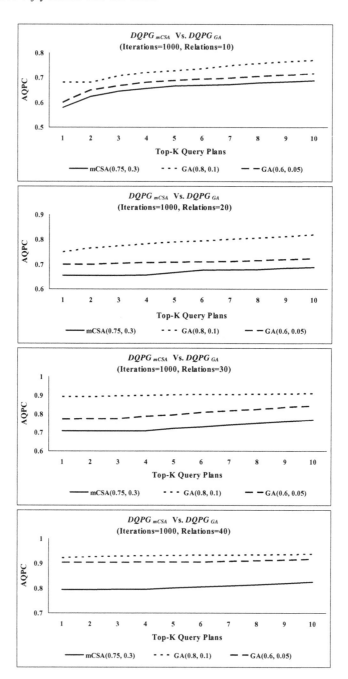

Fig. 14. $DQPG_{mCSA}$ Vs. $DQPG_{GA}$ – $AQPC$ Vs. $Top\text{-}K$ Query Plans (10, 20, 30 and 40 relations after 1000 iterations)

the corresponding graphs are shown in Fig. 13. The values of p_α and p_β considered are (0.25, 0.5, 0.75) [6, 20] and (0.2, 0.25, 0.3) [6] respectively. From these graphs and tables, the appropriate values of p_α and p_β obtained for 10, 20, 30 and 40 relations are $DQPG_{mCSA}(p_\alpha = 0.75, p_\beta = 0.3)$, $DQPG_{mCSA}(p_\alpha = 0.75, p_\beta = 0.25)$, $DQPG_{mCSA}(p_\alpha = 0.75, p_\beta = 0.2)$ and $DQPG_{mCSA}(p_\alpha = 0.75, p_\beta = 0.25)$ respectively. These observed values are used for comparing $DQPG_{mCSA}$ with $DQPG_{GA}$ (($P_c = 0.6$, $P_m = 0.05$), ($P_c = 0.8$, $P_m = 0.1$) [15, 16] with respect to the $AQPC$ of the $Top-K$ ($K = 1$ to 10) query plans generated by them. The comparison graphs for 10, 20, 30 and 40 relations are shown in Fig. 14. It can clearly be inferred from the graphs that the proposed algorithms $DQPG_{mCSAA}$ are able to generate $Top-K$ query plans at a comparatively lower $AQPC$ than those generated using $DQPG_{GA}$.

5 Conclusion

In this paper, a modified CSA based $DQPG$ algorithm $DQPG_{mCSA}$ has been proposed to address the $DQPG$ problem given in [15, 16]. $DQPG_{mCSA}$ attempts to generate the $Top-K$ 'close' query plans for a given distributed query that incur a lower total query processing cost. Experiments are carried out to determine the appropriate values of p_α and p_β for which $DQPG_{mCSA}$ is able to generate the $Top-K$ query plans having a minimum $AQPC$. Thereafter, these observed values are used for comparing $DQPG_{mCSA}$ with $DQPG_{GA}$. The experimental results show that $DQPG_{mCSA}$ is able to generate $Top-K$ query plans that have a comparatively lower $AQPC$ than those generated using $DQPG_{GA}$ [15, 16]. The performance of $DQPG_{mCSA}$ becomes better with increase in the number of relations accessed by the distributed query. Thus, the query plans generated using $DQPG_{mCSA}$ have a comparatively better query response times and thus would provide greater assistance in the decision making process.

References

1. Burnwal, S., Deb, S.: Scheduling optimization of flexible manufacturing system using cuckoo search-based approach. Int. J. Adv. Manufact. Technol. **64**, 951–959 (2013)
2. Ceri, S., Pelagati, G.: Distributed Database: Principles and Systems. Tata McGraw-Hill Publishing Company, New Delhi (2008)
3. Chifu, V.R., Pop, C.B., Salomie, I., Suia, D.S., Niculici, A.N.: Optimizing the semantic web service composition process using cuckoo search. In: Brazier, F.M.T., Nieuwenhuis, K., Pavlin, G., Warnier, M., Badica, C. (eds.) Intelligent Distributed Computing V. Studies in Computational Intelligence, vol. 382, pp. 93–102. Springer, Heidelberg (2012)
4. Kosmann, D.: The state of the art in distributed query processing. ACM Comput. Surv. **32**(4), 422–469 (2000)
5. Ouaarab, A., Ahiod, B., Yang, X.S.: Discrete cuckoo search algorithm for the travelling salesman problem. Neural Comput. Appl. **24**, 1659–1669 (2013)
6. Ouyang, X., Zhou, Y., Luo, Q., Chen, H.: A novel discrete cuckoo search algorithm for spherical traveling salesman problem. Appl. Math. Inf. Sci. **7**(2), 777–784 (2013)

7. Ozsu, M., Valduriez, P.: Principles of Distributed Database Systems. Pearson Education, India (2011)
8. Payne, R.B., Sorenson, M.D., Klitz, K.: The Cuckoos. Oxford University Press, Oxford (2005)
9. Shlesinger, M., Zaslavsky, G.M., Frisch, U.: Lévy Flights and Related Topics in Physics. Springer, Heidelberg (1995)
10. Srivastava, P.R., Chis, M., Deb, S., Yang, X.S.: An efficient optimization algorithm for structural software testing. Int. J. Artif. Intell. **9**(S12), 68–77 (2012)
11. Vijay Kumar, T.V., Arun, B., Kumar, L.: Distributed query plan generation using HBMO. In: Ramanna, S., Lingras, P., Sombattheera, C., Krishna, A. (eds.) MIWAI 2013. LNCS (LNAI), vol. 8271, pp. 293–304. Springer, Heidelberg (2013). doi:10.1007/978-3-642-44949-9_27
12. Vijay Kumar, T.V., Kumar, A., Singh, R.: Distributed query plan generation using particle swarm optimization. Int. J. Swarm Intell. Res. **4**(3), 58–82 (2013)
13. Vijay Kumar, T.V., Kumar, L., Arun, B.: Distributed query plan generation using BCO. Int. J. Swarm Intell. (IJSI) **1**(4), 358–377 (2015)
14. Vijay Kumar, T.V., Singh, R., Kumar, A.: Distributed query plan generation using ant colony optimization. Int. J. Appl. Metaheuristic Comput. (IJAMC) **6**(1), 1–22 (2015)
15. Vijay Kumar, T.V., Singh, V., Verma, A.K.: Distributed query processing plans generation using genetic algorithm. Int. J. Comput. Theor. Eng. **3**(1), 38–45 (2011)
16. Vijay Kumar, T.V., Singh, V. Verma, A.K.: Generating distributed query processing plans using genetic algorithm. In: International Conference on Data Storage and Data Engineering, pp. 173–177 (2010)
17. Walton, S., Hassan, O., Morgan, K., Brown, M.R.: Modified cuckoo search: a new gradient free optimization algorithm. Chaos, Solutions Fractals **44**(9), 710–718 (2011)
18. Yadav, M., Vijay Kumar, T.V.: Optimization using cuckoo search algorithms, communications in dependability and quality management. Int. J. **18**(4), 52–64 (2015)
19. Yang, X.S., Deb, S.: Cuckoo search via levy flights. In: World Congress on Nature & Biologically Inspired Computing (NaBIC 2009), pp. 210–214. IEEE Publications, India (2009)
20. Yang, X.S., Deb, S.: Multi objective cuckoo search for design optimization. Comput. Oper. Res. **40**(6), 1616–1624 (2013)
21. Yu, C.T., Chang, C.C.: Distributed query processing. ACM Comput. Surv. **16**(4), 399–433 (1984)
22. Zheng, H., Zhou, Y., He, S., Ouyang, X.: A discrete cuckoo search algorithm for solving knapsack problems. Adv. Inf. Sci. Serv. Sci. (AISS) **4**(18), 331–339 (2012)

Locally Informed Shuffled Frog Leaping Algorithm

Pragya Sharma$^{(\boxtimes)}$, Nirmala Sharma, and Harish Sharma

Rajasthan Technical University, Kota, India
pragya.gecj@gmail.com

Abstract. Shuffled Frog-Leaping Algorithm (SFLA) is a memetic meta-heuristic approach for solving complex optimization problems. Like other evolutionary algorithms, it may also suffer from the problem of slow convergence. To elevate the convergence property of the algorithm, locally informed search strategy is incorporated with SFLA. To improve the intensification and diversification capabilities of SFLA, locally informed search strategy is embedded by calculating the mean of local best and one randomly selected neighbour solution of memeplex while updating the position of worst solution in local best updating phase. Similarly, mean of global best and a randomly selected neighbour solution is used to improve the position of worst solution while updating the position of worst solution in global best updating phase. The proposed algorithm is named as Locally Informed Shuffled Frog-Leaping Algorithm (LISFLA). The modified algorithm LISFLA is analysed over 15 distinct benchmark test problems and compared with conventional SFLA, its recent variant, namely Binomial Crossover Embedded Shuffled Frog-Leaping Algorithm (BC-SFLA) and three other nature inspired algorithms, namely Gravitational Search Algorithm (GSA), Differential Evolution (DE) and Biogeography-Based Optimization Algorithm (BBO). The results manifest that LISFLA is an antagonist variant of SFLA.

Keywords: Meta-heuristic optimization techniques · Swarm intelligence · Shuffled frog leaping algorithm · Locally informed

1 Introduction

Nature-inspired algorithms (NIAs), that take inspiration from nature and its foundation is biological components of nature i.e. human and nature. The main objective of developing such algorithm is to solve distinct complex real world problems whose absolute solution doesn't exist and is to optimize engineering problems [1]. Swarm intelligence based algorithms [5] are based on mimicking collective behavior of natural swarm's e.g. particle swarm optimization (PSO) [2], artificial bee colony algorithm (ABC) [8], shuffled frog-leaping algorithm (SFLA) [7] and bacterial foraging algorithm (BFO) [9] etc.

© Springer Nature Singapore Pte Ltd. 2017
K. Deep et al. (eds.), *Proceedings of Sixth International Conference on Soft Computing for Problem Solving*, Advances in Intelligent Systems and Computing 546, DOI 10.1007/978-981-10-3322-3_13

SFLA takes inspiration from the grubbing behavior of frogs that replicate contagious information pattern with the natural and social behavior of species. In SFLA, population (frogs) is partitioned into several memeplexes. Frogs exchange their memes with other frogs using memetic evolution procedure which helps to improve the performance of individual frog towards its global optimum solution. There is always a presence of odds with all the evens, basic SFLA converges slowly at the last stage and easily falls into local minima. To elevate the performance of the conventional SFLA algorithm researchers are continuously working on this algorithm [6, 10, 15].

To improve the convergence, intensification and diversification proficiency of basic SFLA, locally informed search strategy is embedded in the local exploration phase of the conventional SFLA to ameliorate the position of the worst solution. In LISFLA, worst solution is take good memes either from local best and one local random solution of the memeplex or global best and a randomly chosen neighbour solution of the entire feasible search space. In this process, worst solution is locally informed through the global best or local best with one randomly selected neighbour to ameliorate the knowledge of worst solution of memeplex. The contemplated algorithm is titled as Locally Informed Shuffled Frog Leaping Algorithm (LISFLA).

The remaining paper is organized as shown: In Sect. 2, a brief overview of standard SFLA is described. Locally Informed Shuffled Frog-Leaping Algorithm (LISFLA) is proposed in Sect. 3. In Sect. 4, performance of LISFLA is tested with several numerical benchmark functions. Finally, Sect. 5 conclude the work.

2   Overview of Shuffled Frog-Leaping Algorithm

Eusuff invented Shuffled Frog Leaping Algorithm (SFLA) in 2003 [7] for solving distinct complex optimization problems. SFLA is a population-based cooperative search metaphor inspired by foraging behaviour of frogs [13]. Memetic evolution is used in SFLA for the purpose of spreading ideas among the solutions in a local exploration which is same as PSO [2]. A shuffling approach helps for exchanging ideas among local searchers that lead them toward a global optimum. SFLA contains elements of global exploration, local exploration and shuffling procedures. In general, a SFLA works as follows. Firstly, the parameters for the SFLA are total number of memeplex (Mmpx), the number of frogs in each memeplex (Fm) and the range of feasible search space are initialized. Therefore, the total population size (N) of swarm is denoted as $N = Mmpx * Fm$. Afterwards, objective value of each frog is calculated. Rank is assigned accordingly to their objective value and sort them in the descending order of their objective values. Then, N frogs are partitioned into memeplexes (M), each containing frogs (F), like that first rank frog goes to first memepelex, second rank frog goes to second memeplex and third rank frog goes to third memeplex and so on. To construct submemeplex, memplexes are divided into the submemeplex with having the goal is that true solution to move towards its optimum solution by elevating their ideas. Submemeplex selection process assigns weights to frogs. The weights are assigned with a triangular probability distribution ($prob_k$) using Eq. 1.

$$prob_k = \frac{2(q+1-k)}{q(q+1)} \tag{1}$$

Here, q is total number of population and $k = 1, 2, 3, ..., q$, represents rank of frogs within the memeplex. The frog with the higher objective value has the higher probability of being selected for the submemeplex. The frogs with the lower objective value has lower probability. The position of best frogs and worst frogs is recorded.

The worst solution is updated their position by using three phases: (1) Local best updating phase (2) Global best updating phase and (3) Randomly initialization of solution in the search space (Censorship).

1. **Local best updating phase**: To improve the position of worst frog. The position update equation for worst solution is defined in Eq. 2.

$$U_{new} = P_W + R(0,1) * (P_{LB} - P_W) \tag{2}$$

Here, U_{new} is the new position of worst frog, P_{LB} and P_W are position of local best frog and worst frog respectively. $R(0,1)$ is a random number in the range [0, 1]. If U_{new} lies in the feasible space, compute the new objective value. Greedy selection strategy is applied for improving the position of worst solution. If the position of worst solution gets better than the previous position then position is updated otherwise it goes in next phase i.e. global best updating phase.

2. **Global best updating phase**: In this phase, the worst frog get chance to update its position with the help of global best frog as defined in Eq. 3.

$$U_{new} = P_W + R(0,1) * (P_{GB} - P_W) \tag{3}$$

Here, P_{GB} represents the global best frog found so far. Again greedy selection strategy is applied for improving the position of worst solution. If worst solution does not update its position then it is randomly initialized in the feasible search space. After this phase memeplexes are updated with the new position of worst frog solution.

3. **Randomly initialization of solution in the search space (Censorship)**: If new position of worst solution is infeasible means worst solution exist outside the range of search space and old position which is calculated by global best solution is not better. Meme of this frog not spread no longer it means that worst frog does not have good meme so, randomly generate a new frog within the range of feasible search space to replace the frog whose new position was not so good to progress.

After the memetic evolutionary steps within each memeplexes are to be shuffled and the population is to be sorted in decreasing order of their objective value. Position of the best frog P_{GB} is get updated. To check convergence, repeat the above procedure until the stopping criteria is met.

3   Locally Informed Shuffled Frog Leaping Algorithm (LISFLA)

In the working of SFLA algorithm, there are enough possibilities for the solutions to get stuck in local optima. It also suffers from the problem of slow convergence. To reduce such problems, locally informed search process is incorporated by taking mean of both local best and one randomly selected local neighbour of memeplex and global best and one random neighbour for improving the position of the worst solution in local best updating phase and global best updating phase of basic SFLA respectively.

As it is clear from the solution's search process of conventional SFLA that the Eqs. (2, 3), the worst solution is updated during each iteration by using three phases: (1) Local best updating phase (2) Global best updating phase and (3) Randomly initialization of solution in the search space. Further, it is to be noted that worst solution is simply influenced by the local best or global best solution, which may lead to trap in local optima and leads to loss of intensification and diversification capability.

To avoid such possibilities (stagnation or converging in local optima), in the proposed strategy, the step size is calculated by taking the mean of both local best or global best solution and one randomly selected neighbour solution of memeplex otherwise it is randomly initialized in the search space. This type of search phenomenon elevates the intensification and diversification proficiency of the algorithm that are chief characteristics of the population-based optimization algorithms. Therefore, to improve the convergence and to maintain the intensification and diversification capability of SFLA, following modifications are proposed.

3.1   Local Best Learning Phase with Random Neighbour

The position of worst solution is updated using the locally informed search process. In the process of local search, the worst solution is get updated (informed) by taking mean of both local best solution and one randomly selected local neighbour solution of particular memeplex . The updated step size and position update equations is defined as Eqs. (4 and 5).

$$Step = R(0,1) * (\frac{P_{LB} + P_{KL}}{2} - P_W) \tag{4}$$

$$U_{new} = P_W + Step \tag{5}$$

Here, U_{new} is the updated position of the worst solution and $Step$ shows the step size. (P_{LB}) and (P_{KL}) represents the local best solution and a local neighbour solution of P_W respectively. $R(0,1)$ is a uniformly distributed random number in the range between $[0, 1]$. If U_{new} lies in the feasible space, compute the new objective value. Greedy selection strategy is applied for improving the position of worst solution in the search space. If the position of worst solution gets better than the previous position then the position of the worst solution is updated otherwise it goes to next phase. In the modified equation a new term is added that is containing the information received from the local best solution and a random neighbour solution of memeplex. In Eqs. (4, 5), worst solution P_W is informed through the both both P_{LB} and P_{KL}. In basic SFLA Eq. (2), sometimes the worst solution (P_W) are not updated through the local best solution that leads to loss of intensification and diversification proficiency of the search space. Therefore, in Eqs. (4, 5) to improve the position of worst solution mean of both P_{LB} and P_{KL} are taken that leads to move P_W toward its optimum solution.

3.2 Global Best Learning Phase with Random Neighbour

In this phase the worst solution is get update its position by taking mean of both global best and a random neighbour solutions of the feasible search space. The position update process of worst solution is defined as Eqs. (6 and 7).

$$Step = R(0,1) * (\frac{P_{GB} + P_{KG}}{2} - P_W) \qquad (6)$$

$$U_{new} = P_W + Step \qquad (7)$$

Here, U_{new} is the updated position of worst solution and $Step$ shows the step size. P_{GB} and P_{KG} represents the global best solution and neighbour solution of P_W respectively. If U_{new} is exist in the feasible space, compute its objective value and apply greedy selection strategy between the new and worst solution. If worst solution does not update its position then it is randomly initialized in the feasible search space. In Eqs. (6, 7) both P_{GB} and P_{KG} has the ability to attract the worst solution P_W. In the modified equation a new term is added that is containing the mean of the information received from the global best solution and a randomly selected neighbour solution of the search space. According to Eqs. (6, 7), in place of only global best solution, mean of both randomly chosen neighbour solution and global best solution is used that commute the worst solution in the direction of the global best solution with neighbour solution that enhance the intensification and diversification potential of SFLA algorithm.

3.3 Censorship

If new position of worst solution is infeasible means worst solution exist outside the range of search space and old position which is calculated by global best solution is not better. Meme of this frog not spread no longer it means that worst frog does not have good meme so, generate a new solution randomly within the range of feasible search space to inplace the frog whose new position was not so good to evolution.

After the memetic evolutionary steps within each memeplex, the memeplexes are to be shuffled and the population is to be sorted in decreasing order of their objective value. Position of best frog P_{GB} is get updated. Then we check the stopping criteria of algorithm if it is satisfied then stop the process. Otherwise, again partition the frogs into memeplexes.

Like SFLA, the LISFLA algorithm is also divided into two phases, namely global exploration phase and local exploration phase. The locally informed search strategy is embedded in the local exploration phase of the algorithm, while global exploration phases are kept same as in the basic SFLA.

4   Results and Discussions

To analyze the validity of LISFLA algorithm, 15 distinct global optimization functions (f_1 to f_{15}) are used here, demonstrated in Table 1.

To accredit the pursuance of the proposed algorithm LISFLA, a comparative experiment is carried out among LISFLA, SFLA [7], BC-SFLA, GSA [11], DE and BBO [14]. LISFLA is tested with the basic SFLA, BC-SFLA, BBO, DE and GSA over considered optimization test functions. Following experimental parameters are adopted:

- The number of simulations/run = 30,
- Total number of memeplexes ($Mmpx$) = 5
- Number of frogs in each memeplex (Fm) = 10 and total population Size (Mmpx * Fm) (N) = 50
- Parameter settings for the algorithms basic SFLA, BC-SFLA, GSA, DE and BBO are imitated from their elementary research papers. [7,11,14]

Table 2 shows the experimental results of the SFLA, BC-SFLA, GSA, DE and BBO algorithms and also furnishes about the standard deviation (SD), average number of function evaluations (AFE), mean error (ME), and success rate (SR). Results in Table 2 reflects that most of the time LISFLA outrun in terms of reliability, robustness, efficiency as well as accuracy in comparison to the SFLA, BC-SFLA, GSA, DE and BBO.

Besides, boxplots [3,12] analysis is carried out for comparing the examined algorithms in the form of combined performance though it can efficiently depict

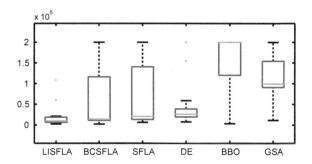

Fig. 1. Boxplots graphs (Average Function Evaluation)

Table 1. TP: Test Problems, OF: Objective Function, SR: Search Range, OV: Optimum Value, D: Dimension, AE: Acceptable Error for LISFLA

Test problem	Objective function	Search range	Optimum value	D	AE
De Jong f4	$f_1(x) = \sum_{i=1}^{n} i.(x_i)^4$	[-5.12 , 5.12]	$f(0) = 0$	30	$1.00E-05$
Griewank	$f_2(x) = 1 + \frac{1}{4000}\sum_{i=1}^{D} x_i^2 - \prod_{i=1}^{D}\cos(\frac{x_i}{\sqrt{i}})$	[-600 , 600]	$f(0) = 0$	30	$1.0E-05$
Ackley	$f_3(x) = -20 + e + exp(-\frac{0.2}{D}\sqrt{\sum_{i=1}^{D} x_i^3}) - exp(\frac{1}{D}\sum_{i=1}^{D}\cos(2\pi.x_i)x_i)$	[-1 , 1]	$f(0) = 0$	30	$1.0E-05$
Cosine mixture	$f_4(x) = \sum_{i=1}^{D} x_i^2 - 0.1(\sum_{i=1}^{D}\cos 5\pi x_i)+0.1D$	[-1 , 1]	$f(0) = -1 \times 0.1$	30	$1.0E-05$
Exponential	$f_5(x) = -(exp(-0.5\sum_{i=1}^{D} x_i^2))+1$	[-1 , 1]	$f(0) = -1$	30	$1.0E-05$
brown3	$f_6(x) = \sum_{i=1}^{D-1}(x_i^{2(x_{i+1}^2+1)} + x_{i+1}^{2x_i^2+1})$	[-1 , 4]	$f(0) = 0$	30	$1.0E-05$
Salomon problem	$f_7(x) = 1 - \cos(2\pi\sqrt{\sum_{i=1}^{D} x_i^2}) + 0.1(\sqrt{\sum_{i=1}^{D} x_i^2})$	[-100 , 100]	$f(0) = 0$	30	$1.0E-01$
Axis parallel hyper-ellipsoid	$f_8(x) = \sum_{i=1}^{D} i.x_i^2$	[-5.12 , 5.12]	$f(0) = 0$	30	$1.0E-05$
Sum of different powers	$f_9(x) = \sum_{i=1}^{D} |x_i|^{i+1}$	[-1 , 1]	$f(0) = 0$	30	$1.0E-05$
Step function	$f_{10}(x) = \sum_{i=1}^{D}([x_i + 0.5])^2$	[-100 , 100]	$f(-0.5 \le x \le 0.5) = 0$	30	$1.0E-05$
Rotated hyper-ellipsoid	$f_{11}(x) = \sum_{i=1}^{D}\sum_{j=1}^{i} x_j^2$	[-65.536 , 65.536]	$f(0) = 0$	30	$1.0E-05$
Levy montalvo 1	$f_{12}(x) = \frac{\Pi}{D}(10\sin^2(\Pi y_1)+\sum_{i=1}^{D-1}(y_i-1)^2 \times (1+10\sin^2(\Pi y_{i+1}))+(y_D-1)^2)$, where $y_i = 1 + \frac{1}{4}(x_i+1)$	[-10 , 10]	$f(-1) = 0$	30	$1.0E-05$
Levy montalvo 2	$f_{13}(x) = 0.1(\sin^2(3\Pi x_1) + \sum_{i=1}^{D-1}(x_i-1)^2 \times (1+\sin^2(3\Pi x_{i+1})) + (x_D-1)^2(1+\sin^2(2\Pi x_D)))$	[-5 , 5]	$f(1) = 0$	30	$1.0E-05$
Ellipsoidal	$f_{14}(x) = \sum_{i=1}^{D}(x_i - i)^2$	[-30 , 30]	$f(1, 2, 3,...,D) = 0$	30	$1.0E-05$
Moved axis parallel hyper-ellipsoid	$f_{15}(x) = \sum_{i=1}^{D} 5i \times x_i^2$	[-5.12 , 5.12]	$f(x) = 0; x(i) = 5 \times i, i = 1 : D$	30	$1.0E-15$

Table 2. Comparative result of TP, TP: Test Problem for LISFLA

TP	Algorithm	SD	ME	AFE	SR
f_1	LISFLA	9.80E−07	8.32E−06	5999.47	30
	BCSFLA	9.67E−07	8.24E−06	8546.30	30
	SFLA	1.12E−06	8.73E−06	12333.10	30
	DE	1.16E−06	8.58E−06	18770.00	30
	BBO	8.01E−07	8.98E−06	46193.33	30
	GSA	1.34E−06	8.03E−06	63223.33	30
f_2	LISFLA	1.03E−06	8.85E−06	8502.20	30
	BCSFLA	1.77E−02	8.05E−01	200000.00	0
	SFLA	2.19E−02	8.25E−01	200000.00	0
	DE	1.03E−16	7.59E−01	200000.00	0
	BBO	1.13E−03	7.63E−01	200000.00	0
	GSA	3.21E−03	7.60E−01	200000.00	0
f_3	LISFLA	5.23E−07	9.38E−06	14840.93	30
	BCSFLA	5.95E−01	2.77E−01	56603.13	24
	SFLA	5.62E−01	2.69E−01	61437.80	24
	DE	4.28E−07	9.49E−06	42043.33	30
	BBO	1.00E−02	4.53E−02	200000.00	0
	GSA	5.84E−07	9.37E−06	161030.00	30
f_4	LISFLA	3.69E−02	9.86E−03	20234.33	28
	BCSFLA	1.61E−01	1.67E−01	136836.07	10
	SFLA	2.47E−01	5.52E−01	193757.90	1
	DE	7.96E−07	9.02E−06	21763.33	30
	BBO	1.87E−01	1.58E−01	200000.00	0
	GSA	8.13E−07	8.66E−06	111176.67	30
f_5	LISFLA	8.41E−07	8.90E−06	5554.03	30
	BCSFLA	8.74E−07	8.96E−06	8074.27	30
	SFLA	7.73E−07	8.92E−06	10541.43	30
	DE	6.91E−07	9.02E−06	17165.00	30
	BBO	3.62E−07	9.54E−06	93826.67	30
	GSA	7.56E−07	8.86E−06	91298.33	30
f_6	LISFLA	9.45E−07	8.87E−06	9168.40	30
	BCSFLA	6.82E−07	9.02E−06	10797.23	30
	SFLA	7.06E−07	8.94E−06	18280.53	30
	DE	6.87E−07	9.21E−06	22221.67	30
	BBO	1.65E−05	4.26E−05	200000.00	0
	GSA	1.07E−06	8.84E−06	99265.00	30

(*continued*)

Table 2. (*continued*)

f_7	LISFLA	6.19E−02	2.50E−01	108844.53	17
	BCSFLA	1.29E−01	5.17E−01	200000.00	0
	SFLA	1.47E−01	3.60E−01	167215.30	7
	DE	3.00E−02	2.10E−01	155318.33	26
	BBO	5.37E−02	4.67E−01	200000.00	0
	GSA	5.82E−02	8.00E−01	200000.00	0
f_8	LISFLA	8.56E−07	8.66E−06	8698.07	30
	BCSFLA	6.53E−07	9.11E−06	12437.80	30
	SFLA	9.55E−07	8.84E−06	16133.77	30
	DE	7.81E−07	9.12E−06	25906.67	30
	BBO	4.33E−04	1.28E−03	200000.00	0
	GSA	1.02E−06	8.91E−06	109881.67	30
f_9	LISFLA	2.73E−06	6.81E−06	2738.93	30
	BCSFLA	1.93E−06	7.11E−06	2266.53	30
	SFLA	1.36E−06	8.09E−06	6626.67	30
	DE	2.28E−06	7.26E−06	7665.00	30
	BBO	2.20E−06	7.48E−06	3241.67	30
	GSA	1.81E−06	6.49E−06	46398.33	30
f_{10}	LISFLA	0.00E+00	0.00E+00	3844.73	30
	BCSFLA	4.76E−01	2.00E−01	38038.30	25
	SFLA	0.00E+00	0.00E+00	9004.93	30
	DE	0.00E+00	0.00E+00	10886.67	30
	BBO	0.00E+00	0.00E+00	5351.67	30
	GSA	0.00E+00	0.00E+00	11583.33	30
f_{11}	LISFLA	1.07E−06	8.58E−06	9897.87	30
	BCSFLA	8.49E−07	8.66E−06	14172.93	30
	SFLA	8.00E−07	8.95E−06	18385.47	30
	DE	9.63E−07	8.95E−06	29205.00	30
	BBO	5.39E−03	1.44E−02	200000.00	0
	GSA	9.47E−07	8.75E−06	95278.33	30
f_{12}	LISFLA	3.96E−02	2.39E−02	59982.17	22
	BCSFLA	6.44E−01	5.53E−01	149901.03	8
	SFLA	3.15E+00	4.85E+00	194013.27	1
	DE	7.17E−07	8.90E−06	26943.33	30
	BBO	7.07E−01	8.54E−01	200000.00	0
	GSA	8.82E−07	8.81E−06	90630.00	30

(*continued*)

Table 2. (*continued*)

f_{13}	LISFLA	1.79E−02	3.32E−03	14040.03	29
	BCSFLA	8.89E−07	8.78E−06	11064.47	30
	SFLA	1.79E−02	3.32E−03	20893.87	29
	DE	1.02E−06	8.88E−06	23088.33	30
	BBO	3.09E−05	1.04E−04	200000.00	0
	GSA	6.13E−07	9.00E−06	95498.33	30
f_{14}	LISFLA	5.87E−07	9.07E−06	12680.57	30
	BCSFLA	1.03E−06	8.59E−06	13154.93	30
	SFLA	6.76E−07	9.16E−06	25357.17	30
	DE	5.87E−07	9.08E−06	26216.67	30
	BBO	6.60E−04	2.33E−03	200000.00	0
	GSA	2.33E−05	7.78E−04	134665.00	28
f_{15}	LISFLA	9.31E−17	8.72E−16	21009.73	30
	BCSFLA	9.95E−17	8.45E−16	28845.97	30
	SFLA	4.86E−17	9.30E−16	37419.80	30
	DE	6.21E−17	9.08E−16	58980.00	30
	BBO	2.02E−03	5.94E−03	200000.00	0
	GSA	1.34E−12	1.34E−11	200000.00	0

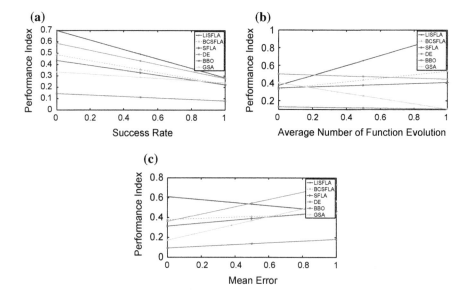

Fig. 2. Performance index for test problems; (a) for weighted importance to SR, (b) for weighted importance to AFE and (c) for weighted importance to ME.

the empirical dispersion of data graphically. The boxplots for LISFLA, SFLA, BC-SFLA, GSA, DE and BBO are displayed in Fig. 1. The results manifest that interquartile range and medians of LISFLA are comparatively low.

Nextly, all considered algorithms are also compared by giving weighted importance to the ME, SR, and AFE. This comparison is measured using the performance indices which is described in [3,4]. The resultant values of PI for the LISFLA, BC-SFLA, SFLA, GSA, DE and BBO are computed and corresponding PIs graphs are demonstrated in Fig. 2.

The graphs belonging to each of the cases i.e. giving weighted importance to AFE, SR and ME (as explained in [3,4]) are depicted in Figs. 2(a), (b), and (c) respectively. In these figures, horizontal axis represents the weights while vertical axis expresses the PI.

It is clear from Fig. 2 that PI of LISFLA are superior than the other considered algorithms in each case. i.e. LISFLA performs better on the considered test problems as compare to the BCSFLA, SFLA, GSA, DE and BBO.

5   Conclusion

In this paper, a new variant of SFLA algorithm is presented, namely Locally Informed Shuffled Frog Leaping Algorithm (LISFLA). In the proposed LISFLA, a new position update strategy for the worst solution is proposed and that is embedded in local exploration phase of primary SFLA. In the proposed locally informed update process, the step size of the worst solution is decided on the basis local best or global best and a local randomly selected neighbour solution of memeplex. In this proposed LISFLA, local best and global best solutions are intensified the search space while, randomly selected neighbour solution is diversified the search area. Further, the proposed algorithm is compared with basic SFLA, its recent variant, namely Binomial Crossover Embedded Shuffled-Frog Leaping Algorithm (BC-SFLA) and three other nature inspired algorithms, namely Gravitational Search Algorithm (GSA), Differential Evolution (DE) and Biogeography-Based Optimization Algorithm (BBO). Experiments over the test functions, depicts that the LISFLA outplays to the considered algorithms.

References

1. Agarwal, P., Mehta, S.: Nature-inspired algorithms: state-of-art, problems and prospects. Int. J. Comput. Appl. **100**(14), 14–21 (2014)
2. Banks, A., Vincent, J., Anyakoha, C.: A review of particle swarm optimization. part ii: hybridisation, combinatorial, multicriteria and constrained optimization, and indicative applications. Nat. Comput. **7**(1), 109–124 (2008)
3. Bansal, J.C., Sharma, H.: Cognitive learning in differential evolution and its application to model order reduction problem for single-input single-output systems. Memetic Comput. **4**(3), 209–229 (2012)
4. Bansal, J.C., Sharma, H., Arya, K.V., Nagar, A.: Memetic search in artificial bee colony algorithm. Soft Comput. **17**(10), 1911–1928 (2013)

5. Blum, C., Li, X.: Swarm intelligence in optimization. In: Blum, C., Merkle, D. (eds.) Swarm Intelligence. Natural Computing Series, pp. 43–85. Springer, Heidelberg (2008)
6. Elbeltagi, E., Hegazy, T., Grierson, D.: A modified shuffled frog-leaping optimization algorithm: applications to project management. Struct. Infrastruct. Eng. **3**(1), 53–60 (2007)
7. Eusuff, M., Lansey, K., Pasha, F.: Shuffled frog-leaping algorithm: a memetic metaheuristic for discrete optimization. Eng. Optim. **38**(2), 129–154 (2006)
8. Karaboga, D.: An idea based on honey bee swarm for numerical optimization. Technical report, Technical report-tr06, Erciyes university, engineering faculty, computer engineering department (2005)
9. Passino, K.M.: Biomimicry of bacterial foraging for distributed optimization and control. IEEE Control Syst. **22**(3), 52–67 (2002)
10. Pourmahmood, M., Akbari, M.E., Mohammadpour, A.: An efficient modified shuffled frog leaping optimization algorithm. Int. J. Comput. Appl. **32**(1), 0975–8887 (2011)
11. Rashedi, E., Nezamabadi-Pour, H., Saryazdi, S.: GSA: a gravitational search algorithm. Inf. Sci. **179**(13), 2232–2248 (2009)
12. Sharma, H., Bansal, J.C., Arya, K.V.: Self balanced differential evolution. J. Comput. Sci. **5**(2), 312–323 (2014)
13. Sharma, S., Sharma, T.K., Pant, M., Rajpurohit, J., Naruka, B.: Centroid mutation embedded shuffled frog-leaping algorithm. Procedia Comput. Sci. **46**, 127–134 (2015)
14. Simon, D.: Biogeography-based optimization. IEEE Trans. Evol. Comput. **12**(6), 702–713 (2008)
15. Zhao, J., Lv, L.: Two-phases learning shuffled frog leaping algorithm. Int. J. Hybrid Inf. Technol. **8**(5), 195–206 (2015)

An Astute Artificial Bee Colony Algorithm

Avadh Kishor[1]([✉]), Manik Chandra[1], and Pramod Kumar Singh[2]

[1] Department of Computer Science and Engineering,
Indian Institute of Engineering and Technology, Roorkee, India
avadhkishor133@gmail.com, manikchandra.cse@gmail.com
[2] Computational Intelligence and Data Mining Research Laboratory,
ABV-Indian Institute of Information Technology and Management, Gwalior, India
pksingh@iiitm.ac.in

Abstract. Artificial bee colony (ABC) algorithm is one of the most popular optimization methods for global optimization over real-valued parameters. Though it has been shown very competitive to other natureinspired methods, it suffers from some challenging problems, e.g., slow convergence speed while solving unimodal problems, local optima stagnation (premature convergence) while dealing with the complex multimodal problems, and scalability problem in case of high dimensional problems. In order to circumvent these problems, we propose a new variant of the ABC, called *Astute Artificial Bee Colony* (AsABC) algorithm, which is able to maintain a better trade-off between two conflicting aspects, exploration and exploitation in the search space. In AsABC, we model a new search behavior of the onlooker bees to foster the solutions towards better region and to make the algorithm scalable. Performance of the AsABC is evaluated on a test suite of 12 benchmark functions of three different categories: unimodal, multimodal, and rotated multimodal. Comprehensive benchmarking and comparison of the AsABC with three other state-of-the-art variants of the ABC demonstrate its superior performance in terms of solution quality, scalability, robustness, and convergence speed.

Keywords: Artificial bee colony · Arithmetic recombination · Neighbor solution · Genetic crossover

1 Introduction

In 2005, Karaboga [1] developed an algorithm called ABC by modeling the intelligent foraging behavior of honey bees. Since then, It has emerged as a potential method to solve many real-world problems [2,3]. In most of the works pertaining to the ABC, generally a new version is designed to ameliorate its search ability. In this context, a lot of work has been done to improve performance of ABC for global optimization. Many researchers worked on the search equation of ABC and proposed various ABC variants. The existing variants of ABC can be grouped in two possible categories: (i) modification in the search equation and (ii) hybridization of ABC with other operators. A brief review of the existing work is delineated in subsequent paragraphs.

© Springer Nature Singapore Pte Ltd. 2017
K. Deep et al. (eds.), *Proceedings of Sixth International Conference on Soft Computing for Problem Solving*, Advances in Intelligent Systems and Computing 546, DOI 10.1007/978-981-10-3322-3_14

– *Modification in the search equation:* Akay et al. [4] observed that the convergence rate of ABC is very slow while solving complex problems. They claimed that the single dimensional perturbation of the solution is responsible for slow convergence. Therefore, they introduced two new parameters to control the frequency and magnitude of perturbation. Zhu et al. [5] pointed out that random neighbor selection strategy of ABC leads to poor exploitation. In this context, they integrated a best solution from the current population as an additional neighbor. Xiang et al. [6] suggested a new version of ABC by modifying search equations of both the onlooker bee and the scout bee. Kiran et al. [7] integrated direction with each solution vector to improve the local search ability. Karaboga et al. [8] proposed quick ABC with improved local search ability by selecting most profitable neighbor in spite of randomly selected neighbor.
– *Hybridization of ABC with other operators:* ABC has been hybridized by incorporating other operators into it. For example, Gao et al. [9] introduced a modified version of ABC, in which the mutation operator of DE is adopted to avoid the problem of premature convergence. Yan et al. [10], incorporated the crossover operator of GA into the ABC. Yang et al. [11] proposed a hybrid ABC-DE. The rationale behind the hybridization is to utilize the advantages of both the ABC and the DE.

Recently, Kishor and Singh [12] made a comprehensive study of ABC to divulge its pitfalls and proclivities. They discerned that ABC suffers from the problems of slow convergence rate, premature convergence and scalability. Search strategy of the ABC is itself responsible for the above mentioned drawbacks. Since, to find a new solution, a bee makes random selection of another bee which is associated with a different solution and communicate with her. Whilst, it is obvious that there is only 50 % chance for random selected bee to be good one. Therefore, this uncertainty entails all the issues.

In this paper, we propose new variant of ABC called AsABC which is geared towards solving complex optimization problems in the continuous search space. In AsABC, search strategies of the employed bees and the onlooker bees are quite different. The search equation of the onlooker bee is replaced with the arithmetic recombination operator of genetic algorithm (GA) as devised in [10]. Thus, the proposed AsABC belongs to the hybridization category as an arithmetic recombination operator of GA is introduced in to it. The rest of the paper is organized as follows. Standard ABC algorithm is briefed in Sect. 2. In Sect. 3, the AsABC is presented in detail. In Sect. 4, benchmark functions, tuning parameter for algorithms, and experimental results are presented and discussed. In last, a brief conclusion of the study is drawn in Sect. 5.

2   Standard ABC

Artificial bee colony (ABC) algorithm is a population-based stochastic optimization technique [1]. According to the foraging behavior of honey bees, the colony of bees is divided into three groups. The members of the first group are known as employed bees. Every employed bee keep a predefined position of a food source in her mind and go for that position to discern the richness (i.e., nectar amount) of that food. After discerning the richness of food source, each employed bee performs waggle dance in the dancing area of the hive. A group of bees who decide their food source by relying on the information disseminated by employed bees, are called onlooker bees. Onlooker bees exploit the food sources, albeit with a probability that corresponds to the qualities of the food sources. The third group of bees is scout bees which explores new food source without paying any heed for other bees.

In classical ABC algorithm, a food source position corresponds to a viable solution, the virtual foraging land represents the search space and the nectar amount of a food source signifies the fitness of that solution in the optimization process.

The step-by-step procedure of the algorithm is outlined in Algorithm 1 and a brief explanation is presented in the subsequent paragraphs.

Similar to other optimization algorithms, the ABC also requires the setting of some fundamental control parameters such as population size (number of food sources or colony size) $\mathcal{N}$, maximum cycle (generation) $\mathcal{G}_{max}$ as the stopping criterion, $limit$ to determine abandoned food source and $\mathcal{D}$ to signify a solution as a $\mathcal{D}$-dimensional parameter vector. In this paper, the ith solution in the population is represented as

$$\mathcal{X}^i = [x_1^i, .., x_j^i, .., x_{\mathcal{D}}^i]$$

where x_j^i denotes the jth dimension of the ith solution.

The x_j^i is initialized by (1).

$$x_j^i = x_l^i + \phi_j^i(x_{u_j}^i - x_{l_j}^i) \tag{1}$$

where ϕ_j^i is a uniformly distributed random number over (0,1). The upper and lower search bounds for a dimension j are denoted as $x_{u_j}^i$ and $x_{l_j}^i$, respectively. Initially, value of $limit$ for each solution is set to zero. The three phases of the ABC, i.e., employed bee phase, onlooker bee phase, and scout bee phase come to play after initialization of the solutions. The detail description of these three phases are as follows.

In the employed bee phase, every employed bee generate a new solution (candidate solution) by mutating one of its parameter (dimension) of the solution vector as shown in (2).

$$v_j^i = x_j^i + \psi_j^i(x_j^i - x_k^i) \tag{2}$$

where $k \in (1, \mathcal{N})$ such that $k \neq i$, j is a randomly chosen index, and ψ_j^i is a random number over $(-1,1)$.

After generating the candidate solution, its fitness is computed and compared with its corresponding original solution. The new solution replaces the old one if it is better than that and its $limit$ counter is set to zero; otherwise the old one remains in the population and its $limit$ value is incremented by one.

After the employed bee phase, an onlooker bee selects an affluent solution by performing stochastic sampling in the current generation. The selection probability p_i of a each solution i is calculated by (3).

$$p_i = \frac{fit_i}{\sum\limits_{i=1}^{N} fit_i} \tag{3}$$

Here, fit_i corresponds to the fitness of ith solution.

After selecting a solution through probability, a candidate solution is generated as in employed bee phase by (2). The replacement policy and setting of the *limit* counter are also similar to the employed bee phase.

A scout bee checks whether the value of *limit* counter of a solution exceeds a predefined threshold. If yes, then she replaces such a solution with a newly generated random solution using (1).

Algorithm 1. The Standard ABC Algorithm

```
1  begin
2     for i = 1 to N do
3        generate random solution by (1);
4        trial(i) :=0;

5     iter := 1;
6     while iter < Gmax do
7        for i = 1 to N do
8           generate candidate solution by (2);
9           apply greedy selection between candidate and original solution;
10          update the trial counter accordingly;

11       Calculate Probabilities for onlooker bees by (3);
12       for i = 1 to N do
13          select a solution according to the probability;
14          perform step 8;
15          perform step 9 and 10;

16       if trial(i) > limit then
17          generate random solution by (1);
18          trial(i) :=0;
```

---

3   Astute Artificial Bee Colony Algorithm

According to Karaboga [1], every employed bee exploits the food source which is already in her mind before leaving from the hive. On the other hand, every onlooker bee exploits a food source on the basis of information shared by employed bees. Thus, theoretically, the food source exploitation policies of employed bees and onlooker bees are absolutely different. But, it is quite noticeable that in the standard ABC, this remarkable difference is not considered and the search strategies of both the employed bees and the onlooker bees are modeled with same mathematical formula (2).

Moreover, to find a better solution in the vicinity of a solution $\mathcal{X}^i$, both employed as well as onlooker choose a companion of $\mathcal{X}^i$, say $\mathcal{X}^k$, randomly from the current population. Thereafter, the information sharing between these two solutions is carried out on random dimension. Though, this search mechanism may be reasonable for exploration, it is inadequate for the exploitation. The rationale behind this is very clear: (a) a bee $\mathcal{X}^i$ learns from its companion $\mathcal{X}^k$, while she herself is not sure that her companion $\mathcal{X}^k$ is good enough to guide the search towards a better region, and (b) Suppose, $\mathcal{X}^k$ is a better solution, yet, for deciding the next potential solution, two bees communicate over single dimension, which seems very poor in case of high dimensional problems.

In recent years, however, some researchers have made attempts to circumvent these inefficiencies. Zhu et al. [5] included the information of the best solution in the search equation of both employed as well as onlooker to improve exploitation and suggested a new version of ABC called gbest-guided ABC (GABC). However, Gao et al. [13] claim that GABC suffers from a problem called oscillation phenomenon which causes slow convergence. Further, Karaboga et al. [8] utilized the best solution for onlooker bees to enhance local search ability and proposed qABC. However, experimental study, clearly manifests that it is not able to avoid premature convergence particularly in case of complex multimodal problems because the information sharing between bees is still single dimensional. Yan et al. [10] suggested a hybridized ABC in which an extra phase is added. This extra phase includes the arithmetic crossover and binary tournament selection operators of GA. Consequently, the extra phase increases the computational complexity of the algorithm, which is unenviable for a good optimizer.

In this paper, without adding any extra phase, we make only two simple modifications in the onlooker bee phase of the original ABC and propose a new variant of ABC. Firstly, instead of choosing a neighbor solution $\mathcal{X}^k$ randomly, a solution with better fitness is selected as a neighbor solution, i.e., a solution whose probability p_i is greater than a random number ϕ (which is drawn from the interval (0,1)) is selected for companionship. Secondly, the search equation (2) is exchanged by (4) which is an arithmetic recombination operator of GA as devised in [10]. This remodeled ABC is named as *Astute Artificial Bee Colony* (AsABC) algorithm.

We do not present separate pseudo code for the AsABC as it is similar to ABC (Algorithm 1) except one minor change that is as follows. In AsABC step 14 of Algorithm 1 reads as follows: generate candidate solution by (4)

$$V^i = \phi_i * \mathcal{X}^i + \phi_k * \mathcal{X}^k \tag{4}$$

where ϕ_i and ϕ_k are uniformly distributed random numbers over (0,1).

4 Experimental Study and Discussion

4.1 Benchmark Functions

To evaluate the performance of AsABC, we do experiment over a test suite of 12 benchmark functions suggested in CEC 2005 [14] which are reported in Table 1.

Table 1. Benchmark functions Here, the global optimal value of each function is "0"; for f_9 to f_{12}, M is an orthogonal matrix and $y_i = M \times x_i$. Rotated is abbreviated as "Rot"

Objective function	Range
Sphere: $f_1(x) = \sum_{i=1}^{D} x_i^2$	$[-5, 5]$
Schwefel 2.22: $f_2(x) = \sum_{i=1}^{D}(\lvert x_i \rvert) + \prod_{i=1}^{D}(\lvert x_i \rvert)$	$[-100, 100]$
Rosenbrock: $f_3(x) = \sum_{i=1}^{D-1}(100(x_i^2 - x_{i+1})^2 + (1 - x_i)^2)$	$[-15, 15]$
Schaffer: $f_4(x) = \dfrac{0.5 + sin^2(\sqrt{\sum_{i=1}^{D}(x_i^2)} - 0.5)}{(1 + 0.001(\sum_{i=1}^{D}(x_i^2)))}$	$[-100, 100]$
Rastrigin: $f_5(x) = \sum_{i=1}^{D}(x_i^2 - \cos(2\pi x_i) + 10)$	$[-15, 15]$
Griewank: $f_6(x) = \frac{1}{4000}(\sum_{i=1}^{D} x_i^2) - (\prod_{i=1}^{D} \cos(\frac{x_i}{\sqrt{i}})) + 1$	$[-600, 600]$
Ackley: $f_7(x) = 20 + e - 20e^{(-0.2\sqrt{\frac{1}{D}\sum_{i=1}^{D} x_i^2})} - e^{(\frac{1}{D}\sum_{i=1}^{D} \cos(2\pi x_i))}$	$[-32, 32]$
Weierstrass: $f_8(x) =$ $\sum_{i=1}^{D}(\sum_{k=0}^{k_{max}}[a^k \cos(2\pi b^k z_i)]) - D \times \sum_{k=0}^{k_{max}}[a^k \cos(2\pi b^k \times 0.5)]$ $a = 0.5, b = 3, k_{max} = 20, z_i = (x_i + 0.5)$	$[-0.5, 0.5]$
Rot Rastrigin: $f_9(x) = \sum_{i=1}^{D}(y_i^2 - \cos(2\pi y_i) + 10)$	$[-15, 15]$
Rot Griewank: $f_{10}(x) = \frac{1}{4000}(\sum_{i=1}^{D} y_i^2) - (\prod_{i=1}^{D} \cos(\frac{y_i}{\sqrt{i}})) + 1$	$[-600, 600]$
Rot Ackley: $f_{11}(x) = 20 + e - 20e^{(-0.2\sqrt{\frac{1}{D}\sum_{i=1}^{D} y_i^2})} - e^{(\frac{1}{D}\sum_{i=1}^{D} \cos(2\pi y_i))}$	$[-32, 32]$
Rot Weierstrass: $f_{12}(x) = \sum_{i=1}^{D}(\sum_{k=0}^{k_{max}}[a^k \cos(2\pi b^k z_i)]) - D \times \sum_{k=0}^{k_{max}}[a^k \cos(2\pi b^k \times 0.5)]$ $z_i = (y_i + 0.5)$	$[-0.5, 0.5]$

The first three functions $f_1 - f_3$ are unimodal functions. The next five functions $f_4 f_8$ are multimodal functions with many local optimal solutions other than the global optimum. Finally, the last four functions $f9 f_{12}$ are rotated multimodal functions introduced in [15].

4.2 Experimental Setting

To analyze relative performance of the AsABC, it is compared with original ABC and two other state-of-the-art variants of ABC, i.e., GABC [5] and ABCDE [11]. The parameter setting adopted for all the algorithms is as follows. The population size is set to 100 (colony size), the number of scout bee in an iteration is at most one [1], and $limit = (\mathcal{N}* \mathcal{D}/2)$ [1]. For each test function, all the algorithms are simulated 30 times independently. The algorithm stops after 1,00,000 number of function evaluations(FEs). For the GABC, C = 1.5 [5] and for the ABC-DE, F = 0.5, CR = 0.9 [11].

4.3 Discussion of the Results

On the Quality of Solution. The mean and standard deviation (std) values yielded by different variants of ABC are summarized in Table 2.

Table 2. Mean and the standard deviation (std) values achieved by algorithms

ABCs		Functions					
		f_1	f_2	f_3	f_4	f_5	f_6
AsABC	mean	7.77E$-$19	1.32E$-$17	4.85E+01	0	0	0
	std	7.25E$-$19	7.32E$-$18	2.76E$-$01	0	0	0
ABC	mean	9.14E$-$15	4.78E$-$06	2.51E$-$01	4.63E$-$01	2.00E$-$01	1.27E$-$07
	std	6.19E$-$15	1.95E$-$06	2.38E$-$01	1.04E$-$02	4.04E$-$01	3.83E$-$07
GABC	mean	9.38E$-$15	4.14E$-$06	2.94E$-$01	4.61E$-$01	3.48E$-$02	2.91E$-$04
	std	4.28E$-$15	1.24E$-$06	2.71E$-$01	1.46E$-$02	1.82E$-$01	1.60E$-$03
ABC-DE	mean	3.61E$-$17	2.74E$-$10	2.78E+01	1.52E$-$02	7.98E$-$01	0
	std	9.62E$-$18	1.01E$-$10	9.29E$-$02	1.23E$-$02	1.09E+00	0
ABCs		Functions					
		f_7	f_8	f_9	f_{10}	f_{11}	f_{12}
AsABC	mean	8.88E$-$16	0	0	0	8.88E$-$16	0
	std	0	0	0	0	0	0
ABC	mean	5.68E$-$06	4.50E$-$04	3.24E+02	1.23E$-$02	7.19E+00	3.70E+01
	std	2.64E$-$06	1.32E$-$04	2.12E+01	1.03E$-$02	4.12E+00	1.15E+00
GABC	mean	6.27E$-$06	4.27E$-$04	3.24E+02	1.17E$-$02	6.65E+00	3.72E+0
	std	3.23E$-$06	1.18E$-$04	2.48E+01	7.86E$-$03	3.73E+00	1.10E+00
ABC-DE	mean	5.43E$-$11	1.21E$-$07	1.66E+02	2.83E$-$10	7.23E$-$11	1.17E$-$03
	std	4.26E$-$11	4.43E$-$08	1.38E+01	2.87E$-$10	1.60E$-$11	2.49E$-$04

Numerical results reported in Table 2, imply that the AsABC is more robust and reliable than other contestants. Furthermore, it is interesting to see that in all the multimodal and rotated multimodal functions, the AsABC yields the global optimal results, while others fail to obtain optimal solution, i.e., 0.0.

On the Convergence Rate. In order to compare the convergence speed of the algorithms, the convergence characteristics of 6 test functions (2 from each category) are illustrated in Fig. 1. From Fig. 1, it can be seen that the convergence speed of the AsABC is far better than the other competitors as it requires comparatively very less number of FEs to converge to global optimum.

On the Scalability of ABC Algorithms. In general, for most of the population based optimization algorithms, it is per se very difficult to solve high dimensional problems efficiently. It is because increase in the dimensionality of the search space increases its hyper-volume also which in turns deteriorates efficacy of the algorithm. Therefore, to check the scalability of the AsABC, we test it on three different functions. From Table 3, it is seen that AsABC is not only scalable but insensitive to dimensionality growth of the problems.

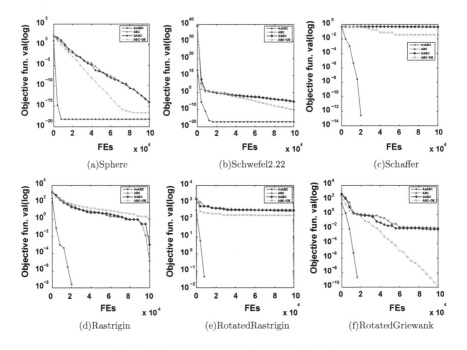

Fig. 1. Convergence graph on test functions

4.4 Time Complexity of RABC

As, we do not introduce any extra phase/operator in the standard ABC, the worse-case running time of the AsABC remains similar to ABC, i.e., $\mathcal{O}(\mathcal{N}.\mathcal{D}.\mathcal{G}_{max})$.

4.5 Effect of "Population Size" on AsABC

For investigating the effect of population size on AsABC, three benchmark functions, i.e., f_3, f_7, and f_9, are chosen from Table 1. Each function has different property. The AsABC is tested on these three function for colony size: 50, 100, 200, and 400. The rest of the parameter setting is as in above experiment. The results are shown in Table 4 which clearly indicate that the performance of the AsABC is not influenced by changing population size. Results for other 9 benchmark functions follow a similar trend and have not been included to save space.

4.6 Effect of "*limit*" on AsABC

Karaboga [1] anticipated that in honey bees colony, scout bees are 5–10 % of the population. They are responsible for exploration and random selection process and the number of scout is controlled by *limit* parameter. Thus, to assess the

Table 3. M = mean, S = std values obtained by AsABC, ABC, GABC, and ABC-DE at $\mathcal{D} = 30,50,100$ in 30 runs

F.		AsABC			ABC			GABC			ABC-DE		
		$\mathcal{D}=30$	$\mathcal{D}=50$	$\mathcal{D}=100$	$\mathcal{D}=30$	$\mathcal{D}=50$	$\mathcal{D}=100$	$\mathcal{D}=30$	$\mathcal{D}=50$	$\mathcal{D}=100$	$\mathcal{D}=30$	$\mathcal{D}=50$	$\mathcal{D}=100$
f_1	M	7E−19	1E−18	1E−18	9E−15	3E−08	3E−04	9E−15	1E−07	3E−04	3E−17	5E−17	1E−14
	S	7E−19	9E−19	1E−18	6E−15	3E−08	2E−04	4E−15	9E−08	3E−04	9E−18	1E−17	1E−14
f_2	M	1E−17	1E−17	1E−17	4E−06	3E−03	3E−01	4E−06	3E−03	3E−01	2E−10	1E−07	3E−05
	S	7E−18	6E−18	7E−18	1E−06	1E−03	8E−02	1E−06	1E−03	7E−02	1E−10	5E−08	2E−05
f_4	M	0	0	0	4E−01	4E−01	5E−01	4E−01	4E−01	5E−01	7E−01	9E−03	3E−02
	S	0	0	0	1E−02	2E−04	3E−06	1E−02	3E−04	2E−06	1E+00	2E−09	1E−10
f_5	M	0	0	0	2E−01	5E+00	6E+01	3E−02	4E+00	7E+01	0	1E+01	5E+01
	S	0	0	0	4E−01	1E+00	1E+01	1E−01	2E+00	1E+01	0	1E+00	1E+01
f_9	M	0	0	0	3E+02	7E+02	2E+03	3E+02	7E+02	2E+03	1E+02	3E+02	7E+02
	S	0	0	0	2E+01	4E+01	1E+02	2E+01	5E+01	1E+02	1E+01	2E+00	2E+01
f_{10}	M	0	0	0	1E−02	7E−02	8E−01	1E−02	5E−02	8E−01	2E−10	4E−08	1E−07
	S	0	0	0	1E−02	4E−02	1E−01	7E−03	2E−02	1E−01	2E−10	3E−08	7E−08

Table 4. Effect of colony size on the AsABC

	f_3				f_7				f_9			
	Colony size				Colony size				Colony size			
	50	100	200	400	50	100	200	400	50	100	200	400
mean	2E+01	4E+01	2E+01	2E+01	8E−16	8E−16	8E−16	8E−16	0	0	0	0
std	5E−01	2E−01	1E−01	1E−01	0	0	0	0	0	0	0	0

Table 5. Effect of *limit* value on the AsABC

	f_3				f_7				f_9			
	limit size				*limit* size				*limit* size			
	100	200	400	800	100	200	400	800	100	200	400	800
mean	2E+01	2E+01	2E+01	4E+01	8E−16	8E−16	8E−16	8E−16	0	0	0	0
std	3E−01	3E−01	3E−01	2E−01	0	0	0	0	0	0	0	0

performance of the AsABC in various *limit* values, we test it on three benchmark functions over different *limit* size. The statistical results are recorded in Table 5. It is interesting to see that the performance of the AsABC is not affected in different setting of *limit* value.

5   Conclusion

In this paper, we proposed a new version of ABC algorithm namely AsABC, which is based on intelligent foraging behavior of honey bees. In the AsABC, the search strategy of onlooker bees is different to the employed bees and is capable to maintain a proper trade-off between exploration and exploitation. In order to judge the efficacy of the AsABC, we compare it with other three variants of the ABC over a test bed of 12 benchmark functions. The superior performance of the AsABC lead us to assert that it is a better alternative for optimization. For future research, it would be very interesting to apply the AsABC in some relevant real-world application and extended it to multi-objective domain.

References

1. Karaboga, D., Basturk, B.: A powerful and efficient algorithm for numerical function optimization: artificial bee colony (ABC) algorithm. J. Global Optim. **39**(3), 459–471 (2007)
2. Kishor, A., Singh, P.K., Prakash, J.: NSABC: non-dominated sorting based multi-objective artificial bee colony algorithm and its application in data clustering. Neurocomputing **216**, 514–533 (2016). doi:10.1016/j.neucom.2016.08.003
3. Bharti, K.K., Singh, P.K.: Chaotic gradient artificial bee colony for text clustering. Soft Comput. **20**(3), 1113–1126 (2016)
4. Akay, B., Karaboga, D.: A modified artificial bee colony algorithm for real-parameter optimization. Inf. Sci. **192**, 120–142 (2012)
5. Zhu, G., Kwong, S.: Gbest-guided artificial bee colony algorithm for numerical function optimization. Appl. Math. Comput. **217**(7), 3166–3173 (2010)
6. Xiang, W.-L., An, M.-Q.: An efficient and robust artificial bee colony algorithm for numerical optimization. Comput. Oper. Res. **40**(5), 1256–1265 (2013)
7. Kıran, M.S., Fındık, O.: A directed artificial bee colony algorithm. Appl. Soft Comput. **26**, 454–462 (2015)
8. Karaboga, D., Gorkemli, B.: A quick artificial bee colony (QABC) algorithm and its performance on optimization problems. Appl. Soft Comput. **23**, 227–238 (2014)
9. Gao, W., Liu, S.: A modified artificial bee colony algorithm. Comput. Oper. Res. **39**(3), 687–697 (2012)
10. Yan, X., Zhu, Y., Zou, W., Wang, L.: A new approach for data clustering using hybrid artificial bee colony algorithm. Neurocomputing **97**, 241–250 (2012)
11. Yang, J., Li, W.-T., Shi, X.-W., Xin, L., Jian-Feng, Y.: A hybrid ABC-DE algorithm and its application for time-modulated arrays pattern synthesis. IEEE Trans. Antennas Propag. **61**(11), 5485–5495 (2013)
12. Kishor, A., Singh, P.K.: Comparative study of artificial bee colony algorithm and real coded genetic algorithm for analysing their performances and development of a new algorithmic framework. In: 2015 Second International Conference on Soft Computing and Machine Intelligence (ISCMI), pp. 15–19. IEEE (2015)
13. Gao, W., Liu, S., Huang, L.: A novel artificial bee colony algorithm based on modified search equation and orthogonal learning. IEEE Trans. Cybern. **43**(3), 1011–1024 (2013)
14. Suganthan, P.N., Hansen, N., Liang, J.J., Deb, K., Chen, Y.-P., Auger, A., Tiwari, S.: Problem definitions and evaluation criteria for the CEC 2005 special session on real-parameter optimization. KanGAL report, 2005005:2005 (2005)
15. Liang, J.J., Qin, A.K., Suganthan, P.N., Baskar, S.: Comprehensive learning particle swarm optimizer for global optimization of multimodal functions. IEEE Trans. Evol. Comput. **10**(3), 281–295 (2006)

Exploitative Gravitational Search Algorithm

Aditi Gupta$^{(\boxtimes)}$, Nirmala Sharma, and Harish Sharma

Rajasthan Technical University, Kota, India
aditi.2404.gupta@gmail.com, harish.sharma0107@gmail.com

Abstract. Gravitational search algorithm (GSA) is a simple well known meta-heuristic search algorithm based on the law of gravity and the law of motion. In this article, a new variant of GSA is introduced, namely Exploitative Gravitational Search Algorithm (EGSA). In the proposed EGSA, two control parameters (Kbest and Gravitational constant) are modified that play an important role in GSA. Gravitation constant G is reduced iteratively to maintain a proper balance between exploration and exploitation of the search space. Further, To enhance the searching speed of algorithm *Kbest* (best individuals) is exponentially decreased. The performance of proposed algorithm is measured in term of reliability, robustness and accuracy through various statistical analyses over 12 complex test problems. To show the competitiveness of the proposed strategy, the reported results are compared with the results of GSA, Fitness Based Gravitational Search Algorithm (FBGSA) and Biogeography Based Optimization (BBO) algorithms.

Keywords: Gravitational search algorithm · Swarm intelligence · Heuristic search algorithm · Elitism · Exponential · Gravitational constant

1 Introduction

Nature is an origin of inspiration for solving hard and complex problems. Nature-inspired algorithms (NIAs) are inspired by nature and used to deal with difficult real-world engineering problems [6]. Swarm intelligence algorithms [1,2] are inspired by any type of collective behaviours of individuals in nature. Gravitational search algorithm (GSA) [6] is a swarm intelligence type algorithm that is inspired by the Newton's physics concept gravitational force and motion of individuals in nature. Individuals fascinate each others by the gravity force and accelerate according to the force applied on individuals. Individuals with heavier masses have high attraction power compare to lower masses individuals. By this attraction power of individuals higher masses individuals move slowly as compare to lower masses individuals. GSA is an optimization algorithm and provides proper balancing between exploitation and exploration capabilities. So in this algorithm, heavier masses individuals are responsible for exploitation whereas lighter masses individuals are responsible for the exploration of the search area.

© Springer Nature Singapore Pte Ltd. 2017
K. Deep et al. (eds.), *Proceedings of Sixth International Conference on Soft Computing for Problem Solving*, Advances in Intelligent Systems and Computing 546, DOI 10.1007/978-981-10-3322-3_15

When searching process start lighter masses (individuals are far from the optimum solutions) individuals move with large step size (exploration) and after this when individuals converge to the optimum solutions i.e. higher masses individuals move with comparative small step size (exploitation). Researchers have been progressively produce new techniques to refine the performance of the algorithm [3,7,8,11].

In this paper, to improve the searching capability and diversification and intensification proficiency of GSA algorithm, a new variant of GSA is designed, namely Exploitative Gravitational Search Algorithm (EGSA). In the proposed EGSA, *Kbest* (best individuals) and Gravitational constant G are modified such that searching efficiency of GSA is increased and the solution explore the search space in early iteration having large step size while exploit the identified search region in later iterations with small step size. To balance the number of individuals that applied force to other individuals as the iteration is increased, Kbest value is exponentially decreased. Gravitational constant is also decreased through iterations that is responsible for the step size of individuals.

The remaining paper is organised as shown: In Sect. 2, a brief overview of GSA is illustrated. Exploitative GSA algorithm (EGSA) is proposed in Sect. 3. In Sect. 4, performance of EGSA is tested with several numerical benchmark functions. Finally, Sect. 5 gives a summary and conclude the work.

2   Gravitational Search Algorithm

E. Rashedi et al. developed Gravitational Search Algorithm (GSA) in 2009 [6]. GSA is a population-based stochastic search algorithm inspired by the Newton's gravitational law and movement of individuals in universe by gravitational force. According to Newton's gravity law *"Every individuals in-universe fascinate each other with force, this force is directly proportional to the product of individuals masses and inversely proportional to the square of the distance between individuals masses* [6]". Force applied to the individuals, by this force individuals are accelerated from their position. Performance of individuals is measure by their mass. Agents with higher mass are good as compare to lighter mass agents.

The GSA algorithm is described as follows: Each individual X_i in search space with I number of individuals is represented as:

$$X_i = (x_i^1, ..., x_i^d, ..., x_i^n) \qquad for\ i = 1, 2, ..., I, \qquad (1)$$

here x_i^d shows the position of ith individual in d dimensional area.

Mass of individuals is a based on the individuals fitness. The fitness of all individuals are calculated and worst and best fitness are identified for calculating the mass of individuals.

– For minimization problems best and worst fitness are:

$$best(g) = minfit_j(g) \qquad j \in 1, \cdots, I \qquad (2)$$

$$worst(g) = maxfit_j(g) \qquad j \in 1, \cdots, I \qquad (3)$$

– For maximization problems best and worst fitness are:

$$best(g) = maxfit_j(g) \qquad j \in 1, \cdots, I \tag{4}$$

$$worst(g) = minfit_j(g) \qquad j \in 1, \cdots, I \tag{5}$$

$maxfit_j(g)$ and $minfit_j(g)$ show the maximum and minimum fitness value of the jth individual at iteration g.

In GSA inertia, active and passive gravitational masses are equal. Individual with heavier masses are more efficient. Heavier masses individuals have higher attraction power and move slowly. Masses in GSA depend on the fitness value of individuals and calculated as follows:

$$M_{aj} = M_{pi} = M_{ii} = M_i, \qquad i = 1, 2, ..., I. \tag{6}$$

$$m_i(g) = \frac{fit_i - worst(g)}{best(g) - worst(g)} \tag{7}$$

$$M_i = \frac{m_i(g)}{\sum_{j=1}^{I} m_j(g)} \tag{8}$$

here M_{ii} and M_{pi} are inertia and passive gravitational masses of ith individual respectively and M_{aj} is active gravitational mass of jth individual. fit_i is the fitness value of ith individual.
$G(g)$ is the gravitational constant computed as Eq. 9.

$$G(g) = G_0 e^{(-\alpha g/MaxIt)} \tag{9}$$

Here, G_0 and α are constant and initialized at the starting. The value of $G(g)$ is reduced exponentially during each iteration for controlling search accuracy. $MaxIt$ is total number of iteration. Acceleration of individuals depends upon the ratio of force and mass of the individual [5] and calculated as follows:

$$a_i^d(g) = F_i^d(g)/M_{ii}(g) \tag{10}$$

$F_i^d(g)$ is the overall force acting on i^{th} individual computed as:

$$F_i^d(g) = \sum_{j \in Kbest, j \neq i} rand_j F_{ij}^d(g) \tag{11}$$

$Kbest$ is computed as follows:

$$Kbest = finalper + (1 - \frac{g}{MaxIt}) \times (N - finalper) \tag{12}$$

$$Kbest = round(N \times \frac{Kbest}{N}) \tag{13}$$

Here, $finalper$ is the constant and N is the total number of individuals in the search space. $Kbest$ is initial N individuals with the best fitness value and highest mass. Kbest will reduce linearly in each iteration and at the final only one individual applying force to the other individuals.

Force on i^{th} individual by j^{th} individuals mass during iteration g is computed using the following Eq. 14:

$$F_{ij}^d(g) = G(g).(M_{pi}(g) \times M_{aj}(g)/R_{ij}(g) + \varepsilon).(x_j^d(g) - x_i^d(g)) \qquad (14)$$

Here, $R_{ij}(g)$ is the Euclidian-distance between two individuals i and j at iteration g. Gravitational constant $G(g)$ is calculated using Eq. 9 while ε is a small constant. The velocity update equation for individuals is defined as:

$$v_i^d(g+1) = rand_i \times v_i^d(g) + a_i^d(g) \qquad (15)$$

here, $rand$ is random variable in interval [0, 1]. $v_i^d(g)$ and $v_i^d(g+1)$ are the velocity of ith individual at the iteration g and $g+1$ subsequently.

The position update equation for individuals is defined as:

$$x_i^d(g+1) = x_i^d(g) + v_i^d(g+1) \qquad (16)$$

here, $x_i^d(g)$ and $x_i^d(g+1)$ are the position of ith individual at the iteration g and $g+1$ subsequently. Velocity of individuals is updated during each iteration. Due to changes in the velocity every individual update its position.

This procedure is carry on until their termination criteria is met or iteration reach their maximum limit.

3　Exploitative Gravitational Search Algorithm

In the population-based algorithms, behavior of agent is measured by the exploitation and exploration capability in the search space. Exploration is for the enlarging the entire search space and exploitation is the finding optimum solution from the previously visited good solutions. During the early iteration of the algorithm, GSA visit the entire search space to find out the optimal solutions. After the lapse of iteration GSA exploit the search space by visiting the previously visited points. For the better performance of any population-based algorithm it is necessary to maintain a proper balance between the exploitation and exploration. Initially, when the individuals are not converged, exploration is needed to find out the good solutions in the whole search space. For the exploration large step size is necessary. After the lapse of iteration, individuals are converged. Hence for finding the optimal solution of the algorithm, individuals needs to exploit the search space (step size is comparatively less). In GSA gravitational constant G affects the step size of individuals. As mentioned in Eq. 14, force is directly proportional to the gravitational constant G and from the Eq. 10, acceleration is depend on the force of individuals. Acceleration in GSA plays a vital role for the step size of the individuals. Therefore in this paper gravitational constant G is modified as shown here:

$$G(g) = G_0 e^{(-\alpha g/MaxIt)}(1 - \frac{g}{MaxIt}) \qquad (17)$$

From the Eq. 17, it is clear that the value of gravitational constant will be high during the initial iteration and value is reduced iteratively. Therefore, the acceleration and step size of the individuals are decreased as the number of iteration increased.

$Kbest$ in Eq. 13 controls the number of individuals that apply the force to other individuals in search space. A large number of $Kbest$ (individuals) means large number of individuals interact with each other and movement between the individuals is high. As the result convergence speed is lower. From Eq. 13 it is clear that $Kbest$ is linearly decreased therefore change in $Kbest$ is very small as the number of iterations increase. Due to this movement and interaction between individuals is comparatively lower but the effect in convergence speed is not much. Therefore in this paper $Kbest$ is modified as shown here:

$$Kbest = round(N \times exp(-\beta g/MaxIt)) \tag{18}$$

Here, N is total number of individuals and β is constant. The value of $Kbest$ is reduced exponentially during each iteration. From the Eq. 18 it is clear that $Kbest$ is exponentially decreased. At initial iteration $Kbest$ is large therefore movement and interaction between the individuals is more that show the exploration of the search space. Whereas the number of iteration is increased $Kbest$ is reduced therefore movement and interaction between the individuals is comparative less that shows the exploitation of the individuals.

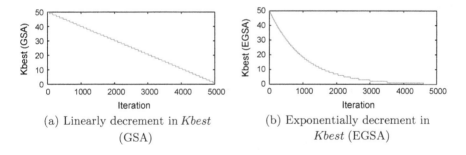

(a) Linearly decrement in $Kbest$ (GSA)

(b) Exponentially decrement in $Kbest$ (EGSA)

Fig. 1. Effect of $Kbest$ in number of individuals

Behaviour of $Kbest$ through iterations is shown in Fig. 1. It is clear from this figure that in EGSA, $Kbest$ is decreases exponentially through iterations whereas in GSA $Kbest$ is decreases linearly through iteration. So initially large number of individuals apply force as the number of iteration is increased comparatively less number of individuals apply force. So, EGSA regulates a proper balance between diversification and intensification proficiency and improve the searching ability as the number of iteration increase. The pseudo-code of EGSA is shown in Algorithm 1.

  – Identification of search area.
  – Creating a randomly dispersed set of individuals;

while *Stopping condition is not satisfied* **do**

  | – Calculate fitness of individuals.
  | – Calculate individuals mass by Eq. 6 and 7.
  | – Evaluate constant (G) by Eq. 17.
  | – Evaluate $Kbest$ by Eq. 18.
  | – Calculate force (F) for each direction by Eq. 14 and acceleration (a) for
  | every individuals by Eq. 10.
  | – Update individuals velocity by Eq. 17 and position by Eq. 18.

end

Algorithm 1. EGSA Algorithm

4   Results and Discussions

To examine the outcome of the EGSA, 12 different benchmark functions (f_1 to f_{12}) are picked as shown in Table 1.

  To certify the prosection of the proposed algorithm EGSA, a comparative analysis is carried out among EGSA, standard GSA, FBGSA [4] and BBO [10]. To authenticate the performance of the considered algorithms over the test problems, the experimental setting is given below:

– The number of simulations/run = 30,
– Number of population (N) = 50,
– $G_0 = 100$, $\alpha = 20$, $\beta = 5$ and $finalper = 2$,
– Experimental settings for the algorithms GSA, FBGSA [4] and BBO [10] are simulated from their primary research papers.

  Table 2 display the experimental results of the examine algorithms. A detailed analysis about the standard deviation (SD), mean error (ME), average number of function evaluations (AFE) along with the success rate (SR) are shown in Table 2. Results in Table 2 replicates, many times EGSA exceeds in terms of reliability, efficiency as well as accuracy as compare to the GSA, FBGSA and BBO.

  Further, Mann-Whitney U rank sum test [9] is performed at 5% level of remarkable ($\alpha = 0.05$) between EGSA - GSA, EGSA - FBGSA and EGSA - BBO. Table 3 display the compared results of mean function evaluation and Mann-Whitney test for 30 simulations. In Mann-Whitney test, we observe the remarkable difference between two data set. If remarkable difference is not seen then = symbol appears and when remarkable difference is observed then comparison is performed in terms of the AFEs. And we use + and - symbol, + represent the EGSA is superior than the examined algorithms and - represent the algorithm is inferior. The last row in Table 3, authorize the excellence of EGSA over GSA, FBGSA and BBO.

Table 1. Test problems, D: Dimension, AE: Acceptable Error

Test problem	Objective function	Search range	Optimum value	D	AE
Alpine	$f_1(x) = \sum_{i=1}^{D} |x_i \sin x_i + 0.1 x_i|$	$[-10\ 10]$	$f(0) = 0$	30	$1.0E-05$
Zakharov	$f_2(x) = \sum_{i=1}^{D} x_i^2 + \left(\sum_{i=1}^{D} \frac{i x_i}{2}\right)^2 + \left(\sum_{i=1}^{D} \frac{i x_i}{2}\right)^4$	$[-5.12\ 5.12]$	$f(0) = 0$	30	$1.0E-02$
Schwefel	$f_3(x) = \sum_{i=1}^{D} |x_i| + \prod_{i=1}^{n} |x_i|$	$[-10\ 10]$	$f(0) = 0$	30	$1.0E-05$
Step function	$f_4(x) = \sum_{i=1}^{D} (\lfloor x_i + 0.5 \rfloor)^2$	$[-100\ 100]$	$f(-0.5 \le x \le 0.5) = 0$	30	$1.0E-05$
Neumaier 3 problem (NF3)	$f_5(x) = \sum_{i=1}^{D}(x_i - 1)^2 - \sum_{i=2}^{n} x_i x_{i-1}$	$[-900\ 900]$	$f(0 = -(D*(D+4)*(D-1))/6.0)$	10	$1.0E-05$
Colville	$f_6(x) = 100[x_2 - x_1^2]^2 + (1-x_1)^2 + 90(x_4 - x_3^2)^2 + (1-x_3)^2 + 10.1[(x_2-1)^2 + (x_4-1)^2] + 19.8(x_2-1)(x_4-1)$	$[-10\ 10]$	$f(1) = 0$	4	$1.0E-05$
Gear train problem	$f_7(x) = ((1/6.931) - ((x(1)*x(2))/(x(3)*x(4))))^2$	$[-4\ 4]$	$f(1) = 4$	4	$1.0E-05$
Six-hump camel back	$f_8(x) = (4 - 2.1x_1^2 + x_1^4/3)x_1^2 + x_1 x_2 + (-4 + 4x_2^2)x_2^2$	$[-5\ 5]$	$f(-0.0898, 0.7126) = -1.0316$	2	$1.0E-05$
Easom's function	$f_9(x) = -\cos x_1 \cos x_2 e^{((-(x_1-\pi)^2 - (x_2-\pi)^2))}$	$[-100\ 100]$	$f(\pi, \pi) = -1$	2	$1.0E-13$
Hosaki problem	$f_{10}(x) = (1 - 8x_1 + 7x_1^2 - 7/3x_1^3 + 1/4x_1^4)x_2^2 \exp(-x_2)$, subject to $0 \le x_1 \le 5, 0 \le x_2 \le 6$	$[0\ 5],\ [0\ 6]$	-2.3458	2	$1.0E-05$
McCormick	$f_{11}(x) = \sin(x_1 + x_2) + (x_1 - x_2)^2 - \frac{3}{2}x_1 + \frac{5}{2}x_2 + 1$	$-1.5 \le x_1 \le 4,\ -3 \le x_2 \le 3$	$f(-0.547, -1.547) = -1.9133$	30	$1.0E-04$
Moved axis parallel hyper-ellipsoid	$f_{12}(x) = \sum_{i=1}^{D} 5i \times x_i^2$	$[-5.12\ 5.12]$	$f(x) = 0; x(i) = 5 \times i, i = 1 : D$	30	$1.0E-12$

Table 2. Comparison of the results of test functions, TP: Test Problem

TP	Algorithm	SD	ME	AFE	SR
f_1	EGSA	4.76E-07	9.44E-06	127540.00	30
	GSA	5.59E-07	9.30E-06	154615.00	30
	FBGSA	5.57E-07	9.25E-06	141046.67	30
	BBO	5.78E-03	1.04E-02	200000.00	0
f_2	EGSA	1.01E-04	9.89E-03	84075.00	30
	GSA	1.98E+00	3.61E+00	200000.00	0
	FBGSA	8.65E+00	4.04E+01	200000.00	0
	BBO	5.28E-03	1.77E-02	200000.00	0
f_3	EGSA	4.65E-07	9.50E-06	151536.67	30
	GSA	4.94E-07	9.42E-06	181821.67	30
	FBGSA	6.47E-07	9.23E-06	166096.67	30
	BBO	5.61E-02	5.61E-02	20000.00	0
f_4	EGSA	1.25E-04	1.54E-01	8996.67	30
	GSA	4.96E-03	2.56E+02	11583.33	30
	FBGSA	5.60E-04	4.15E-02	12635.00	30
	BBO	4.40E+01	4.80E-05	5351.67	30
f_5	EGSA	1.49E-06	8.45E-06	67220.00	30
	GSA	1.55E-06	7.73E-06	86376.67	30
	FBGSA	1.86E-06	7.45E-06	81458.33	30
	BBO	5.49E-02	1.42E-01	200000.00	0
f_6	EGSA	1.60E-04	9.02E-04	74060.00	30
	GSA	5.23E-02	1.88E-02	142220.00	26
	FBGSA	3.73E-01	9.92E-02	200000.00	0
	BBO	1.41E+00	2.73E-01	191210.00	3
f_7	EGSA	7.92E-13	1.83E-12	14473.33	30
	GSA	8.88E-13	1.85E-12	21290.00	30
	FBGSA	7.81E-13	1.89E-12	21995.00	30
	BBO	8.25E-13	1.79E-12	1090.00	30
f_8	EGSA	1.15E-05	1.14E-05	36455.00	30
	GSA	1.16E-05	1.17E-05	49801.67	30
	FBGSA	1.20E-05	1.22E-05	47368.33	30
	BBO	4.00E-01	3.26E-01	83296.67	18
f_9	EGSA	2.70E-14	3.90E-14	127580.00	30
	GSA	6.47E-02	6.67E-02	160010.00	28
	FBGSA	2.70E-14	5.34E-14	155450.00	30
	BBO	2.58E-14	3.10E-14	18711.67	30
f_{10}	EGSA	6.74E-06	6.33E-06	29600.00	30
	GSA	5.95E-06	5.50E-06	39686.67	30
	FBGSA	6.24E-06	6.29E-06	36780.00	30
	BBO	3.34E-16	1.22E+00	200000.00	0
f_{11}	EGSA	7.48E-06	8.88E-05	30623.33	30
	GSA	6.00E-06	8.96E-05	43526.67	30
	FBGSA	5.66E-06	8.89E-05	41330.00	30
	BBO	6.55E-06	8.96E-05	2600.00	30
f_{12}	EGSA	4.13E-14	9.53E-13	179236.67	30
	GSA	1.34E-12	1.34E-11	200000.00	0
	FBGSA	9.24E-14	8.73E-13	192680.00	30
	BBO	2.02E-03	5.94E-03	200000.00	0

Table 3. Comparison based on Mann-Whitney U rank test at significance level $\alpha = 0.05$ and mean function evaluations

Test problems	EGSA Vs GSA	EGSA Vs FBGSA	EGSA Vs BBO
f_1	+	+	+
f_2	+	+	+
f_3	+	+	+
f_4	+	+	-
f_5	+	+	+
f_6	+	+	+
f_7	+	+	-
f_8	+	+	+
f_9	+	+	-
f_{10}	+	+	+
f_{11}	+	+	-
f_{12}	+	+	+
Total number of + sign	12	12	08

Moreover, for comparison of examined algorithms, in form of consolidated achievement boxplots [2] study of AFE is carried out. Boxplot study efficiently describe the empirical circulation of data graphically. The boxplots for EGSA, GSA, FBGSA and BBO are depicted in Fig. 2. The results clearly show that interquartile range and medians of EGSA are relatively low.

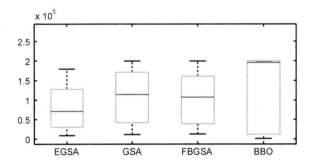

Fig. 2. Boxplots graphs (Average number of function evaluation)

To calculate the convergence speed of modified algorithm we use Acceleration Rate (AR) [9] which is represent as shown below:

$$AR = \frac{AFE_{compareAlgo}}{AFE_{EGSA}} \qquad (19)$$

Table 4. Test Problems: TP, Acceleration Rate (AR) of *EGSA* compare to the Standard *GSA*, *FBGSA* and *BBO*

TP	GSA	FBGSA	BBO
f_1	1.212286342	1.105901417	1.568135487
f_2	2.378828427	2.378828427	2.378828427
f_3	1.199852621	1.096082356	1.319812587
f_4	1.287513894	1.40440904	0.594849944
f_5	1.284984628	1.21181692	2.975304969
f_6	1.920334864	2.700513097	2.581825547
f_7	1.470981115	1.519691386	0.075310917
f_8	1.366113473	1.299364513	2.284917478
f_9	1.254193447	1.218451168	0.146666144
f_{10}	1.340765766	1.242567568	6.756756757
f_{11}	1.421356264	1.349624469	0.08490258
f_{12}	1.115843112	1.075003255	1.115843112

here, $compareAlgo \in$ (GSA, FBGSA and BBO) and $AR > 1$ means EGSA is faster than compared algorithms. For investigate the AR of modified algorithm it compared with standard GSA, FBGSA and BBO, results of Table 2 are analyzed and the value of AR is calculated using Eq. 19. It is cleared from the Table 4 that convergence speed of EGSA is faster than other examined algorithms.

5   Conclusion

This paper presents a variant of GSA algorithm, known as Exploitative Gravitational Search Algorithm (EGSA). In the modified version, two control parameters gravitational constant G and *Kbest* are modified. *Kbest* is number of best individuals that exponentially decreases with the number of iterations increases and it increase the searching speed of algorithm. Gravitational constant G is also deceased with number of iterations increased and it enhance the convergence speed. This methodology is reliable and efficient and maintain the proper balance between the exploitation and exploration proficiency of the algorithm. The proposed algorithm is compared with GSA, FBGSA and BBO over different benchmark functions. The obtained results state that EGSA is a competitive variant of GSA and also a good choice for solving the continuous optimization problems. In future, the newly developed algorithm may be used to solve various real world optimization problems of continuous nature.

References

1. Bansal, J.C., Sharma, H., Arya, K.V., Deep, K., Pant, M.: Self-adaptive artificial bee colony. Optimization **63**(10), 1513–1532 (2014)
2. Bansal, J.C., Sharma, H., Arya, K.V., Nagar, A.: Memetic search in artificial bee colony algorithm. Soft Comput. **17**(10), 1911–1928 (2013)
3. Guo, Z.: A hybrid optimization algorithm based on artificial bee colony and gravitational search algorithm. Int. J. Digital Content Technol. Appl. **6**(17) (2012)
4. Gupta, A., Sharma, N., Sharma, H.: Fitness based gravitational search algorithm. In: Proceedings of IEEE International Conference on Computing Communication and Automation. IEEE (Accepted 2016)
5. Holliday, D., Resnick, R., Walker, J.: Fundamentals of physics (1993)
6. Rashedi, E., Nezamabadi-Pour, H., Saryazdi, S.: Gsa: a gravitational search algorithm. Inf. Sci. **179**(13), 2232–2248 (2009)
7. Rashedi, E., Nezamabadi-Pour, H., Saryazdi, S.: Bgsa: binary gravitational search algorithm. Nat. Comput. **9**(3), 727–745 (2010)
8. Sarafrazi, S., Nezamabadi-Pour, H., Saryazdi, S.: Disruption: a new operator in gravitational search algorithm. Scientia Iranica **18**(3), 539–548 (2011)
9. Sharma, K., Gupta, P.C., Sharma, H.: Fully informed artificial bee colony algorithm. J. Exp. Theor. Artif. Intell. 1–14 (2015)
10. Simon, D.: Biogeography-based optimization. IEEE Trans. Evol. Comput. **12**(6), 702–713 (2008)
11. Sudin, S., Nawawi, S.W., Faiz, A., Abidin, Z., Rahim, M.A.A., Khalil, K., Ibrahim, Z., Md Yusof, Z.: A modified gravitational search algorithm for discrete optimization problem

A Systematic Review of Software Testing Using Evolutionary Techniques

Deepti Bala Mishra[1], Rajashree Mishra[2(✉)], Kedar Nath Das[3], and Arup Abhinna Acharya[1]

[1] School of Computer Engineering, KIIT University,
Bhubaneswar 751024, India
dbm2980@gmail.com, aacharyafcs@kiit.ac.in
[2] School of Applied Sciences, KIIT University, Bhubaneswar 751024, India
rajashreemishra011@gmail.com
[3] Department of Mathematics, NIT Silchar, Assam, India
kedar.iitr@gmail.com

Abstract. A best solution for decreasing software cost and reducing the cycle time during software development is automatic software testing and it has been seen by various organization. User specifications and requirements can be fully achieved by software testing. A number of issues are underlying in the field of software testing such as prioritization of test cases and automatic and effective test case generation are to be handled properly and they mostly depends on duration, cost and effort during the testing process. Testing can be done in two different ways such as manual testing and automatic testing by using different testing tools. Manual testing are very time consuming and this can be overcome by automatic testing by generating test cases automatically. Several types of evolutionary techniques like Genetic Algorithm, Particle Swarm Optimization and Bee Colony Optimization have been used for software testing. In this research paper, a survey of different evolutionary techniques used in software testing have been presented by taking the various issues in to account.

Keywords: Test data generation · Software testing · Genetic algorithm (GA) · Particle swarm optimization (PSO) · Bee Colony Optimization (BCO)

1 Introduction

Now-a-days automated software testing and developing of high quality test cases are two main objectives in the software industry. To support a high quality assurance of software, to create reliable, robust and trust worthy software or to deliver error free software, testing is performed by gathering required information of the software. It is also defined by the process of verification and validation, which meets the technical and business requirements [1, 2] in software development process. Testing is a most time consuming task which takes approximately 60% work load of the total software development time. If the testing is performed using automated testing then it will lead to reduce in software development cost by a significant margin [3–5].

© Springer Nature Singapore Pte Ltd. 2017
K. Deep et al. (eds.), *Proceedings of Sixth International Conference on Soft Computing for Problem Solving*, Advances in Intelligent Systems and Computing 546, DOI 10.1007/978-981-10-3322-3_16

A best solution for decreasing software cost and reducing the cycle time during software development is automatic software testing and it has been seen by various organization [6]. By using different software tools software can tested either manually or automatically. It is proved that automated software testing is better than manual testing as manual testing is a very time consuming and expensive task [7, 8]. Various types of techniques have been proposed by researchers and a lot of work has been done for software testing using soft computing techniques such as GA, Neural Network, genetic programming, fuzzy logic and evolutionary computing by providing high quality test data [8–10]. These techniques can be applied for test data generation to optimized problems.

This paper presents a survey of how different types of evolutionary techniques such as GA, PSO and BCO have been efficiently used in software testing and have been applied extensively for automated test data generation. Further the paper is partitioned into 4 sections. Section 1 presents the Introduction to software testing, Sect. 2 presents related work in the field of software testing using different types of evolutionary techniques, Sect. 3 contains a brief description about the working of GA, PSO and BCO and Sect. 4 gives a conclusion followed by our future work.

2 Related Work

This section provides a survey on different evolutionary techniques like GA, PSO and BCO used in software testing field for generating best test cases.

Last et al. [11], proposed a hybrid fuzzy based GA, which is an age extension of GA (FAexGA) to generate test cases for mutation testing. They found a very minimal set of test cases. The faults in test cases are exposed by the use of mutated versions of the original method. The proposed method uses a FLC (Fuzzy Logic Controller) for obtaining the probability of crossover. The probability of crossover differs according to the age intervals allocated during lifetime. The life time and age of chromosomes (parents) are defined by the FLC state variables. The truth value for obtaining Young-age, middle-age and old-age are shown in Table 1. Where,

Table 1. Fuzzy rule for cross over probability [14]

		Parent 1		
		Young-age	Middle-age	Old-age
Parent 2	Young-age	Low	Medium	High
	Middle-age	Medium	High	Medium
	Old-age	Low	Medium	Low

Age ∈ [Young-age, Middle-age, Old-age]
Crossover Probability ∈ [Low, Medium, High]

In their work an effective set of test cases are generated for a Boolean expression of 100 Boolean attributes by using three logical operators AND, OR, and NOT. An external application generates the correct expression randomly and one simple function is evaluated for each test case to generate an erroneous expression. Here a 100-bit lengthen binary strings of one dimension are generated as chromosomes.

Hla et al. [12], proposed a particle swarm optimization (PSO) algorithm based on modified software units for embedded real time software regression testing. The proposed algorithm prioritize the test cases automatically so that new higher priority test cases are selected for regression testing. The PSO algorithm successfully applied to the prioritization problem by taking solution as particle space and from which the best new positions of test cases, based on software unit can be found. Their results shows that the PSO algorithm can prioritized the test cases in the test suites by new best positions effectively and efficiently.

McCaffrey [13], developed a simulated BCO algorithm by which pair wise test sets can be generated to reduce the test set size as all systems are not supported for exhaustive testing with all possible inputs. The technique is a combinatorial NP-hard technique and it takes more time to generate test sets, which are far better than the test sets generated by deterministic approach.

Nachiyappan et al. [14], proposed a model based on genetic algorithm to decrease the cost of regression testing. Their proposed model creates population by taking the test history, the fitness value is calculated depending on the block based coverage value and run time of test case and the genetic operators are used for successive generations till the test cases with optimum value is found. They used Average Percentage of Faults Detected (APFD) metric to calculate the fitness function of individual test cases. The APFD can determine the effectiveness about cost, coverage, runtime and ordering of the new test case. The test cases are rejected which violates the specified time constraints. The model shows a good optimal sized test set by reduced test suite technique and the method is very highly adaptive as test case reduction is more effective when the fitness granularity is increased.

Kaur and Goyal [15], presented a BCO algorithm for fault coverage to a maximum limit. The authors have mapped the farmer bee's scenarios to prioritize the test suite. They explained their work by taking two examples like "college program for admission in courses" and "Hotel Reservation". In their work values have been compared using APFD (Average Percentage of Fault Detection) metrics and the proposed algorithm has been implemented in CPP compiler.

Ferrer et al. [16], presented two search based approaches as GTSG (Genetic Test Sequence Generator) and ACOts (Ant Colony Optimization approach for Test Sequence) for test sequence generation in functional testing with shortest valid path, which covers full transition and class. They used one CIT (Combinatorial Interaction Testing) approach, the classification tree method for test planning and design in functional testing. The authors defined the entire model as an extended classification tree to generate test sequences for a SUT (Software Under Test), which is needed for both industry and academia. Their first approach is GTSG with memory operator to preserve the memory for population evaluation as well as faster computation to get the solution. The second one is ACOts, which deals with large construction graphs. Test sequence can be generated with near-optimal solutions, where search spaces are separated. The authors performed the experiments using 12 software models by comparing their proposed approaches with greedy algorithm and they found their approaches can generate test sequences with shortest valid path, which covers full transition and class in functional testing.

Ankur and Srivastav [17], used GA to generate test data automatically for branch testing. They developed an improved approach which focuses on branch ordering, memory and elitism. The authors have discussed about DFS (Depth First Strategy), BFS (Breadth First Strategy) and PPS (Path Prefix Strategy) for ordering the branches, which are to be covered for testing. For improve test data they used elitism and memory with branch orderings. They compared each strategy with RAN and RNS and found best results with a mean number of generations and longer populations.

Andalib and Babamir [18], used PSO in discrete space for generating test data where there is no data dependency between program lines in a software. They proposed a method that produced minimum numbers of Test Case (TC) automatically with highest covering of codes in a program. In their method Mc Cabe theory was used to find the independent paths by reducing the number of paths in a program for selecting the best test case. Investing the motion of all the particles (birds/fish), the fitness function was taken for an optimal solution. They executed an integrated sorting program and with only one TC, they found 75% of the regions and 50% of independent paths are covered. The authors has compared their proposed algorithm with GA to covered 100% of independent paths and found more efficient result.

Dixit and Tomar [19], developed and implemented a hybrid algorithm GPSHA (Genetic Particle Swarm Hybrid Algorithm) combining the power of GA and PSO and they found a less number of generations and less number of test cases which covers around 100% of a program. Their results confirmed the effectiveness of the GPSHA over GA and PSO after performing in real world problems.

Sharma et al. [20], implemented GA in software testing to increase the efficiency and process time of testing. They generate test cases by using GA in Ruby, C++ and Matlab. It is found that the best fitness function is evaluated to a population of 50 and maximum generation 500. When the stopping condition is satisfied the iterative generation is stopped by providing an optimized and unique solution.

Yang et al. [21], developed a new intelligent search based algorithm RGA (Regenerate Genetic Algorithm) to increase test coverage, search efficiency, restrain population aging and produce less number of test cases for coverage oriented software testing. They found RGA can give better optimized solution for large scale, highly complex problems and solve the population aging problem. After comparing with GA and random test method, authors found RGA is more efficient for required coverage criteria of test cases and achieving greater test coverage with fewer iterations and test cases.

Shahbazi and Miller [22], used a multi objective optimization in black box string test case generation for random testing and adaptive random testing. The authors performed their experiments by taking six different types of string distance functions such as Levenstein, Hamming, Cosine, Manhattan, Cartesian and Locality-Sensitive Hashing, to find effectiveness and run time of test cases. They introduce two objectives for effective string test cases such as the length distribution of the string test cases and the diversity control of the test cases within a test set. They used one diversity- based fitness function to generate optimized test sets to reveal faults more effectively and found superior test cases are produced by applying the objectives.

Zhenga et al. [23], developed a decomposition based multi-objective evolutionary algorithm (MOEA/D) for regression testing of programs from SIR repository. The experiments are evaluated in four approaches such as NSGA-II (non-dominated sorting genetic algorithm, MOEA/D (parameter c, used in normalization is fixed), MOEA/D (c is chosen from tuning) and classic greedy algorithm. The authors compared their work with Yoo and Harman [24] multi objective approaches where they used greedy algorithm and two versions of NSGA-II. They found among all the approaches MOEA/D with varying c is most effective and it produce the lowest HV(Hyper Volume) values with cheapest test suite. The two variants of MOEA/D have superior performance in comparison to NSGA-II and greedy algorithm.

After an extensive study of different evolutionary techniques used in software testing, we came to learn GA, PSO and BCO are used efficiently for generating test cases and solving many complex problems. Table 2 shows a brief summary of different evolutionary algorithm used in software testing and the results found in different related work has been already done.

Table 2. A brief summary of different evolutionary algorithm used in software testing and results found.

Authors	Problem discussed and solved	Algorithm used	Work done in particular area	Results
Mark Last et al. [11]	Generate test cases for mutation testing.	GA	Used a FLC (Fuzzy Logic Controller) for obtaining the probability of crossover	FAexGA is efficient as the rate of finding error is very fast and number of solution is distinct.
Hla et al. [12]	Prioritize test cases to increase effectiveness in regression testing for embedded real time software.	PSO	Focused on coverage based prioritization of test suite.	PSO algorithm can prioritized the test cases in very effectively and efficiently
McCaffrey [13]	Reducing test set in pair wise testing	BCO	Combinatorial NP hard and Pair wise Testing	Test cases are far better than the test sets generated by deterministic approach.
Nachiyappan et al. [14]	Decrease the cost of regression testing by reducing the test suite.	GA	APFD is used to determine the effectiveness of test cases.	The method is very highly adaptive as test case reduction is more effective with increase of fitness granularity.

(*continued*)

Table 2. (*continued*)

Authors	Problem discussed and solved	Algorithm used	Work done in particular area	Results
Kaur and Goyal [15]	Fault based test suit prioritization.	BCO	Used APFD (Average Percentage of Fault Detection) metrics and CPP compiler.	Maximum numbers of faults are covered in regression testing.
Ferrer et al. [16]	Test sequence generation with shortest valid path to cover transition and class.	GA & ABC	CIT (Combinatorial Interaction Testing), the extended classification tree method.	Generate test sequences with shortest valid path, which covers full transition and class in functional testing.
Ankur and Srivastav. [17]	Generate test data automatically for branch testing.	GA	Focused on branch ordering, memory and elitism.	Generate best results with a mean number of generations and longer populations.
Andalib and Babamir [18]	Generating minimum number of Test Case (TC) automatically with highest covering of codes in a program.	PSO	Used Mc Cabe theory to find the independent paths for selecting the best test case.	Covered 100% of independent paths and found more efficient result.
Dixit and Tomar [19]	Generation of Less and unique numbers of test cases.	GA & PSO	Combining the power of GA and PSO	GPSHA results a less number of generations and less number of test cases and covers around 100% of a program.
Sharma et al. [20]	Increase the efficiency and process time of testing.	GA	Generate test cases by using GA in Ruby, C ++ and Matlab.	Providing an optimized and unique solution for testing.

(*continued*)

Table 2. (*continued*)

Authors	Problem discussed and solved	Algorithm used	Work done in particular area	Results
Yang et al. [21]	Judging the population aging process.	GA	Used population regeneration strategy	RGA is more efficient by reducing the number of test cases and achieving greater test coverage with fewer iterations and test cases.
Shahbazi and Miller [22]	Generate effective set of black box string test cases through multi objective optimization.	GA & MOGA	Used several string distance functions to find effectiveness and run time of test cases.	Superior test cases are produced by using multi objective optimization technique.
Zhenga et al. [23]	To achieve full coverage for regression testing	MOEA & GA	Used MOEA/D with a normalization parameter c to solve multi objective optimization problem	MOEA/D have superior performance in comparison to NSGA-II and greedy algorithm.

3   An Introduction to Evolutionary Algorithm

Evolutionary algorithm based on biological behavior or evolution of population, which can be used to solve many complex and real life problems by producing high quality test data automatically [15, 23]. This algorithm is based on the principle of survival of the fittest and models some natural phenomena like genetic inheritance and Darwinian strife for survival, constitute an interesting category of modern heuristic search [9, 19]. Figure 1 shows the work flow of evolutionary technique.

3.1   Genetic Algorithm (GA)

GA has emerged as a practical, robust optimization technique and search method and it is inspired by the way nature evolves species using natural selection of the fittest individuals.

The algorithm was developed by John Holland in United States [14]. The solution to a specific problem can be solved by a population of chromosomes. A chromosome is

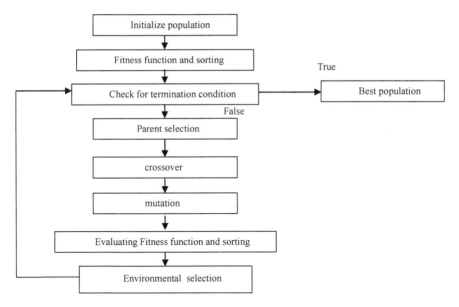

Fig. 1. Work flow of evolutionary technique

a string of binary digits and each digit is called a gene and population can be created randomly. It is a best way to solve optimization problems by searching for good genes and applying the different genetic operators like selection, crossover, mutation and Elitism [11, 17, 22].

Selection: A selection operation is performed to determine the individuals that meets the fitness function where fitness function is a specific function depending upon the criteria which returns a number indicating the acceptability of the program. This function is used in the selection process to determine the optimum point and the variants survive to the next iteration [8, 21]. Selection methods are of six different types such as roulette wheel, stochastic universal sampling, linear rank, exponential rank, binary tournament and truncation.

Crossover or Recombination: After selection, the crossover operation is applied to the selected chromosomes, which swaps genes or sequence of bits in the string between two individuals. For binary encoding different types of crossover operators are used like one point, two point, uniform and arithmetic. Cross over process is repeated with different parent individuals. Finally the mutation operator is applied to a randomly selected subset of the population [17, 20].

Mutation: It is used to maintain genetic diversity in the population by altering chromosomes to introduce new good traits. Basically six types of mutation operators are used in Genetic algorithm such as Bit string, flip bit, boundary, uniform, non uniform and Gaussian [17, 21].

Elitism: Elitism process involves copying a small proportion of the fittest candidates into the next generation, which are related to the best solution found [17].

The basic process of Genetic algorithms mainly involves creating an initial set of random solutions (population) and evaluating them [21, 23], by using the GA operators, in which the better solutions are identified (parents) and are then used to generate new solutions (children). These values can be used to replace with other population. This new population (generation), is then reevaluated and the process for generating new values continues until a final solution is found based on a specified condition of the fitness function [14]. Finally the function minimization is applied to the fitness function for test data generation.

3.2 Particle Swarm Optimization Algorithm (PSO)

PSO is a search based optimization technique that studies the social behavior of bird flocking or fish schooling. This algorithm mainly based on the movement and intelligence of swarms [18, 19]. The best solution can be found by a number of particles constituting a swarm, moving around in a particular search space of N-dimensional and adjusting their flying according to own and other's flying experience. Particles are always keeping track for personal best solution, denoted by *p-best* and the best value of any particle, denoted by *g-best*. Simultaneously the speed is adjusted dynamically of each particle depending on flying experiences. The velocity of each particle can be changed by considering the parameters like current position and velocity, distance between current position and its *p-best* as well as the distance between current position and its *g-best* [12, 19].

3.3 Bee Colony Optimization Algorithm (BCO)

Bee Colony Optimization (BCO) is a special type of Swarm Intelligence (SI), where the honey bees are the agents of the group. They communicate with each other by "Waggle Dance" principle to exchange information about the location for rich food source. In this system there is a well coordinated interaction between bees of a particular colony, organized team work and simultaneous task performance [13, 15].

In a bee colony different types of bees are present like a queen bee, many male drone bees and thousands of worker bees where the Queen is responsible to lay eggs for creating new colonies the male drones are responsible to mate with the Queen. At the time of downfall, male drones are discarded from the colony. The females of the hive are the worker bees. They are main responsible to build blocks of the hive as well as to comb, clean, maintain, guard the hive, search and collect rich food to feed the queen and drones. The worker bees are of two types such as forager bees and scout bees. The scout bees search food sources randomly and after finishing their distance limits they return back to the hive to give the information to foragers by "Waggle Dance" principle. Finally after observing the direction and information regarding location of rich food sources the foragers start flying to collect food [13, 24]. BCO algorithms are used to solve diverse domains problems, bench mark problems like routing problems, NP-hard problems and Travelling Salesman Problems [15].

4 Conclusion and Future Work

In this review paper we analyzed how different types of evolutionary techniques such as GA, PSO, ABCO and BCO have been efficiently used in software testing and have been applied extensively for automated test data generation. The results and performance of testing can be improved by these techniques. The evolutionary generation of test cases is proved to be very efficient and cost effective than manual testing. In future, we planned to combine the power of GA, PSO, ABCO and BCO in such a way that the new hybridized algorithm can produce a less number of test generations from which best test cases can be achieved for software testing. It is also planned to develop a new algorithm to generate test cases randomly and further optimize to find the best test cases.

References

1. Chauhan, N.: Software Testing: Principles and Practices. Oxford University Press, Oxford (2010)
2. Jogersen, P.C.: Software Testing: A Craftsman Approach, 3rd edn. CRC Presses, Boca Raton (2008)
3. Srivastava, P.R., Kim, T.H.: Application of genetic algorithm in software testing. Int. J. Softw. Eng. Appl. **3**(4), 87–96 (2009)
4. Berndt, D.J, Watkins, A.: High volume software testing using genetic algorithms. In: Proceedings of the 38th Annual Hawaii International Conference on System Sciences – Volume 09, vol. 9, pp. 318–326. IEEE Computer Society, Washington, DC (2005)
5. Wang, J., Changan, W., Shouda, J.: Test data generation algorithm of combinatorial testing based on differential evolution. In: Third International Conference on IEEE Instrumentation, Measurement, Computer, Communication and Control (IMCCC) (2013)
6. Vahid, G., Mäntylä, M.K.: When and what to automate in software testing? A Multi-Vocal Lit. Rev., Inf. Softw. Technol. **76**, 92–117 (2016)
7. Vudatha, C.P., Nalliboena, S., Jammalamadaka, S.K., Duvvuri, B.K.K., Reddy, L.: Automated generation of test cases from output domain of an embedded system using genetic algorithms. In: 3rd International Conference on Electronics Computer Technology (ICECT), vol. 5. IEEE (2011)
8. Sharma, C., Sabharwal, S., Sibal, R.: A survey on software testing techniques using genetic algorithm. arXiv preprint arXiv, pp. 1411–1154 (2014)
9. Wappler, S., Lammermann, F.: Using evolutionary algorithms for unit testing of object oriented software. In: GECCO, pp. 1925–1932. ACM (2005)
10. Goldberg, D.E: Genetic Algorithms: In Search, Optimization and Machine Learning. Addison Wesley, MA (1989)
11. Last, M., Eyal, S., Kandel, A.: Effective black-box testing with genetic algorithms. In: Ur, S., Bin, E., Wolfsthal, Y. (eds.) HVC 2005. LNCS, vol. 3875, pp. 134–148. Springer, Heidelberg (2006). doi:10.1007/11678779_10
12. Hla, K.H.S., Choi, Y., Park, J.S.: Applying particle swarm optimization to prioritizing test cases for embedded real time software retesting. In: IEEE 8th International Conference on Computer and Information Technology Workshops, CIT Workshops 2008, pp. 527–532. IEEE, July 2008

13. McCaffrey, J.D.: Generation of pair wise test sets using a simulated bee colony algorithm. In: IEEE International Conference on Information Reuse and Integration, IRI 2009. IEEE (2009)
14. Nachiyappan, S., Vimaladevi, A., Selva Lakshmi, C.B.: An evolutionary algorithm for regression test suite reduction. In: 2010 International Conference on Communication and Computational Intelligence (INCOCCI), pp. 503–508. IEEE, December 2010
15. Kaur, A., Goyal, S.: A survey on the applications of bee colony optimization techniques. Int. J. Comput. Sci. Eng. **3**(8), 30–37 (2011)
16. Ferrer, J., Kruse, P.M., Chicano, F., Enrique Alba, E.: Evolutionary algorithm for prioritized pairwise test data generation. In: Proceedings of the 14th Annual Conference on Genetic and Evolutionary Computation, pp. 1213–1220. ACM (2012)
17. Ankur, P., Srivastav, G.: Automated test data generation for branch testing using genetic algorithm: an improved approach using branch ordering, memory and elitism. J. Syst. Softw. **86**(5), 1191–1208 (2013)
18. Andalib, A., Babamir, S.M.: A new approach for test case generation by discrete particle swarm optimization algorithm. In: The 22nd Iranian Conference on Electrical Engineering (ICEE), May 20–22. Shahid Beheshti University (2014)
19. Dixit, S., Tomar, P.: Automated test data generation using computational intelligence, Reliability. In: 4th International Conference on Infocom Technologies and Optimization (ICRITO) (Trends and Future Directions). IEEE (2015)
20. Sharma, A., Rishon, P., Aggarwal, A.: Software testing using genetic algorithms. Int. J. Comput. Sci. Eng. Surv. (IJCSES) **7**(2), 21–33 (2016). doi:10.5121/ijcses
21. Yang, S., Man, T., Xu, J., Zeng, F., Li, K.: RGA: a lightweight and effective regeneration genetic algorithm for coverage-oriented software test data generation. Inf. Softw. Technol. **76**, 19–30 (2016)
22. Shahbazi, A., Miller, J.: Black-box string test case generation through a multi-objective optimization. IEEE Trans. Softw. Eng. **42**(4), 361–378 (2016)
23. Zheng, W., Hierons, R.M., Li, M., Liu, X., Vinciotti, V.: Multi-objective optimisation for regression testing. Inf. Sci. **334**, 1–16 (2016)
24. Yoo, S., Harman, M.: Regression testing minimization, selection and prioritization: a survey. Softw. Test. Verification Reliab. **22**(2), 67–120 (2012)

On the Hybridization of Spider Monkey Optimization and Genetic Algorithms

Anivesh Agrawal[1], Pushpa Farswan[2(✉)], Vani Agrawal[3], D.C. Tiwari[4], and Jagdish Chand Bansal[2]

[1] Indian Institute of Technology Kanpur, Kanpur, India
anivesh.agrawal@gmail.com
[2] South Asian University, New Delhi, India
pushpafarswan6@gmail.com, jcbansal@gmail.com
[3] Prestige Institute of Management, Gwalior, India
vaniagrawal.mca@gmail.com
[4] Jiwaji University, Gwalior, India
dctiwari2001@yahoo.com

Abstract. Genetic algorithm (GA) is adaptive heuristic search evolutionary algorithm. GA has had a great measure of success in the optimization process. Spider monkey optimization (SMO) is the relatively new swarm intelligence algorithm. SMO inspired by food foraging behavior of spider monkeys. We introduce a new idea that integrates swarm intelligence and evolutionary technique into the optimization process. In this article, we propose two hybridization methodologies for SMO and GA, namely SMOGA (SMO followed by GA) and GASMO (GA followed by SMO) for the numerical optimization problems. These algorithms effectiveness have been tested here on both its "ancestors", SMO and GA for various benchmark problems.

Keywords: Spider monkey optimization · Genetic algorithms · Hybridization · Metaheuristic

1 Introduction

Metaheuristic algorithms are solving tool for real-world optimization problems using stochastic techniques. Each algorithm has an own skill to solve optimization problems. In the literature various metaheuristic algorithms such as genetic algorithm (GA) [7], ant colony optimization (ACO) [3], particle swarm optimization (PSO) [4], differential evolution (DE) [20], bacterial foraging optimization (BFO) [16], artificial bee colony (ABC) [12], biogeography-based optimization (BBO) [19], harmony search algorithm (HSA) [24] and spider monkey optimization (SMO) [2] etc. are available. These metaheuristic algorithms developed according to the demand of the problems. In view of no free lunch theorem given by Wolpert and Macready [22], there is no such algorithm which can beat the other algorithm for solving all optimization problems motivates researchers to

© Springer Nature Singapore Pte Ltd. 2017
K. Deep et al. (eds.), *Proceedings of Sixth International Conference on Soft Computing for Problem Solving*, Advances in Intelligent Systems and Computing 546, DOI 10.1007/978-981-10-3322-3_17

develop better algorithms. Therefore hybridization is the new view to perform better in the optimization process. Hybridization is the agreement or correlation of different schemes in time. In this paper, we considered an evolutionary algorithm (real coded GA) with swarm intelligence technique (SMO) to solve numerical optimization problems. Some hybridized schemes with classical GA are as follows: In [9], Juang et al. developed hybrid genetic algorithm and particle swarm optimization (HGAPSO). In HGAPSO, new individuals generation function mimics the social behavior of animals, breeding and survival of the fittest and applied to recurrent neural/fuzzy network design. Lee et al. [13] proposed a hybrid search algorithm with the advantage of genetic algorithm and ant colony optimization that performs better exploration and exploitation skills. Hwang et al. [8] proposed a hybrid genetic algorithm called a novel adaptive real-parameter simulated annealing genetic algorithm (ARSAGA). In ARSAGA, the crossover is maintained by GA, and mutation is utilized by GA and SA. Harada et al. [6] developed a scheme with hybridization of GA and local search (LS) and applied in multi objective optimization problem. In this hybridized scheme the advantages of both GA and LS exploits maximally. Grimaccia et al. [5] presented genetical swarm optimization (GSO) combined by PSO and GA has a characteristic to set the parameter as hybridization coefficient (hc) in order to adjust itself to any specific problem. GA-PSO developed by Kao et al. [11] is the hybridized optimization algorithm incorporates the concept of GA and PSO. In GA-PSO, individuals in each iteration improved based on social interaction and their personal experience. Wahed et al. [1] introduced a hybrid approach combined by PSO and GA for solving nonlinear optimization problems. Pan et al. [15] developed an improves multi-agent genetic algorithm (IMAGA) and Tian et al. [21] presented a hybrid adaptive genetic algorithm with chaos searching technique for numerical optimization.

This paper proposed two hybridized algorithms using genetic algorithm (GA) and spider monkey optimization (SMO) technique. These proposed algorithms are spider monkey optimization followed by genetic algorithm (SMOGA) and genetic algorithm followed by spider monkey optimization (GASMO). Further, these proposed algorithms are tested on various benchmark test problems.

The remaining content of the paper is organized as follows: In Sect. 2, a brief explanation of spider monkey optimization and genetic algorithm have given. Details of proposed hybridization are given in Sect. 3. Section 4 describes the numerical experiments and discussion. The paper is concluded in Sect. 5.

2   Spider Monkey Optimization and Genetic Algorithm

Swarm intelligence and evolutionary algorithm both are stochastic techniques. Spider monkey optimization (SMO) is swarm intelligence algorithm and genetic algorithm (GA) is an evolutionary algorithm. In this section, a brief discussion of SMO and GA are introduced.

2.1 Spider Monkey Optimization

Spider Monkey Optimization Algorithm (SMO) is a relatively new swarm intelligence algorithm based on the food foraging behavior of spider monkeys. SMO is fast emerging swarm intelligence technique which incorporates meta-heuristic approach to solve global optimization problems. The algorithm is inspired by foraging behavior of intelligent animals including spider monkeys which use fission-fusion social structure (FFSS) to efficiently locate food resources. The algorithm comprises 7 major phases:

Initializing population: In this phase, N potential solutions (each of dimension D) are uniformly seeded across the search space. Each spider monkey initialized as follows:

$$sm_{ij} = sm_{minj} + u(0,1) \times (sm_{maxj} - sm_{minj}) \qquad (1)$$

Where sm represents the spider monkey. Here $i \in \{1, 2, ..., N\}$ and $j \in \{1, 2, ..., D\}$. u(0,1) is the uniformly distributed random number between 0 and 1. sm_{minj} is the lower bound of sm_{ij} and sm_{maxj} is the upper bound of sm_{ij}.

Local Leader Phase (LLP): In local leader phase, each spider monkey updates its position based on the experience of the local group leader and fellow group members. The fitness value of the updated position is computed and the monkey updates its position if the value is greater than its previous value. In this phase spider monkey updated as follows:

$$sm(new)_{ij} = sm_{ij} + u(0,1) \times (ll_{kj} - sm_{ij}) + u(-1,1) \times (sm_{rj} - sm_{ij}) \qquad (2)$$

Where ll represents the local group leader. u(-1,1) is the uniformly distributed random number between -1 and 1. Here $r \neq i$, $r \in \{1, 2, ..., N\}$, $k \in$ local group.

Global Leader Phase (GLP): After the completion of local leader phase, global leader phase starts. In this phase, spider monkeys update their position based on the experience of global leader and fellow group members. The position update in this phase is based on the probability values which is computed using fitness. Hence, the one with higher fitness has more chances to update its position and reach the optimum. The new fitness value is calculated and position is updated in case of better fitness value. The new position of spider monkey is updated in this phase as follows:

$$sm(new)_{ij} = sm_{ij} + u(0,1) \times (gl_j - sm_{ij}) + u(-1,1) \times (sm_{rj} - sm_{ij}) \qquad (3)$$

Where gl represents the global leader.

Global Leader Learning Phase (GLL): In this phase, the position of the global leader is updated making a greedy selection amongst the whole population. Hence, the monkey with best fitness value is chosen as the new global leader. In case the global leader does not update its position, the global limit count is increased by 1.

Local Leader Learning Phase (LLL): In this phase, the monkey with the best fitness in each group is chosen to be the local leader. If the local leader does not update its position, the local limit count is increased by 1.

Local Leader Decision Phase (LLD): If the local leader does not update its position in subsequent previously defined local leader limit, all members of the group re-initialize their position either randomly or based on the combined experience of local and global leaders. The local leader count is made 0. The updated position of spider monkeys' as follows:

$$sm(new)_{ij} = sm_{ij} + u(0,1) \times (gl_j - sm_{ij}) + u(0,1) \times (sm_{ij} - ll_{kj}) \quad (4)$$

Global Leader Decision Phase (GLD): If the global leader does not update its position in a subsequent pre-decided number called the global leader limit, the global leader splits the population into smaller groups until it reaches maximum group number. The global leader decision phase is always followed by local leader learning phase to elect a leader in each newly formed group. If the global leader's position is not updated despite the maximum number of groups, the global leader recombines all groups to form a single group and the complete process restarts.

2.2   Genetic Algorithm

Genetic algorithm (GA) is inspired by natural evolution and developed by John Holland in the 1960s [7]. GA is an optimization method and optimization is based on the development of the population. The population of candidate solutions are called individuals to an optimization problem. Here in real coded GA, chromosome is a vector of real numbers represents the solution of the optimization problem. Each chromosome includes the set of genes and each gene represents the variable of the problem. Initially, a finite number of individuals are generated randomly, each is associated with fitness. Fitness is usually a value of the objective function in the optimization problem being solved. Based on fitness value and genetic operators, new population generated iteratively.

Reproduction, crossover and mutation are three genetic operators used in the development of genetic algorithm as follows:

Reproduction/selection: This is the first operator applied on population. Parents (chromosomes) are selected from the population and produce offspring by crossover. According to Darwin's theory of evolution, only fittest candidate survive. Thus reproduction operator is also known as selection operator. During this phase, in each successive generation a proportion of the existing population is selected to breed a new generation. There are various selecting methods are available as: Roulette-wheel selection, Boltzmann selection, Tournament selection, Rank selection and Steady-state selection. But in this paper randomized selection is applied.

Crossover: Crossover operator is applied after completing reproduction process. This operator applied to the mating pool to create a better individual. Generally in crossover operator, randomly pick two individuals from the mating pool and exchange some gene (variable) between the individuals. There are numbers of crossover have been defined for real coded GA. Some commonly used crossover in real coded GA are Flat crossover [17], Arithmetic crossover [14], BLX-α crossover (Blend crossover) [18], Linear BGA (Breeder Genetic Algorithm) crossover [18] and Wright's heuristic crossover [23]. In this paper, we have used Arithmetic crossover. Short review of Arithmetic crossover is:

Let us assume two parents (chromosomes) are P_1 and P_2 are selected for crossover as follows:

$$
\begin{aligned}
P^1 &= (p_1^1, p_2^1, ...,p_n^1)\\
P^2 &= (p_1^2, p_2^2, ...,p_n^2)
\end{aligned}
\tag{5}
$$

Where n represents the number of genes (variables) in each chromosome.

Similar representation for two offspring O^1 and O^2 produced by two parents are as follows:

$$
\begin{aligned}
O^1 &= (o_1^1, o_2^1, ...,o_n^1)\\
O^2 &= (o_1^2, o_2^2, ...,o_n^2)
\end{aligned}
\tag{6}
$$

The offspring are in Arithmetic crossover defined by Kaelo et al. [10] as:

$$
o_i^1 = \alpha_i p_i^1 + (1 - \alpha_i)p_i^2 \tag{7}
$$
$$
o_i^2 = \alpha_i p_i^2 + (1 - \alpha_i)p_i^1 \tag{8}
$$

Where i is a position $\in \{1, 2, ..., n\}$ and α_i is uniform random number [14].

Mutation: Mutation operator preserves and introduces population diversity. After crossover, chromosomes are subjected to mutation. Mutation operator is applied according to user definable mutation probability. Mutation probability is set low because the search will be covert into primitive random search due to high mutation probability.

3 Hybridization of SMO and GA

Numerical optimization requires a balance between exploration and exploitation. This paper discusses the hybridization of one meta-heuristic approach to other, in order to provide a better balance between exploration and exploitation. The two algorithms discussed here are genetic algorithm (GA) and spider monkey optimization (SMO). The proposed structure has two broad phases, each consisting of one of the two classical (real coded GA and SMO) algorithms. The first phase deals with randomly seeding potential solutions throughout the search space followed by optimizing the test function by using one algorithm (real coded GA or SMO). The optimized output of the first phase is fed as input to the second phase which makes use of the second algorithm (SMO or GA) to optimize the test function further.

3.1 Spider Monkey Optimization Followed by Genetic Algorithm (SMOGA)

In this algorithm, the first broad case consists of spider monkey optimization algorithm followed by genetic algorithm (SMOGA) in the subsequent phase. SMOGA algorithm is developed to combine the properties of two algorithms SMO and GA. Hybrid technique SMOGA maintains the integration of two techniques for the entire run. Initially individual are generated randomly and updated by SMO procedure until (no progress > 100 iterations). Updating procedures are followed by GA until (no progress > 100 iterations). Working procedure of SMOGA is depicted in Fig. 1. The termination criteria for SMOGA is as follows:

Continue SMO
Until (no progress > 100 iterations)
Continue GA
Until (no progress > 100 iterations)
The first phase is further divided into 6 sub-divisions:

- Local Leader Phase
- Global Leader Phase
- Global Leader Learning Phase
- Local Leader Learning Phase
- Local Leader Decision Phase
- Global Leader Decision Phase

There are four control parameters in SMO algorithm: LocalLeaderLimit, GlobalLeaderLimit, Maximum Groups (MG) and Perturbation Rate (pr).

The second phase is subdivided as in 5 sub-divisions:

- Initialization
- Selection
- Crossover
- Mutation
- Termination

In the initialization step optimized solutions of the first phase are initialized as input to the second phase. Selection, crossover and mutation are processed. The algorithm is terminated if the best solution after the iterations gets over is presented as the output of the algorithm.

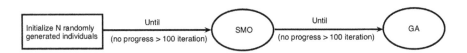

Fig. 1. Schematic representation of SMOGA

3.2 Genetic Algorithm Followed by Spider Monkey Optimization (GASMO)

In genetic algorithm followed by spider monkey optimization (GASMO), genetic algorithm and spider monkey optimization are used in the subsequent phase. GASMO incorporates the properties of both algorithms GA and SMO. Hybrid technique GASMO also maintains the integration of two techniques for the entire run. Here the procedure starts with randomly generated individuals and updated by GA procedure until (no progress > 100 iterations). Updating procedures are followed by SMO until (no progress > 100 iterations). Working procedure of GASMO is depicted in Fig. 2. The termination criteria for GASMO is as follows:

Continue SMO
Until (no progress > 100 iterations)
Continue GA
Until (no progress > 100 iterations)

Fig. 2. Schematic representation of GASMO

4   Numerical Experiments and Discussion

In this section, we compare basic SMO, real coded GA, SMOGA and GASMO. The parameters settings adopted for experiment are:

- Population Size: 100
- LocalLeaderLimit: 3000
- GlobalLeaderLimit: 100
- Perturbation Rate: $[0.1, 0.4]$ linearly increasing over iterations
- Maximum Groups: 10
- Mutation Rate: 0.1
- Number of Simulations: 30
- Max Iterations: 2000

In order to compare the effect of proposed algorithm, 20 different unconstrained continuous minimization benchmark functions are selected with their optimum values given in Table 1. Comparison results of GA, SMO, SMOGA and GASMO based on best fitness and average fitness over 30 runs are reported in Table 2. In Table 2, column 1 illustrates the benchmark function, columns 2, 4, 6 and 8 report the best fitness of 30 runs using GA, SMO, SMOGA and GASMO algorithms, respectively. Columns 3, 5, 7 and 9 report the average fitness over 30

Table 1. Test problems (TP: Test problem, D: Dimensions)

TP	Objective function	Search range	Optimum value	D
Parabola sphere	$f_1(x) = \sum_{i=1}^{D} x_i^2$	[−5.12,5.12]	0	30
Step function	$f_2(x) = \sum_{i=1}^{D} \lfloor x_i + 0.5 \rfloor^2$	[−100,100]	0	30
Ackley	$f_3(x) = -20 \exp\left(-0.2\sqrt{\frac{1}{D}\sum_{i=1}^{D} x_i^2}\right) - \exp(\frac{1}{D}\sum_{i=1}^{D}\cos(2\pi x_i)) + 20 + e$	[−32.768,32.768]	0	30
Griewank	$f_4(x) = \sum_{i=1}^{D}\frac{x_i^2}{4000} - \prod_{i=1}^{D}\cos(\frac{x_i}{\sqrt{i}}) + 1$	[−600,600]	0	30
Axis parallel hyper ellipsoid	$f_5(x) = \sum_{i=1}^{D} i x_i^2$	[−5.12,5.12]	0	30
Levy	$f_6(x) = sin^2(\pi w_1)\sum_{i=1}^{d-1}(w_i - 1)^2[1 + 10sin^2(\pi w_i + 1)] + (w_d - 1)^2[1 + sin^2(2\pi w_d)], Where\ w_i = 1 + \frac{x_i - 1}{4}, i = 1,....,d$	[−10,10]	0	30
Rastrigin	$f_7(x) = \sum_{i=1}^{D}(x_i^2 - 10\cos(2\pi x_i) + 10D)$	[−5.12,5.12]	0	30
Rosenbrock	$f_8(x) = \sum_{i=1}^{D}[100(x_i^2 - x_{i+1})^2 + (x_i - 1)^2]$	[−5,10]	0	30
Schewefel	$f_9(x) = -\sum_{i=1}^{D} x_i sin(\sqrt{|x_i|})$	[−500,500]	−12569.487	30
Schewefel1.2	$f_{10}(x) = -\sum_{i=1}^{D}(\sum_{j=1}^{i} x_j)^2$	[−100,100]	0	30
Sum of different power	$f_{11}(x) = \sum_{i=1}^{D} |x_i|^{i+1}$	[−1,1]	0	30
Dixon price	$f_{12}(x) = (x_1 - 1)^2 + \sum_{i=2}^{D} i(2x_i^2 - x_{i-1})^2$	[−10,10]	0	30
Easom	$f_{13}(x) = -(-1)^n(\prod_{i=1}^{D} cos^2(x_i))\exp[-\sum_{i=1}^{D}(x_i - \pi)^2]$	[−2π, 2π]	0	30
Michalewicz	$f_{14}(x) = -\sum_{i=1}^{D} sin(x_i)\left[\frac{sini(x_i)^2}{\pi}\right]^{20}$	[0,π]	−9.66015	30
Perm	$f_{15}(x) = \sum_{i=1}^{D}(\sum_{j=1}^{D}(j + \beta)(x_j^i - \frac{1}{j^i}))^2$	[−30,30]	0	30
Rotated hyper ellipsoid	$f_{16}(x) = \sum_{i=1}^{D}\sum_{j=1}^{i} x_j^2$	[−65.536,65.536]	0	30
Styblinski Tang	$f_{17}(x) = \frac{1}{2}\sum_{i=1}^{D}(x_i^4 - 16x_i^2 + 5x_i)$	[−5,5]	−1174.9797	30
Trid Function	$f_{18}(x) = \sum_{i=1}^{D}(x_i - 1)^2 - \sum_{i=2}^{D} x_i x_{i-1}$	[−900,900]	$f_{min} = -\frac{(D*(D+4)*(D-1))}{6}$	30
Xin She	$f_{19}(x) = (\sum_{i=1}^{D} |x_i|)\exp[-\sum_{i=1}^{D} sin(x_i^2)]$	[−2π, 2π]	0	30
Zakharov's	$f_{20}(x) = \sum_{i=1}^{D} x_i^2 + \left(\sum_{i=1}^{D}\frac{i}{2}x_i\right)^2 + \left(\sum_{i=1}^{D}\frac{i}{2}x_i\right)^4$	[−5,10]	0	30

runs using GA, SMO, SMOGA and GASMO algorithms, respectively. According to the average fitness over 30 runs, GASMO outperforms on 2 test problems (f_{10}, f_{13}), SMO outperforms on 5 test problems (f_8, f_{14}, f_{18}, f_{19}, f_{20}) and SMOGA outperforms on 10 test problems (f_1, f_3, f_4, f_5, f_6, f_9, f_{11}, f_{15}, f_{16}, f_{17}). SMO, SMOGA and GASMO algorithms are equally better than GA algorithm for f_{12}. Also for 2 test problems (f_2, f_7), SMO and SMOGA obtained the optimum value. Comparing according to the best value, SMO and SMOGA outperform over GA and GASMO. Thus overall comparison shows that SMOGA algorithm outperforms over GA, SMO and GASMO algorithms.

Here some statistical analysis based on boxplot and Mann–Whitney U rank sum test are also presented. The boxplot for average fitness over 30 runs for all considered algorithms (SMO, GA, SMOGA and GASMO) have been depicted in Fig. 3. From the Fig. 3, it is clearly seen that SMOGA is the better candidate as compare to SMO, GA and GASMO.

Further Table 3, shows the Mann–Whitney U rank sum test to see the significant difference in the performance of considered algorithms. In this paper, this test is performed on average fitness over 30 runs to see the difference SMO Vs SMOGA, GA Vs SMOGA, SMO Vs GASMO and GA Vs GASMO. In Table 3, column 1 illustrates the test problems. Columns 2, 3, 4 and 5 report the significance difference between SMO Vs SMOGA, GA Vs SMOGA, SMO Vs GASMO and GA Vs GASMO, respectively. In column 2, '+' sign indicates SMOGA is significantly better than SMO, '=' sign indicates there is no significance difference between SMO and SMOGA and '−' sign indicates SMOGA performs worse than SMO. Here '+' sign indicates in column 3 (SMOGA is better than GA), in column 4 (GASMO is better than SMO) and in column 5 (GASMO is better than GA). Similarly '=' and '−' sign indicates as discussed above. In Table 3, total number of '+' signs are 8 in SMO Vs SMOGA, 18 in GA Vs SMOGA, 4 in SMO Vs GASMO and 12 in GA Vs GASMO. It is clear that SMOGA and GASMO outperform over GA.

Table 2. Comparison of GA, SMO, SMOGA and GASMO based on average fitness over 30 runs

TP	GA		SMO		SMOGA		GASMO	
	Best	Average	Best	Average	Best	Average	Best	Average
f_1	9.97E−66	8.79E−08	5.45E−17	1.03E−16	9.24E−53	1.49E−22	1.47E−34	2.21E−13
f_2	0.00E+00	2.33E−01	0.00E+00	0.00E+00	0.00E+00	0.00E+00	0.00E+00	1.00E−01
f_3	2.22E−14	2.67E−01	7.99E−15	2.10E−01	7.99E−15	9.94E−06	7.99E−15	6.20E−02
f_4	0.00E+00	7.01E−03	0.00E+00	3.61E−03	0.00E+00	8.21E−04	0.00E+00	1.81E−03
f_5	5.73E−65	5.65E−07	6.26E−17	1.11E−16	6.83E−53	2.71E−27	4.06E−34	7.70E−12
f_6	1.50E−32	2.26E−05	4.95E−17	8.99E−17	1.50E−32	6.32E−20	1.50E−32	6.11E−14
f_7	5.97E+00	9.37E+00	0.00E+00	0.00E+00	0.00E+00	0.00E+00	8.95E+00	1.44E+01
f_8	2.64E+01	8.12E+01	4.97E−07	9.52E+00	7.63E−03	2.28E+01	3.72E−03	2.09E+01
f_9	−1.15E+04	−1.10E+04	−1.26E+04	−1.26E+04	−1.26E+04	−1.26E+04	−1.17E+04	−1.11E+04
f_{10}	1.46E+02	3.60E+02	2.31E−12	3.49E−03	3.53E−03	1.25E+01	2.53E−08	1.71E−03
f_{11}	1.44E−34	2.61E−20	2.09E−18	3.25E−17	9.13E−36	1.34E−21	1.20E−35	5.22E−18
f_{12}	6.67E−01	1.60E+00	6.67E−01	6.67E−01	6.67E−01	6.67E−01	6.67E−01	6.67E−01
f_{13}	−1.00E+00	−1.00E+00	−7.85E−139	−2.62E−140	−1.00E+00	−1.00E+00	−1.00E+00	−1.00E+00
f_{14}	−2.78E+01	−2.68E+01	−2.96E+01	−2.94E+01	−2.95E+01	−2.92E+01	−2.81E+01	−2.62E+01
f_{15}	7.17E+00	6.85E+02	1.30E−04	2.38E+01	2.05E−04	1.55E+01	1.81E−04	2.20E+02
f_{16}	4.72E−61	3.87E−01	5.26E−17	9.32E−17	1.10E−52	1.88E−24	2.19E−32	6.58E−12
f_{17}	−1.16E+03	−1.10E+03	−1.17E+03	−1.17E+03	−1.17E+03	−1.17E+03	−1.16E+03	−1.10E+03
f_{18}	3.60E+02	2.66E+03	1.51E+01	1.08E+02	1.50E+01	1.99E+02	1.50E+01	1.32E+02
f_{19}	5.89E−12	7.79E−12	3.51E−12	3.52E−12	3.51E−12	3.66E−12	5.18E−12	8.10E−12
f_{20}	1.18E−01	5.39E−01	1.17E−12	6.56E−10	1.07E−07	4.78E−05	3.98E−10	1.29E−08

Table 3. Results of Mann–Whitney U rank sum test

TP	SMO Vs SMOGA	GA Vs SMOGA	SMO Vs GASMO	GA Vs GASMO
f_1	+	+	−	+
f_2	=	+	=	=
f_3	+	+	+	+
f_4	+	+	=	=
f_5	+	+	−	+
f_6	+	=	−	+
f_7	=	+	−	−
f_8	−	+	−	+
f_9	=	+	−	=
f_{10}	−	+	+	+
f_{11}	+	+	+	−
f_{12}	=	+	=	+
f_{13}	+	=	+	+
f_{14}	−	+	−	−
f_{15}	=	+	=	+
f_{16}	+	+	−	+
f_{17}	=	+	−	=
f_{18}	−	+	=	+
f_{19}	−	+	−	=
f_{20}	−	+	−	+
Total number of '+' signs	8	18	4	12

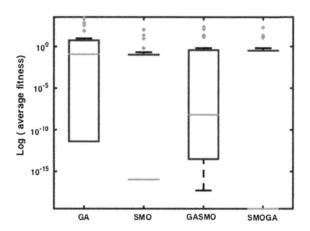

Fig. 3. Boxplot for average fitness

5   Conclusion

In this paper, two hybrid optimization algorithms, spider monkey optimization followed by genetic algorithm (SMOGA) and genetic algorithm followed by spider monkey optimization (GASMO) have been presented. Both hybridized methodologies (SMOGA and GASMO) which yielded better results in once instance while worse in the other. Hence, hybridization has the potential to improve the performances of various algorithms and their combinations provided it is used efficiently. These proposed algorithms have been tested over most known benchmark problems and outperform over classical procedures.

This paper opens possibilities to hybridize varied combinations of optimization algorithms for better performance and their application to various real world optimization problems.

References

1. Abd-El-Wahed, W.F., Mousa, A.A., El-Shorbagy, M.A.: Integrating particle swarm optimization with genetic algorithms for solving nonlinear optimization problems. J. Comput. Appl. Math. **235**(5), 1446–1453 (2011)
2. Bansal, J.C., Sharma, H., Jadon, S.S., Clerc, M.: Spider monkey optimization algorithm for numerical optimization. Memetic Comput. **6**(1), 31–47 (2014)
3. Dorigo, M.: Optimization, learning and natural algorithms. Ph.D. Thesis, Politecnico di Milano, Italy (1992)
4. Eberhart, R.C., Kennedy, J., et al.: A new optimizer using particle swarm theory. In: Proceedings of the Sixth International Symposium on Micro Machine and Human Science, vol. 1, pp. 39–43. New York (1995)
5. Grimaccia, F., Mussetta, M., Zich, R.E.: Genetical swarm optimization: self-adaptive hybrid evolutionary algorithm for electromagnetics. IEEE Trans. Antennas Propag. **55**(3), 781–785 (2007)
6. Harada, K., Ikeda, K., Kobayashi, S.: Hybridization of genetic algorithm and local search in multiobjective function optimization: recommendation of GA then LS. In: Proceedings of the 8th Annual Conference on Genetic and Evolutionary Computation, pp. 667–674. ACM (2006)
7. Holland, J.H.: Adaptation in Natural and Artificial Systems: An Introductory Analysis with Applications to Biology, Control, and Artificial Intelligence. University of Michigan Press, Ann Arbor (1975)
8. Hwang, S.-F., He, R.-S.: A hybrid real-parameter genetic algorithm for function optimization. Adv. Eng. Inform. **20**(1), 7–21 (2006)
9. Juang, C.-F.: A hybrid of genetic algorithm and particle swarm optimization for recurrent network design. IEEE Trans. Syst. Man Cybern. Part B (Cybern.) **34**(2), 997–1006 (2004)
10. Kaelo, P., Ali, M.M.: Integrated crossover rules in real coded genetic algorithms. Eur. J. Oper. Res. **176**(1), 60–76 (2007)
11. Kao, Y.-T., Zahara, E.: A hybrid genetic algorithm and particle swarm optimization for multimodal functions. Appl. Soft Comput. **8**(2), 849–857 (2008)
12. Karaboga, D.: An idea based on honey bee swarm for numerical optimization. Technical report, Technical report-TR06, Erciyes University, Engineering Faculty, Computer Engineering Department (2005)
13. Lee, Z.-J., Lee, C.-Y.: A hybrid search algorithm with heuristics for resource allocation problem. Inf. Sci. **173**(1), 155–167 (2005)
14. Michalewics, Z.: Genetic Algorithms + Data Structures = Evolution Programs. Springer, Heidelberg (1996)
15. Pan, X., Jiao, L., Liu, F.: An improved multi-agent genetic algorithm for numerical optimization. Nat. Comput. **10**(1), 487–506 (2011)

16. Passino, K.M.: Biomimicry of bacterial foraging for distributed optimization and control. IEEE Control Syst. **22**(3), 52–67 (2002)
17. Radcliffe, N.J.: Equivalence class analysis of genetic algorithms. Complex Syst. **5**(2), 183–205 (1991)
18. Schlierkamp-Voosen, D.: Strategy adaptation by competition. In: Proceedings of the Second European Congress on Intelligent Techniques and Soft Computing, pp. 1270–1274 (1994)
19. Simon, D.: Biogeography-based optimization. IEEE Trans. Evol. Comput. **12**(6), 702–713 (2008)
20. Storn, R., Price, K.: Differential evolution-A simple and efficient heuristic for global optimization over continuous spaces. J. Global Optim. **11**(4), 341–359 (1997)
21. Tian, D.: Hybridizing adaptive genetic algorithm with chaos searching technique for numerical optimization. Int. J. Grid. Distrib. Comput. **9**(2), 131–144 (2016)
22. Wolpert, D.H., Macready, W.G.: No free lunch theorems for optimization. IEEE Trans. Evol. Comput. **1**(1), 67–82 (1997)
23. Wright, A.H., et al.: Genetic algorithms for real parameter optimization. Found. Genet. Algorithms **1**, 205–218 (1991)
24. Yang, X.-S.: Harmony search as a metaheuristic algorithm. In: Geem, Z.W. (ed.) Music-Inspired Harmony Search Algorithm. SCI, vol. 191, pp. 1–14. Springer, Heidelberg (2009)

An Analysis of Modeling and Optimization Production Cost Through Fuzzy Linear Programming Problem with Symmetric and Right Angle Triangular Fuzzy Number

Rajesh Kumar Chandrawat[1](✉), Rakesh Kumar[1](✉), B.P. Garg[2](✉),
Gaurav Dhiman[3](✉), and Sumit Kumar[1](✉)

[1] Department of Mathematics, Lovely Professional University, Jalandhar, India
Rajesh.16786@lpu.co.in, rakeshmalhan23@gmail.com, siwachsumit27@gmail.com
[2] Department of Mathematics,
I.K. Gujral Punjab Technical University, Jalandhar, India
bkgarg2007@gmail.com
[3] Department of Computer Science and Engineering,
Thapar University, Patiala, India
gaurav.dhiman@thapar.edu

Abstract. The main objective of this paper is to do the modeling and optimization of production cost of RCF kapurthala using TFLPP-(s, l, r) and triangular (Right angle) fuzzy linear programming problem. The total costs of the different constrains are vacillating or uncertain, so to minimize the production cost, fuzzy LPP (right angle triangular) and TFPP- (s, l, r) model are used. Owing to probabilistic increments in the availability of different constrains, the actual cost of production is to leading the destruction. Here the situational based Fuzzy model is being expressed to mitigate the destruction in the cost optimization and examining the credibility of optimized value. The data of RCF Kapurthala constitutes the production cost of different coaches from the year 2009–10. The total cost has been targeted to optimize with respect to the constraints of Labor cost, Material cost, Administrative overhead charges, Factory overhead charges, Township overhead charges, Shop overhead charges and Performa charges. The lower and upper bound have been calculated using TFLPP-(s, l, r), TFLPP-(s, l), TFLPP-(s, r) and TFLPP-(s) for the objective function of the optimized fuzzy LPP. This optimized fuzzy LPP will provide the membership grade for the optimized production cost.

Keywords: Fuzzy linear programming · Ranking · Trapezoidal fuzzy number · Optimization

1 Introduction

Operation research has become increasingly important in the face of fast moving technology and increasing complexities in business and industry. Business and

© Springer Nature Singapore Pte Ltd. 2017
K. Deep et al. (eds.), *Proceedings of Sixth International Conference on Soft Computing for Problem Solving*, Advances in Intelligent Systems and Computing 546, DOI 10.1007/978-981-10-3322-3_18

economic situation are concerned with planning activity. It can be maximum production, minimum cost, and maximum profit under limited resource constraints.
Such problems are referred to as the problem of constraints optimization.

A linear programming is a technique for determining an optimum schedule of interdependent activities in view of accessible resources. A problem thus
obtained, known as linear programming problem. Linear programming also called
linear optimization is a technique to achieve the best outcome in a mathematical
model. This new approach to systematic and scientific study of the operation of
the system was called the operation research.

With the help of linear programming problem, the optimal solution and the
best sense of efficiency can be emphasized.

In the mathematical model of LPP, the requirements are represented by linear
relation. The representation of linear programming problem is as follows:

$$
\begin{aligned}
&\text{Maximize/Minimize} &&c^T x \\
&\text{subject to} &&Ax \le b \\
&\text{and} &&x \ge 0
\end{aligned}
\tag{1}
$$

Standard Form of Linear Programming problem can be written as:

$$\text{Maximize/Minimize} = c_1 x_1 + c_2 x_2 + \ldots + c_n x_n$$

subject to

$$
\begin{aligned}
&a_{11} x_1 + a_{12} x_2 + \ldots + a_{1n} x_n (\le = \ge) b_1 \\
&a_{21} x_1 + a_{22} x_2 + \ldots + a_{2n} x_n (\le = \ge) b_2 \\
&a_{m1} x_1 + a_{m2} x_2 + \ldots + a_{mn} x_n (\le = \ge) b_m
\end{aligned}
\tag{2}
$$

OR

$$\sum_{i=1, j=1}^{mn} a_{ij} x_j \le \text{ or } \ge b_i$$

These linear equations are the constraints for the objective function. Here
are the decision variables and represents the availability of m constraints Unfor-
tunately, some times, the actual practical situations are often not deterministic.
There exist certain types of dubieties in social, industrial and economic systems,
such as randomness of occurrence of events can lead to improper optimization.
Such types of dubieties (Feasible uncertainties) are associated with the difficulty
of making sharp or precise decision. Feasible uncertainties deal with the situation where the information cannot be valued sharply or cannot be described
clearly in linguistic term, such as preference related information. At a certain
point of time, the availabilities of m constraints can be fluctuated in term of
probabilistic increment, probabilistic decrement or in the both directions then
general LPP cannot explicit the proper optimization. In these situations fuzzy
lpp can provide the better optimization.

If the fluctuation is available in terms of increment or decrement then the
use of triangular (right angle) fuzzy linear programming problem benefits in

introducing the credibility for the increase or decrease in the . This credibility fulfills the necessities to find out the lower and upper bounds for the initial LPP. If the fluctuation is available in the both directions then triangular (s, l, r) fuzzy linear programming problem can be proposed to achieve the required optimization. In this project we are proposing the triangular (s, l, r) fuzzy LPP to achieve realistic optimization. The triangular(s, l, r) Fuzzy LPP in which only the right hand side numbers B_i are fuzzy number can be expressed as:

$$\text{Max} \sum_{j=1}^{n} c_j x_j$$

$$\text{such that} \sum_{j=1}^{n} a_{ij}(s,l,r)x_j \leq B_i(s,l,r) \tag{3}$$

$$\forall x_j \geq 0 \text{ and } j \in N_n$$

where a_{ij} and B_i terms are fuzzy number. This model has an appropriate and reasonable interpretation of situational based optimization and it can fill the gap between the vagueness of constrains and standard optimization.

2   Fuzzy Set

Fuzzy sets [17] are those sets which allows partial membership i.e. between 0 and 1. A fuzzy set S can be defined on the universe of discourse U as follows:

$$S = \{(x, \mu_S(x))|x \epsilon U\} \tag{4}$$

where μ_S is the membership function of fuzzy set S within range [0,1] and $\mu_S(x)$ indicates the degree of membership of x in S lies in range [0,1].

2.1   Convex Fuzzy Set

If the membership value of any membership function are monotonically increase and decrease for some element in universe then those fuzzy set S in universe of discourse U is called a convex fuzzy set [9].

2.2   Normal Fuzzy Set

A fuzzy set [9] is said to be normal fuzzy set if there exists at least one element $x \epsilon U$ such that $\mu_S(x) = 1$ where no membership function has its value equal to 1 is called sub-normal fuzzy set.

2.3   Fuzzy Number

A fuzzy number [9] is a regular number in which the value corresponding to element between 0 and 1, called membership functions, instead of one single value.

2.4   Defuzzification

The process of converting the fuzzy number output to a crisp value is called defuzzification. In order to make decisions to maintenance the actions it is necessary to convert the fuzzy number output into a crisp value.

3   Literature Review

The fuzzy logic idea was first presented by Loft Zadeh, professor at the University of California at Berkley. This fuzzy logic when applied to linear decision making then fuzzy linear programming came in existence. Because of the continuous efforts of the researchers the fuzzy linear programming now days is broadly applicable to many fields. With the assistance of fuzzy programming we can calculate the variation in some objective function when there is variation in the constraints of the objective function. There are numerous real life applications of fuzzy linear programming similar in the analysis of future performance of organizations and factories.

The basic arithmetic operations for two generalized positive parabolic fuzzy numbers [7] by using the concept of the distribution functions. There is no need to compute the -cut of the fuzzy number which becomes more powerful than the standard method. A newly generalized improved score function [6] has been presented to incorporating the idea of weighted average in fuzzy set environment. The method for solving the multi-criteria decision making (MCDM) problem has also been presented for unknown attribute weights. Singh [16] proposed a method to reduce the large data-set using soft computing techniques, such as fuzzy sets and artificial neural network, which can decrease the dimensionality of data-set. Garg [5] proposed a method to quantify the uncertainties, generic, extensible for the application domain and sensitivity of system performance which investigates the various reliability parameters in terms of membership and non-membership functions by using -cut and the weakest t-norm based arithmetic operations on triangular intuitionistic fuzzy sets. Rani, Gulati, and Garg [14] demonstrated a method for solving multi-objective optimization problem under the optimistic and pessimistic view point. This problem considered as the parabolic multi-objective non-linear optimization programming problem (PMONLOPP) such as linear/non-linear membership functions corresponding to each objective has been taken.

Weldon A. Lodwick and Katherine A. Bachman [10] concentrated on solving large scale fuzzy and possibilistic optimization problems. They took an optimization problem in radiation therapy with many orders of complexity from 100 to 62,250 constraints for fuzzy and possibilistic linear and non-linear programming implementations possessing fuzzy inequalities, fuzzy right-hand side values and possibilistic right-hand side is used to show that fuzzy and possibilistic optimization are useful. In this project he concentrated on the uncertainty in the right side of limitations which arises in the context of the radiation therapy problem. The result shows that fuzzy and possibilistic optimization is a natural and effective way to model of various type of optimization under uncertainty problems.

P.K. De and D. Das [1] proposed a new ranking procedure for trapezoidal intuitionistic fuzzy number(TRIFN). To serve this purpose, the value and ambiguity index of TRIFNs have been defined. In order to define the rank of TRIFNs, they proposed a ranking function by taking the sum of value and ambiguity index.

Wan [13] proposed a technique on multi-attribute group decision making problems (MAGDM) in which attribute values are expressed with (TrIFNs), which are further solved by developing a new decision method based on the power average operators of (TrIFNs). Hereby the power average operator of real numbers is extended to four kinds of power operators of (TrIFNs) such as power average operator of (TrIFNs), the weighted power average operator of (TrIFNs), the power ordered weighted average operator of (TrIFNs), and the power hybrid average operator of (TrIFNs).

Ganesan and Veeramani [4] proposed fuzzy linear programming problem which involve symmetric trapezoidal fuzzy numbers. Some interesting and important results are obtained, to a solution of fuzzy linear programming problems without converting them to crisp linear programming problems.

Pandey [11] proposed four new aggregation operators based on the geometric and arithmetic means of L- and R- or right side and left side angles of apex for triangular and trapezoidal fuzzy numbers respectively. In this technique, a new aggregation operator for TFNs in which the L- and R- membership function of lines of the aggregate (TFN) in which slopes are the arithmetic means of the corresponding L- and R- slopes of the individual (TFNs).

Hassan Mishmast Nehi and Hamid Reza Maleki [12] worked on Intuitionistic fuzzy numbers and its applications in fuzzy optimization problem. He introduces the trapezoidal intuitionistic fuzzy numbers and proved some operation for them. He also introduces the intuitionistic fuzzy optimization problem by use of the membership and non-membership functions. Frank Rogers, J. Neggers and Younbae Jun [15] demonstrated method for optimizing linear problems with uncertain constraints. They have focused on linear fuzzy programing problem. When they were solving the problems they found that optimizing fuzzy constraints and objective that consist of triplet and appears like triangular fuzzy numbers but they differ in that way that they are a hybrid fuzzy number that has characteristics that are both fuzzy and crisp.

Ali Ebrahimnejad and Madjid Tavana [3] worked on method for solving linear programing problems with symmetric trapezoidal fuzzy numbers. They proposed a new method for solving fuzzy linear programming problem in which the coefficient of the objective function and the values and the of the right hand side are symbolized by symmetric trapezoidal fuzzy number while the elements of the coefficients matrix are represented as real numbers. Then they converted the fuzzy linear programming problem into an equivalent crisp Linear programming problem and solved the crisp problem with the general primal simplex method. They showed that the method they were using is simpler and computationally more efficient that two competing fuzzy linear programming technique commonly used in the literature.

Yenilmez and Gasimov [8] they concentrate on linear programming problem with only fuzzy technological coefficients. Only the case of fuzzy numbers with linear membership functions is being considered and the "modified sub gradient method" for solving these types of problems have proposed. They also compared this method with well known "fuzzy decisive set method".

4   Methodology

4.1   Method of Calculation:

The general form of triangular Fuzzy LPP is (s, l, r) fuzzy LPP in which a_{ij} and B_i are fuzzy number is:

$$\text{Max} \sum_{j=1}^{n} c_j x_j$$

$$\text{such that} \sum_{j=1}^{n} a_{ij}(s,l,r)x_j \leq B_i(s,l,r) \tag{5}$$

$$\forall x_j \geq 0 \text{ and } j \in N_n$$

where a_{ij} and B_i terms are fuzzy number.

Any triangular fuzzy number A can be represented by three real number (s, l, r) whose meaning is defined in the below Fig. 1.

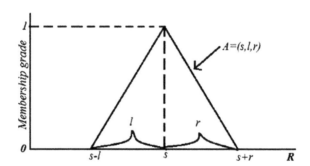

Fig. 1. Triangular (s,l,r) fuzzy number

Using this representation, we can write A= (s, l, r). Now according to D.K.J. Dipankar [2].

$$\text{Max} \sum_{j=1}^{n} c_j x_j$$

$$\text{such that} \sum_{j=1}^{n} (s_{ij}, l_{ij}, r_{ij})x_{ij} \leq (t_i, u_i, v_i)(i \in N_m) \tag{6}$$

$$\forall x_j \geq 0, (j \in N_m)$$

where $a_{ij} = (s_{ij}, l_{ij}, r_{ij})$ and $B_i = (t_i, u_i, v_i)$ are fuzzy number. The general structure of Eq. (3) is defined as follows:

$$\text{Max} \sum_{j=1}^{n} c_j x_j$$

$$\text{such that} \sum_{j=1}^{n} s_{ij} x_j \leq t_i$$

$$\sum_{j=1}^{n} (s_{ij} - l_{ij}) x_j \leq t_i - u_i \tag{7}$$

$$\sum_{j=1}^{n} (s_{ij} + r_{ij}) x_j \leq t_i + v_i (i \in N) \, \forall x_j \geq 0 \text{ and } j \in N$$

Here the problem is of second type i.e. the fuzzy linear programming problem with fuzzy right hand side numbers.

$$\text{Max } z = \sum_{j=1}^{n} c_j x_j$$

$$\text{subject to} \sum_{j=1}^{n} a_{ij} x_j \leq \tilde{b}_i \tag{8}$$

$$\text{where} \quad 1 \leq i \leq m, 1 \leq j \leq n, x_j \geq 0$$

$\tilde{b}_i$ represents the availability of constraints the accent symbol shows that this quantity is fuzzy means there is increase or decrease in this quantity after some time. But in this project work the optimization is with respect to increase in the availability of constraints. This means the problem will be converted into the LPP.

$$\text{Max } z = \sum_{j=1}^{n} c_j x_j$$

$$\text{subject to} \sum_{j=1}^{n} a_{ij} x_j \leq b_i + p_i \tag{9}$$

$$\text{where} \quad 1 \leq i \leq m, 1 \leq j \leq n, x_j \geq 0$$

In this LPP is the probabilistic increase in the availability of constraints. The main task is to optimize the problem when there is an increase in the availability of constraints.

In the above kind of problem, the membership grades can be introduced with respect to the increase in the availability of constraints. The membership grades for will be as follows:

$$B_i = \begin{cases} 1 & \text{when} \quad x \leq b_i, \\ \frac{b_i + p_i - x}{p_i} & \text{when} \quad b_i \leq x \leq b_i + p_i, \\ 0 & \text{when} \quad x \geq b_i + p_i \end{cases} \tag{10}$$

These are the membership grades for the right hand side coefficient i.e. the availability of constraints. Here x is the variable and x R. For the optimization of this type of problem we have to calculate the lower and upper bounds of the optimal values. The value for lower bound (Z_l) will be:

$$
\begin{aligned}
&\text{Max}\, z = \sum_{j=1}^{n} c_j x_j \\
&\text{subject to} \sum_{j=1}^{n} a_{ij} x_j \le b_i \\
&\text{where}\quad 1 \le i \le m,\, 1 \le j \le n,\, x_j \ge 0
\end{aligned}
\tag{11}
$$

The LPP with the initial value of the right hand side coefficient will be the lower bound for the problem.

Now the value for the upper bound (Z_u) will be-

$$
\begin{aligned}
&\text{Max}\quad z = \sum_{j=1}^{n} c_j x_j \\
&\text{subject to} \sum_{j=1}^{n} a_{ij} x_j \le b_i + p_i \\
&\text{where}\quad 1 \le i \le m,\, 1 \le j \le n,\, x_j \ge 0
\end{aligned}
\tag{12}
$$

Here the right hand coefficient will be total probabilistic increase in the availability of constraints.

These LPPs for lower and upper bounds can be solved by using the Simplex method which is a technique for solving LPP. These lower and upper bounds will be used to get the optimized fuzzy LPP.

Optimized fuzzy LPP:

$$
\begin{aligned}
&\text{Max}\quad z = \lambda \\
&\text{subject to}\quad \lambda(Z_u - Z_l) - cx \le -Z_l \\
&\lambda(p_i) + \sum_{j=1}^{n} a_{ij} x_j \le b_i + p_i \\
&\text{where}\quad x \ge 0 \quad and \quad 0 \le \lambda \le 1
\end{aligned}
\tag{13}
$$

This fuzzy optimized LPP will give the membership grade for our initial LPP. Here represents the membership grade and Z_u and Z_l are the upper and lower bounds. is the objective function of the initial LPP. The term with summation sign represents the constraints of given LPP and is the probabilistic increase in the availability of the constraints.

5 Data and Problem Identification

The data given below in Table 1 is of the railway industry, Kapurthala of the year 2009–10. This data shows the manufacturing cost of different types constrains of coaches. Kapurthala railway was established in 1986, it is a coach manufacturing unit of Indian railway and manufactured more than 30000 passenger coaches of different types in Table 1 where LAB= labor, MAT= Material, AOH= Administrative overhead charge, FOH= Factory overhead charges, TOH= Township overhead charges, SOH= Shop overhead charges, PROF. CHAR= Performa charges.

Table 1. Production cost of different coaches

COACH TYPE	LAB.	MAT.	FOH	AOH	TOH	SOH	TOTAL O/Heads	PROF. CHAR	TOTAL COST
SCN/AB	4.11	46.15	6.20	5.49	1.10	0.18	12.97	2.91	66.14
SCN/AB(CBC)	4.27	62.28	6.75	5.09	1.11	0.59	13.54	3.69	83.78
SLR/AB	3.82	39.95	5.76	5.10	1.02	0.15	12.03	2.57	58.37
SLR/AB(CBC)	3.98	38.03	6.28	4.74	1.04	0.36	12.42	2.51	56.94
GS/AB	3.78	41.28	5.70	5.05	1.01	0.16	11.92	2.63	59.61
GS/AB(CBC)	3.92	56.65	6.20	4.67	1.02	0.54	12.43	3.37	76.37
MEMU/MC	9.33	205.01	14.13	12.51	2.49	0.78	29.91	11.26	255.51
MEMU/TC	3.88	45.77	5.85	5.18	1.03	0.17	12.23	2.85	64.73
ACCN/SG	7.11	103.36	11.24	8.48	1.86	0.96	22.54	6.13	139.14
ACCN/SG(CBC)	7.11	116.88	11.24	8.48	1.86	1.11	22.69	6.76	153.44
WACCNH	5.91	63.32	10.22	7.27	1.69	0.37	19.55	4.09	92.87
WACCNH (H.HEIGHT)	5.91	63.32	10.22	7.27	1.69	0.37	19.55	4.09	92.87
WRRMDAC	6.70	179.46	10.13	8.97	1.79	0.68	21.57	9.58	217.31
VPH	2.96	34.16	4.45	3.94	0.79	0.13	9.31	2.14	48.57
VPU	2.99	34.22	4.73	3.57	0.78	0.33	9.41	2.15	48.77
EOG/LBH/ACCB	9.82	178.21	15.13	11.41	2.50	1.69	30.73	8.57	227.33
EOG/LBH/WLRRM	9.94	285.52	17.82	12.05	2.68	1.73	34.28	12.92	342.66
EOG/LBH/ACCW	10.63	177.81	16.41	12.37	2.71	1.69	33.18	8.68	230.30
EOG/LBH/ACCN	10.73	205.56	16.17	14.32	2.85	0.78	34.12	8.58	258.99
LGS(LC)	4.82	62.2	8.39	5.97	1.39	0.42	16.17	3.83	87.02
TOTAL	121.72	2039.140	193.020	151.930	32.410	13.190	356.270	109.310	2660.720

In the year 2009–2010 the total cost of different coaches is taken as an objective function which is to be minimized with respect to the cost constraints. As per the given data the total availability of cost constraints is LAB, MAT, FOH, AOH, TOTAL O/HEAD, and PROF. CHAR: - 121.72, 2039.14, 193.02, 151.93, 32.41, 13.19, 356.27, 109.31 respectively. But they can be increased and decreased as per requirement. So, in this situation we are proposing a Triangular Fuzzy LPP (s, l, r) and Right Angle Triangular Fuzzy LPP to minimize the production cost. However, Table 2 shows the quantities of increments and decrements in the basic production cost.

Table 2. Total basic availability of cost parameter with probabilistic increments and decrements

Cost parameter	Total basic cost	l(Decrement in the cost)	r(Increment in the cost)
1	121.72	6.086	6.086
2	2039.14	101.957	101.957
3	193.02	9.651	9.651
4	151.93	7.5965	7.5965
5	32.41	1.6205	1.6205
6	13.19	0.6595	0.6595
7	356.27	17.8135	17.8135
8	109.31	5.4655	5.4655

5.1   Modeling and Optimization:

The total cost is minimized using the real time data as follows

Minimize $Z = 66.14x_1 + 83.78x_2 + 58.37x_3 + 56.94x_4 + 59.61x_5 + 76.37x_6 + 255.51x_7 + 64.73x_8 + 139.14x_9 + 153.44x_{10} + 92.87x_{11} + 92.87x_{12} + 217.31x_{13} + 48.57x_{14} + 48.77x_{15} + 227.33x_{16} + 342.66x_{17} + 230.30x_{18} + 258.99x_{19} + 87.02x_{20}$

$4.11x_1 + 4.27x_2 + 3.82x_3 + 3.98x_4 + 3.78x_5 + 3.92x_6 + 9.33x_7 + 3.88x_8 + 7.11x_9 + 7.11x_{10} + 5.91x_{11} + 5.91x_{12} + 6.70x_{13} + 2.96x_{14} + 2.99x_{15} + 9.82x_{16} + 9.94x_{17} + 10.63x_{18} + 10.73x_{19} + 4.82x_{20} \leq (121.72, 6.086, 6.086)$

$46.15x_1 + 62.28x_2 + 39.95x_3 + 38.03x_4 + 41.28x_5 + 56.65x_6 + 205.01x_7 + 45.77x_8 + 103.36x_9 + 116.88x_{10} + 63.32x_{11} + 63.32x_{12} + 179.46x_{13} + 34.16x_{14} + 34.22x_{15} + 178.21x_{16} + 285.52x_{17} + 177.81x_{18} + 205.56x_{19} + 62.2x_{20} \leq (2039.14, 101.957, 101.957)$

$6.2x_1 + 6.75x_2 + 5.76x_3 + 6.28x_4 + 5.70x_5 + 6.20x_6 + 14.13x_7 + 5.85x_8 + 11.24x_9 + 11.24x_{10} + 10.22x_{11} + 10.22x_{12} + 10.13x_{13} + 4.45x_{14} + 4.73x_{15} + 15.13x_{16} + 17.82x_{17} + 16.41x_{18} + 16.17x_{19} + 8.39x_{20} \leq (193.02, 9.651, 9.651)$

$5.49x_1 + 5.09x_2 + 5.10x_3 + 4.74x_4 + 5.05x_5 + 4.67x_6 + 12.51x_7 + 5.18x_8 + 8.48x_9 + 8.48x_{10} + 7.27x_{11} + 7.27x_{12} + 8.97x_{13} + 3.94x_{14} + 3.57x_{15} + 11.41x_{16} + 12.05x_{17} + 12.37x_{18} + 14.32x_{19} + 5.97x_{20} \leq (32.41, 7.5965, 7.5965)$

$1.10x_1 + 1.11x_2 + 1.02x_3 + 1.04x_4 + 1.01x_5 + 1.02x_6 + 2.49x_7 + 1.03x_8 + 1.86x_9 + 1.86x_{10} + 1.69x_{11} + 1.69x_{12} + 1.79x_{13} + 0.79x_{14} + 0.78x_{15} + 2.50x_{16} + 2.68x_{17} + 2.71x_{18} + 2.85x_{19} + 1.39x_{20} \leq (32.41, 1.6205, 1.6205)$

$0.18x_1 + 0.59x_2 + 0.15x_3 + 0.36x_4 + 0.16x_5 + 0.54x_6 + 0.78x_7 + 0.17x_8 + 0.96x_9 + 1.11x_{10} + 0.37x_{11} + 0.37x_{12} + 0.68x_{13} + 0.13x_{14} + 0.33x_{15} + 1.69x_{16} + 1.73x_{17} + 1.69x_{18} + 0.78x_{19} + 0.42x_{20} \leq (13.19, 0.6595, 0.6595)$

$12.97x_1 + 13.54x_2 + 12.03x_3 + 12.42x_4 + 11.92x_5 + 12.43x_6 + 29.91x_7 + 12.23x_8 + 22.54x_9 + 22.69x_{10} + 19.55x_{11} + 19.55x_{12} + 21.57x_{13} + 9.31x_{14} + 9.41x_{15} + 30.73x_{16} + 34.28x_{17} + 33.18x_{18} + 34.12x_{19} + 16.17x_{20} \leq (356.27, 17.8135, 17.8135)$

$2.91x_1 + 3.69x_2 + 2.57x_3 + 2.51x_4 + 2.63x_5 + 3.37x_6 + 11.26x_7 + 2.85x_8 + 6.13x_9 + 6.76x_{10} + 4.09x_{11} + 4.09x_{12} + 9.58x_{13} + 2.14x_{14} + 2.15x_{15} + 8.57x_{16} + 12.92x_{17} + 8.68x_{18} + 8.58x_{19} + 3.83x_{20} \leq (109.31, 5.4655, 5.4655)$

6   Results and Discussion

Sometimes classical optimization techniques fail to deliver the targeted result due to uncertainty of data. We can apply the fuzzy optimization techniques in these situations to mitigate the distortion of the result due to uncertainty of data. If the constraints are uncertain and have uncertain increment and decrement then Triangular Fuzzy linear programming problem (s, l, r) help to get the required outcome. Here we have proposed a TFLPP (s, l, r) and triangular fuzzy lpp model to optimize the cost of production of different coaches of RCF Kapurthala. The minimized cost with (s, l, r) fuzzy LPP is Z = 2792.887 . If the cost is minimized using increment, then the minimized cost with (s, r) fuzzy LPP is Z = 2808.6939934732. If the cost is minimized using decrement. The minimized cost with (s, l) is Z = 2541.1993274281. The minimized cost without the increments and decrement is Z = 2674.9466604506. Now these optimal solutions can be categorized into three different cases to achieve the desired membership grade with respect to optimal solution.

6.1   Case-I

The optimized fuzzy lpp (Right angle triangle) for membership grade has been constructed using the optimal solution of (s, l, r)lpp as a lower bound and (s, r)lpp as an upper bound and then the membership grade has been derived. The following graph is representing the membership grade function of right angle triangular fuzzy LPP (Fig. 2).

The optimized membership grade is derived that is 0.013

$$B(X) = \begin{cases} 1 & \text{when} \quad x \leq 2792.887, \\ \frac{2808.693 - x}{15.806} & \text{when} \quad 2792.887 \leq x \leq 2808.693, \\ 0 & \text{when} \quad x \geq 2808.693 \end{cases} \quad (14)$$

The final optimal solution for Case-I is obtained by using the membership grade (0.013) and that is x=2808.483.

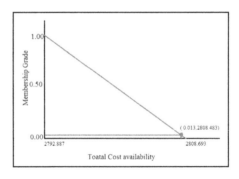

Fig. 2. Membership function of optimized cost using optimal solution of (s, l r) and (s, r) LPP

6.2   Case-II

The optimized fuzzy lpp (Right angle triangle) for membership grade has been constructed using the optimal solution of (s, l) as a lower bound and (s, l, r) as an upper bound and then the membership grade has been derived. The following graph is representing the membership grade function of right angle triangular fuzzy LPP (Fig. 3).

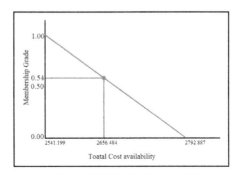

Fig. 3. Membership function of optimized cost using optimal solution of (s, l) and (s, l, r)

The optimized membership grade is derived that is 0.541

$$B(X) = \begin{cases} 1 & \text{when} \quad x \leq 2541.199, \\ \frac{2792.887-x}{251.687} & \text{when} \quad 2541.199 \leq x \leq 2792.887, \\ 0 & \text{when} \quad x \geq 2792.887 \end{cases} \tag{15}$$

The final optimal solution for Case-II is obtained by using the membership grade (0.541) and that is x=2656.484.

6.3   Case-III

The optimized fuzzy lpp (Right angle triangle) for membership grade has been constructed using the optimal solution of (s)lpp as a lower bound and (s, l, r)lpp as an upper bound and then the membership grade has been derived. The following graph is representing the membership grade function of right angle triangular fuzzy LPP (Fig. 4).

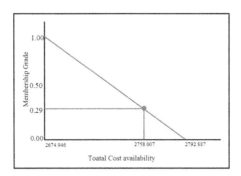

Fig. 4. Membership function of optimized cost using optimal solution of (s, l) and (s, l, r)

The optimized membership grade is derived that is 0.295

$$B(X) = \begin{cases} 1 & \text{when} \quad x \le 2674.946, \\ \frac{2792.887-x}{117.940} & \text{when} \quad 2674.946 \le x \le 2792.887, \\ 0 & \text{when} \quad x \ge 2792.887 \end{cases} \qquad (16)$$

The final optimal solution for Case-III is obtained by using the membership grade (0.295) and that is x=2758.007.

7   Conclusion and Future Scope

The modeling and optimization of production cost of RCF kapurthala has been done using the Triangular (s, l, r)triangular (Right angle) fuzzy linear programming problem. Owing to probabilistic increments in the availability of different constrains, the actual costs of production were vacillating or uncertain, the situational based Fuzzy models have been expressed to mitigate the destruction in the cost optimization and examined the credibility of optimized value. The production costs of different coaches from the year 2009–10 were considered as input. The total cost has been targeted in order to optimize. The lower and upper bound have been calculated using TFLPP-(s, l, r), TFLPP-(s, l), TFLPP-(s, r) and TFLPP-(s) for the objective function of the optimized fuzzy LPP. The minimized cost with (s, l, r) fuzzy LPP is Z = 2792.887. The cost is minimized using the TFLPP-(s, r) and that is Z = 2808. 6939934732.The cost is mini-mized using TFLPP-(s, l) and that is Z = 2541.1993274281 and the minimized cost without the increments and decrement is Z = 2674.9466604506. The fol-lowing results have been made from described modelling. The optimized fuzzy lpp (Right angle triangle) for membership grade has been constructed using the optimal solution of TFLPP (s, l, r) as a lower bound and TFLPP (s, r) as an upper bound and then the membership grade has been derived. The final optimal solution for this case is obtained and that is x=2808.483 with membership grade 0.013 The optimized fuzzy lpp (Right angle triangle) for membership grade has been constructed using the optimal solution of TFLPP (s, l) as a lower bound and TFLPP (s, l, r) as an upper bound and then the membership grade has been derived. The final optimal solution for this case is obtained and that is x=2656.484 with membership grade 0.541 The optimized fuzzy lpp (Right angle triangle) for membership grade has been constructed using the optimal solution of TFLPP(s) as a lower bound and TFLPP (s, l, r) as an upper bound and then the membership grade has been derived. The final optimal solution for this case is obtained and that is x=2758.007 with membership grade 0.295.

The validity of the method has been evaluated by solving some problems to analysis and optimize the production cost with symmetric and right angle Triangular fuzzy number through fuzzy linear programming problem. Further, the proposed approach can be applied to engineering and mathematical science problems which can be taken for further research.

References

1. De, P.K., Das, D.: Ranking of trapezoidal intuitionistic fuzzy numbers. In: 2012 12th International Conference on Intelligent Systems Design and Applications (ISDA), pp. 184–188, November 2012
2. Dipankar Chakraborty, D.K.J., Roy, T.K.: A new approach to solve intuitionistic fuzzy optimization problem using possibility, necessity, and credibility measures. Int. J. Eng. Math. **2014**, 12 pages (2014)
3. Ebrahimnejad, A., Tavana, M.: A novel method for solving linear program-ming problems with symmetric trapezoidal fuzzy numbers. Appl. Math. Model. **38**(1718), 4388–4395 (2014)
4. Ganesan, K., Veeramani, P.: Fuzzy linear programs with trapezoidal fuzzy num-bers. Ann. Oper. Res. **143**(1), 305–315 (2006)
5. Garg, H.: A novel approach for analyzing the behavior of industrial systems using weakest t-norm and intuitionistic fuzzy set theory. ISA Trans. **53**(4), 1199–1208 (2014)
6. Garg, H.: A new generalized improved score function of interval-valued intuition-istic fuzzy sets and applications in expert systems. Appl. Soft Comput. **38**(C), 988–999 (2016)
7. Garg, H., Ansha: Arithmetic operations on generalized parabolic fuzzy numbers and its application. Proc. Natl. Acad. Sci., India Sect. A: Phys. Sci., 1–12 (2016)
8. Gasimov, R.N., Yenilmez, K.: Solving fuzzy linear programming problems with linear membership functions. Turk. J. Math. **26**, 375–396 (2002)
9. Klir, G.J.: Fuzzy arithmetic with requisite constraints. Fuzzy Sets and Syst. **91**(2), 165–175 (1997)

10. Lodwick, W.A., Bachman, K.A.: Solving large-scale fuzzy and possibilistic optimization problems. Fuzzy Optim. Decis. Making **4**(4), 257–278 (2005)
11. Manju Pandey, D.S.S., Khare, N.: New aggregation operator for triangular fuzzy numbers based on the arithmetic means of the slopes of the l-, r- membership functions. Int. J. Comput. Sci. Inf. Technol. **3**(2), 3775–3777 (2012)
12. Nehi, H.M., Maleki, H.R., Mashinchi, M.: A canonical representation for the solution of fuzzy linear system and fuzzy linear programming problem. J. Appl. Math. Comput. **20**(1), 345–354 (2006)
13. Ping Wan, S.: Power average operators of trapezoidal intuitionistic fuzzy numbers and application to multi-attribute group decision making. Appl. Math. Model. **37**(6), 4112–4126 (2013)
14. Rani, D., Gulati, T., Garg, H.: Multi-objective non-linear programming problem in intuitionistic fuzzy environment: optimistic and pessimistic view point. Expert Syst. Appl. **64**, 228–238 (2016)
15. Rogers, F., Jun, Y.: Fuzzy nonlinear optimization for the linear fuzzy real number system. Int. Math. Forum **4**(12), 589–596 (2009)
16. Singh, P.: Big data time series forecasting model: a fuzzy-neuro hybridize approach. In: Acharjya, D.P., Dehuri, S., Sanyal, S. (eds.) Computational Intelligence for Big Data Analysis. ALO, vol. 19, pp. 55–72. Springer, Cham (2015). doi:10.1007/978-3-319-16598-1_2
17. Zadeh, L.: Fuzzy sets. Inf. Control **8**(3), 338–353 (1965)

A New Intuitionistic Fuzzy Entropy of Order-α with Applications in Multiple Attribute Decision Making

Rajesh Joshi$^{(\boxtimes)}$ and Satish Kumar

Department of Mathematics, Maharishi Markandeshwar University,
Mullana-Ambala 133207, India
aprajeshjoshi@gmail.com, drsatish74@resdiffmail.com

Abstract. In this paper, we have introduced a new intuitionistic fuzzy (IF) entropy called intuitionistic fuzzy entropy of order-α in the settings of intuitionistic fuzzy set theory. It considers both the uncertainty and hesitancy degree of IF sets. Also, we have shown that the entropy suggested by Vlachos and Sergiadis is the particular case of the proposed entropy which does not satisfies the maximality property. Further we have proved the validity of the proposed intuitionistic fuzzy entropy. Some of the properties of the proposed entropy are also discussed and proved that the maximum and minimum values of the proposed entropy are independent of α. At last, application of the proposed entropy is given in multiple attribute decision making (MADM) problem. For this purpose, we have taken a case study on insurance companies.

Keywords: Intuitionistic fuzzy entropy · Intuitionistic fuzzy entropy of order-α · Multiple attribute decision making (MADM) · Insurance services

MS Classifications: 94A15 · 94A24 · 26D15

1 Introduction

Zadeh [12], firstly proposed the notion of fuzzy sets to model the vague phenomenan. Since then the theory of fuzzy sets became an thursting area of research in different disciplines such as engineering, medical science, signal processing etc. Fuzziness, a feature of uncertainty, lacks from the fact that a particular element is a member of the set or not. A measure of fuzziness was also first introduced by [13]. [5] first axiomatized the fuzzy entropy and defined a new entropy measure of a fuzzy set based on [8].

Concept of intuitionistic fuzzy set (IFS) was firstly introduced by [1] as a generalization of notion of fuzzy set. The distinguishing fact of intuitionistic fuzzy set is that it considers membership degree, non-membership degree and the hesitancy degree as well. Atanassov and many other researchers studied

© Springer Nature Singapore Pte Ltd. 2017
K. Deep et al. (eds.), *Proceedings of Sixth International Conference on Soft Computing for Problem Solving*, Advances in Intelligent Systems and Computing 546, DOI 10.1007/978-981-10-3322-3_19

various properties of IFSs in decision making problems particularly in madical analysis and sales analysis etc. In the present communication, a new entropy called "intuitionistic fuzzy entropy of order-α" is proposed for intuitionistic fuzzy sets.

This paper is managed as follows. After the introductory section, Sect. 2 is devoted to some needed basic concepts and definitions related to fuzzy sets and intuitionistic fuzzy sets. In Sect. 3, we have introduced a new entropy called "intuitionistic fuzzy entropy of order-α" and established its validity. Some of the mathematical properties of the proposed intuitionistic fuzzy entropy are also discussed in this section. Application of the proposed entropy in multiple attribute decision making is given in Sect. 4. At last paper is concluded with "Concluding Remarks" in Sect. 5.

2 Preliminaries

In this section, some needed basic concepts and definitions of fuzzy sets and intuitionistic fuzzy sets are introduced.

Definition 1 (See [1]). Let $X = (z_1, z_2, \ldots, z_n)$ be a finite universe of discourse. A fuzzy set P is given by

$$P = \{\langle z_i, \mu_P(z_i)\rangle / z_i \in X\}, \tag{1}$$

where $\mu_P : X \to [0, 1]$ is the membership function of P. The number $\mu_P(z_i)$ is the degree of belongingness of $z_i \in X$ in P. [5] defined fuzzy entropy for a fuzzy set P corresponding to Shannon's entropy [8] as:

$$H(P) = -\frac{1}{n} \sum_{i=1}^{n} [\mu_P(z_i) \log(\mu_P(z_i)) + (1 - \mu_P(z_i)) \log(1 - \mu_P(z_i))]. \tag{2}$$

Later, Bhandari and Pal [4] surveyed the information measures on fuzzy sets and proposed some new measures of fuzzy entropy. Corresponding to Renyi's entropy [6], they defined a new measure as:

$$H_\alpha(P) = \frac{1}{n(1 - \alpha)} \sum_{i=1}^{n} \log \left[\mu_P^\alpha(z_i) + (1 - \mu_P(z_i))^\alpha\right]; \alpha \neq 1, \alpha > 0. \tag{3}$$

Attanassov [1,2] as mentioned earlier, generalized Zadeh's [12] idea of fuzzy sets to intuitionistic fuzzy sets, defined as:

Definition 2 (See [7]). An intuitionistic fuzzy set P in a finite universe of discourse $X = (z_1, z_2, \ldots, z_n)$ is given by

$$P = \{\langle z_i, \mu_P(z_i), \nu_P(z_i)\rangle / z_i \in X\}, \tag{4}$$

where $\mu_P : X \to [0, 1]$, $\nu_P : X \to [0, 1]$ satisfying $0 \leq \mu_P(z_i) + \nu_P(z_i) \leq 1$, $\forall z_i \in X$. Here $\mu_P(z_i)$ and $\nu_P(z_i)$, respectively, denotes the *membership degree* and *non-membership degree* of $z_i \in X$ to the set P. For each IFS P in X, $\pi_P(z_i) = 1 - \mu_P(z_i) - \nu_P(z_i), z_i \in X$ represents the hesitancy degree of $z_i \in X$ to the set P and is also called *intuitionistic index*. Obviously, if $\pi_P(z_i) = 0$ then intuitionistic fuzzy set becomes fuzzy set. Thus, the fuzzy sets are special cases of intuitionistic fuzzy sets.

Definition 3 (See [3]**).** Let $IFS(X)$ be the the set of all intuitionistic fuzzy sets in the universe X and let $P, Q \in IFS(X)$ be given by

$$P = \{\langle z_i, \mu_P(z_i), \nu_P(z_i)\rangle / z_i \in X\},$$
$$Q = \{\langle z_i, \mu_Q(z_i), \nu_Q(z_i)\rangle / z_i \in X\}. \tag{5}$$

Then usual set operations and relations are defined as follows:

(i). $P \subseteq Q$ if and only if $\mu_P(z_i) \le \mu_Q(z_i)$ and $\nu_P(z_i) \ge \nu_Q(z_i)$ for all $z_i \in X$;
(ii). $P = Q$ if and only if $P \subseteq Q$ and $Q \subseteq P$;
(iii). $P^c = \{\langle z_i, \nu_P(z_i), \mu_P(z_i)\rangle / z_i \in X\}$;
(iv). $P \cap Q = \{\langle \mu_P(z_i) \wedge \mu_Q(z_i) \text{ and } \nu_P(z_i) \vee \nu_Q(z_i)\rangle / z_i \in X\}$;
(v). $P \cup Q = \{\langle \mu_P(z_i) \vee \mu_Q(z_i) \text{ and } \nu_P(z_i) \wedge \nu_Q(z_i)\rangle / z_i \in X\}$.

[7] generalized the axioms of [5] and proposed a new entropy measure in settings of intuitionistic fuzzy sets.

Definition 4 (See [7]**).** An entropy on $IFS(X)$ is a real-valued function $E : IFS(X) \to [0, 1]$, satisfying the following four axioms:

E1 (Sharpness): $E(P) = 0$ iff P is a crisp set; i.e., $\mu_P(z_i) = 0$, $\nu_P(z_i) = 1$ or $\mu_P(z_i) = 1$, $\nu_P(z_i) = 0$ for all $z_i \in X$.
E2 (Maximality): $E(P) = 1$ if and only if $\mu_P(z_i) = \nu_P(z_i)$ for all $z_i \in X$.
E3 (Resolution): $E(P) \le E(Q)$ if and only if $P \subseteq Q$, i.e., if $\mu_P(z_i) \le \mu_Q(z_i)$ and $\nu_P(z_i) \ge \nu_Q(z_i)$ for $\mu_Q(z_i) \le \nu_Q(z_i)$, or if $\mu_P(z_i) \ge \mu_Q(z_i)$ and $\nu_P(z_i) \le \nu_Q(z_i)$, for $\mu_Q(z_i) \ge \nu_Q(z_i)$ for any $z_i \in X$.
E4 (Symmetry): $E(P) = E(P^c)$.

Throughout this paper, set of all intuitionistic fuzzy sets and fuzzy sets will be denoted by $IFS(X)$ and $FS(X)$, respectively.

With these concepts in mind, in the next section, we introduce a new entropy called "intuitionistic fuzzy entropy of order α" for intuitionistic fuzzy sets with α as a parameter.

3   Intuitionistic Fuzzy Entropy of Order-α

Definition 5. The intuitionistic fuzzy entropy of order-α for intuitionistic fuzzy set P is defined as

$$E_\alpha(P) =$$

$$\frac{1}{n(2^{1-\alpha} - 1)} \sum_{i=1}^{n} \left[(\mu_P^\alpha(z_i) + \nu_P^\alpha(z_i)) \times (\mu_P(z_i) + \nu_P(z_i))^{1-\alpha} + 2^{1-\alpha} \pi_P(z_i) - 1 \right]; \alpha > 0 (\ne 1). \tag{6}$$

Limiting and Particular Cases:

1. When $\alpha \to 1$ (6) becomes

$$E_1(P) = -\frac{1}{n}\sum_{i=1}^{n}\left[(\mu_P(z_i)\log(\mu_P(z_i) + \nu_P(z_i)\log(\nu_P(z_i))\right.$$

$$\left. -(1 - \pi_P(z_i))\log(1 - \pi_P(z_i)) - \pi_P(z_i)\right] \tag{7}$$

which is studied by Vlachos and Sergiadis [10] and does not satisfy the maximality property.

2. If $\pi_P(z_i) = 0$ then (6) becomes

$$E_\alpha(P) = \frac{1}{n(2^{1-\alpha} - 1)}\sum_{i=1}^{n}\left[(\mu_P^\alpha(z_i) + (1 - \mu_P(z_i))^\alpha) - 1\right]; \alpha > 0(\neq 1). \tag{8}$$

Now, we establish the validity of the proposed entropy.

Theorem 1. The intuitionistic fuzzy entropy of order-α, i.e., $E_\alpha(P)$ defined in (6) is a valid entropy for IFSs, i.e., it satisfies all the axioms given in Definition 4.

Proof **(E1).** Let P be the crisp set having membership values either 0 or 1 for all $z_i \in X$. Then from (6), we have $E_\alpha(P) = 0$.
    Conversely, if $E_\alpha(P) = 0$, then

$$\frac{1}{n(2^{1-\alpha} - 1)}\sum_{i=1}^{n}\left[(\mu_P^\alpha(z_i) + \nu_P^\alpha(z_i)) \times (\mu_P(z_i) + \nu_P(z_i))^{1-\alpha} + 2^{1-\alpha}\pi_P(z_i) - 1\right] = 0,$$

Since $\alpha \neq 1$, therefore

$$(\mu_P^\alpha(z_i) + \nu_P^\alpha(z_i)) \times (\mu_P(z_i) + \nu_P(z_i))^{1-\alpha} + 2^{1-\alpha}\pi_P(z_i) - 1 = 0,$$
$$\Rightarrow (\mu_A^\alpha(z_i) + \nu_A^\alpha(z_i)) \times (\mu_A(z_i) + \nu_A(z_i))^{1-\alpha} + 2^{1-\alpha}(1 - \mu_A(z_i) - \nu_A(z_i)) = 1,$$
$$\Rightarrow (\mu_P^\alpha(z_i) + \nu_P^\alpha(z_i)) \times (\mu_P(z_i) + \nu_P(z_i))^{1-\alpha} - 2^{1-\alpha}(\mu_P(z_i) + \nu_P(z_i)) = 1 - 2^{1-\alpha},$$
$$\Rightarrow (\mu_P(z_i) + \nu_P(z_i))\left[\frac{(\mu_P^\alpha(z_i) + \nu_P^\alpha(z_i))}{(\mu_P(z_i) + \nu_P(z_i))^\alpha} - 2^{1-\alpha}\right] = 1 - 2^{1-\alpha} \tag{9}$$

Since, $\alpha > 0(\neq 1)$, therefore (9) will hold only if $\mu_P(z_i) = 0, \nu_P(z_i) = 1$ or $\mu_P(z_i) = 1, \nu_P(z_i) = 0$ for all $z_i \in X$.
    Hence $E_\alpha(P) = 0$ if and only if $\mu_P(z_i) = 0, \nu_P(z_i) = 1$ or $\mu_P(z_i) = 1, \nu_P(z_i) = 0$ for all $z_i \in X$. This proves (E1).

(E2)

$$E_\alpha(P) = 1 \quad \text{iff} \quad \mu_P(z_i) = \nu_P(z_i) \quad \forall z_i \in X.$$

Let

$$E_\alpha(P) = 1$$
$$\Rightarrow \frac{1}{n(2^{1-\alpha}-1)}\sum_{i=1}^{n}\left[(\mu_P^\alpha(z_i) + \nu_P^\alpha(z_i)) \times (\mu_P(z_i) + \nu_P(z_i))^{1-\alpha} + 2^{1-\alpha}\pi_P(z_i) - 1\right] = 1$$
$$\Rightarrow \left[(\mu_P^\alpha(z_i) + \nu_P^\alpha(z_i)) \times (\mu_P(z_i) + \nu_P(z_i))^{1-\alpha} + 2^{1-\alpha}\pi_P(z_i) - 1\right] = 2^{1-\alpha} - 1$$
$$\Rightarrow (\mu_A^\alpha(z_i) + \nu_A^\alpha(z_i)) \times (\mu_A(z_i) + \nu_A(z_i))^{1-\alpha} + 2^{1-\alpha}(1 - \mu_A(z_i) - \nu_A(z_i)) = 2^{1-\alpha},$$
$$\Rightarrow (\mu_P(z_i) + \nu_P(z_i))\left[\frac{(\mu_P^\alpha(z_i) + \nu_P^\alpha(z_i))}{(\mu_P(z_i) + \nu_P(z_i))^\alpha} - 2^{1-\alpha}\right] = 0. \tag{10}$$

(10) will hold only if

$$\text{Either } \mu_P(z_i) + \nu_P(z_i) = 0 \Rightarrow \mu_P(z_i) = \nu_P(z_i) = 0 \quad \forall z_i \in X,$$

$$\text{or } \frac{(\mu_P^\alpha(z_i) + \nu_P^\alpha(z_i))}{(\mu_P(z_i) + \nu_P(z_i))^\alpha} - 2^{1-\alpha} = 0,$$

which is possible only if $\mu_P(z_i) = \nu_P(z_i), \forall z_i \in X$. Therefore

$$E_\alpha(P) = 1 \quad \text{iff} \quad \mu_P(z_i) = \nu_P(z_i) \quad \forall z_i \in X.$$

(E3). $E(P) \leq E(Q)$ if and only if $P \subseteq Q$, i.e., if $\mu_P(z_i) \leq \mu_Q(z_i)$ and $\nu_P(z_i) \geq \nu_Q(z_i)$ for $\mu_Q(z_i) \leq \nu_Q(z_i)$, or if $\mu_P(z_i) \geq \mu_Q(z_i)$ and $\nu_P(z_i) \leq \nu_Q(z_i)$, for $\mu_Q(z_i) \geq \nu_Q(z_i)$ for any $z_i \in X$.

In order to show that (6) satisfies (E3), it is sufficient to prove that the function

$$g(s,t) = \frac{1}{2^{1-\alpha} - 1} \left[(s^\alpha + t^\alpha)(s+t)^{1-\alpha} + 2^{1-\alpha}(1 - s - t) - 1 \right] \tag{11}$$

where $s, t \in [0,1]$ is an increasing function of s and decreasing function of t. Differentiating g partially with respect to s and t, respectively, we get

$$\frac{\partial g(s,t)}{\partial s} = \frac{1}{(2^{1-\alpha} - 1)} \left[(1-\alpha)(s^\alpha + t^\alpha)(s+t)^{-\alpha} + \alpha(s+t)^{1-\alpha}s^{\alpha-1} - 2^{1-\alpha} \right] \tag{12}$$

and

$$\frac{\partial g(s,t)}{\partial t} = \frac{1}{(2^{1-\alpha} - 1)} \left[(1-\alpha)(s^\alpha + t^\alpha)(s+t)^{-\alpha} + \alpha(s+t)^{1-\alpha}t^{\alpha-1} - 2^{1-\alpha} \right] \tag{13}$$

For critical points of g, we put $\partial g(s,t)/\partial s = 0$ and $\partial g(s,t)/\partial t = 0$. This gives

$$s = t \tag{14}$$

From (12) and (14), we get

$$\frac{\partial g(s,t)}{\partial s} \geq 0 \quad \text{when } s \leq t, 0 < \alpha < 1 \text{ and also for } \alpha > 1,$$

$$\frac{\partial g(s,t)}{\partial s} \leq 0 \quad \text{when } s \geq t, 0 < \alpha < 1 \text{ and also for } \alpha > 1. \tag{15}$$

for all $s, t \in [0,1]$. Thus $g(s,t)$ is increasing function of s for $s \leq t$ and decreasing function of s when $s \geq t$.

Similarly, we can prove that

$$\frac{\partial g(s,t)}{\partial t} \leq 0 \quad \text{when } s \leq t, 0 < \alpha < 1 \text{ and also for } \alpha > 1,$$

$$\frac{\partial g(s,t)}{\partial t} \geq 0 \quad \text{when } s \geq t, 0 < \alpha < 1 \text{ and also for } \alpha > 1. \tag{16}$$

Suppose there are two sets $P, Q \in IFS(X)$ such that $P \subseteq Q$. Let $X = \{z_1, z_2, \ldots, z_n\}$, the finite universe of discourse, be divided into two disjoint sets X_1 and X_2 with $X = X_1 \cup X_2$.

Further, suppose that all $z_i \in X_1$ obey the condition

$$\mu_P(z_i) \le \mu_Q(z_i) \le \nu_Q(z_i) \le \nu_P(z_i), \tag{17}$$

and all $z_i \in X_2$,

$$\mu_P(z_i) \ge \mu_Q(z_i) \ge \nu_Q(z_i) \ge \nu_P(z_i). \tag{18}$$

Thus, from the monotonic behaviour of the function g and equation (6), we obtain that $E_\alpha(P) \le E_\alpha(Q)$ when $P \subseteq Q$.

(E4). $E_\alpha(P) = E_\alpha(P^c)$

We know that $P^c = \{\langle z_i, \nu_P(z_i), \mu_P(z_i)\rangle / z_i \in X\}$ for $z_i \in X$; i.e.,

$$\mu_{P^c}(z_i) = \nu_P(z_i), \qquad\qquad \nu_{P^c}(z_i) = \mu_P(z_i). \tag{19}$$

Thus from (6), we have

$$E_\alpha(P) = E_\alpha(P^c). \tag{20}$$

Hence, $E_\alpha(P)$ is a valid intuitionistic fuzzy entropy.

Thus the theorem is proved.   ∎

The proposed entropy (6) also satisfies the following additional properties

Theorem 2 Let P and Q be two IFSs defined in $X = \{z_1, z_2, \ldots, z_n\}$, where $P = \{\langle z_i, \mu_P(z_i), \nu_P(z_i)\rangle / z_i \in X\}$, $Q = \{\langle z_i, \mu_Q(z_i), \nu_Q(z_i)\rangle / z_i \in X\}$, such that for all $z_i \in X$ either $P \subseteq Q$ or $Q \subseteq P$; then

$$E_\alpha(P \cup Q) + E_\alpha(P \cap Q) = E_\alpha(P) + E_\alpha(Q). \tag{21}$$

Corollary: For any $P \in IFS(X)$ and its complement P^c,

$$E_\alpha(P) = E_\alpha(P^c) = E_\alpha(P \cup P^c) = E_\alpha(P \cap P^c). \tag{22}$$

Theorem 3. The value of $E_\alpha(P)$ is maximum when the set is most intuitionistic fuzzy set and is minimum when the set is a crisp set; moreover, maximum and minimum values are free of α.

4   Application of Proposed Entropy in Multiple Attribute Decision Making (MADM) Using TOPSIS Method

An MADM problem is a problem in which there are so many factors need to be considered simultaneously and it becomes very difficult to arrive at a conclusion. So, by using proposed entropy we can compile the information and can take the decision. Now, we apply the proposed entropy to solve an MADM problem in the form of case study.

A Case Study of Insurance Services Quality

In this section, we consider a case study of insurance companies. Let there be n insurance companies $I_1, I_2, \ldots, I_n$ to be evaluated by their customers on the basis of m criteria $Z_1, Z_2, \ldots, Z_m$ through questionnaires. (In current example we take n=m=4. This example is adopted from Toloie *et al.* [9]). The questionnaires consists of four evaluation criteria ,i.e., Z_1 (Confidence), Z_2 (Responsiveness), Z_3 (Reliability), Z_4 (Tangibles). The perception of customers regarding the quality of services is compiled through questionnaires.

We use TOPSIS (Technique for Order Preference by Similarity to Ideal Solution) method for ranking the alternatives. In this method, we take two ideal solutions: One is the best solution and other is the worst solution. Then we determine the solution which is nearest to the best solution and away from the worst solution. This solution will be the best solution among all the solutions. Computational procedure for ranking the performance of companies is as follows:

1. Construction of the IF values decision Matrix (Table 1).

Table 1. The IF values decision matrix.

	I_1	I_2	I_3	I_4
Z_1	(0.449, 0.370)	(0.719, 0.188)	(0.546, 0.192)	(0.520, 0.337)
Z_2	(0.565, 0.162)	(0.630, 0.232)	(0.727, 0.182)	(0.630, 0.100)
Z_3	(0.705, 0.232)	(0.448, 0.378)	(0.641, 0.322)	(0.539, 0.271)
Z_4	(0.730, 0.170)	(0.557, 0.160)	(0.399, 0.200)	(0.679, 0.188)

2. Weight vector of the criteria is obtained by using the principle of minimum entropy suggested by Wang and Wang [11].

Using $min \quad E = \sum_{i=1}^{m} \sum_{j=1}^{n} u_j E_{\alpha(=15)}(\tilde{x}_{ij})$, the weight vector of the evaluated criteria is obtained as

$$min\,(E) = 0.9909u_1 + 0.9660u_2 + 0.9958u_3 + 0.9804u_4, \qquad (23)$$

such that

$$0 \leq u_1 \leq 0.3; 0.1 \leq u_2 \leq 0.2; 0.2 \leq u_3 \leq 0.5; 0.1 \leq u_4 \leq 0.3, \qquad (24)$$

with the condition $u_1 + u_2 + u_3 + u_4 = 1$. Solving this linear programming problem using MATLAB software, the weights obtained are $u_1 = 0.3, u_2 = 0.2, u_3 = 0.2, u_4 = 0.3$.

3. The Best Solution (Z^+) and Worst Solution (Z^-) are respectively given as:

$$Z^+ = ((\alpha_1^+, \beta_1^+), (\alpha_2^+, \beta_2^+), (\alpha_3^+, \beta_3^+)(\alpha_4^+, \beta_4^+)) = ((1,0), (1,0), (1,0), (1,0)),$$
$$Z^- = ((\alpha_1^-, \beta_1^-), (\alpha_2^-, \beta_2^-), (\alpha_3^-, \beta_3^-), (\alpha_4^-, \beta_4^-)) = ((0,1), (0,1), (0,1), (0,1)).$$

4. The distance measures of Z_i's $(i = 1, 2, 3, 4)$ from Z^+ and Z^- is:

$$s(Z_1, Z^+) = 0.4272, s(Z_2, Z^+) = 0.3215, s(Z_3, Z^+) = 0.4282, s(Z_4, Z^+) = 0.4854,$$
$$s(Z_1, Z^-) = 0.7653, s(Z_2, Z^-) = 0.8326, s(Z_3, Z^-) = 0.6821, s(Z_4, Z^-) = 0.8116.$$

5. The calculated relative degrees of closeness are:

$$D_1 = 0.6418, D_2 = 0.7214, D_3 = 0.6143, D_4 = 0.6258.$$

6. Ranking the options in descending order as per the values of D_i's. Based on D_i values, the sequence of alternatives obtained is: $I_2 \succ I_1 \succ I_4 \succ I_3$ and I_2 is the best choice in view of service qualities.

5 Concluding Remarks

In this paper, we have proposed a new intuitionistic fuzzy entropy of order α. It considers membership degree, non-membership degree and hesitancy degree. It is generalized form of entropy studied by Vlachos and Sergiadis [10]. Besides proving the validity of proposed entropy, we have also discussed some of its properties. At last, a case study regarding the selection of insurance company is given based on the TOPSIS method.

References

1. Atanassov, K.T.: Intuitionistic fuzzy sets. Fuzzy Sets Syst. **20**, 87–96 (1986)
2. Atanassov, K.T.: Intuitionistic Fuzzy Sets. Springer, New York (1999)
3. Atanassov, K.T.: Intuitionistic Fuzzy Sets. Physica, Heidelberg (1999)
4. Bhandari, D., Pal, N.R.: Some new information measures for fuzzy sets. Inf. Sci. **67**, 209–228 (1993)
5. De Luca, A., Termini, S.: A definition of non-probabilistic entropy in the settings of fuzzy set theory. Inf. Control. **20**, 301–312 (1972)
6. Renyi, A.: On measures of entropy and information. In: Proceedings of the 4th Berkeley Symposium on Mathematical Statistics and Probability, vol. 1, pp. 547–561. University of California Press, Berkeley (1961)
7. Szmidt, E., Kacprzyk, J.: Using intuitionistic fuzzy sets in group decision-making. Control Cybern. **31**, 1037–1054 (2002)
8. Shannon, C.E.: The mathematical theory of communication. Bell Syst. Tech. J. **27**(379–423), 623–656 (1948)
9. Toloie, A., Nasimi, M.A., Poorebrahimi, A.: Assessing quality of insurance companies using multiple criteria decision making. Eur. J. Sci. Res. **54**, 448–457 (2011)
10. Vlachos, I.K., Sergiadis, G.D.: Intuitionistic fuzzy information- applications to pattern recognition. Pattern Recogn. Lett. **28**, 197–206 (2007)
11. Wang, J., Wang, P.: Intuitionistic linguistic fuzzy multi-critria decision-making method based on intutionistic fuzzy entropy. Control. Decis. **27**, 1694–1698 (2012)
12. Zadeh, L.A.: Fuzzy sets. Inf. Comput. **8**, 338–353 (1965)
13. Zadeh, L.A.: Probability measures of fuzzy events. J. Math. Anal. Appl. **23**, 421–427 (1968)

The Relationship Between Intuitionistic Fuzzy Programming and Goal Programming

Sandeep Kumar[✉]

Department of Mathematics, Chaudhary Charan Singh University,
Meerut 250004, Uttar Pradesh, India
drsandeepmath@gmail.com

Abstract. One of the generalizations of fuzzy programming (FP) is intuitionistic fuzzy programming (IFP). IFP and goal programming (GP) are two important techniques for determining the solution (optimal) of multi-objective optimization problem by transforming it to a single objective one. The main purpose of this article is to introduce the similarities between IFP and GP. In this work, the max and min-operator are considered to transform the IFP to a deterministic program. One example is given to show the applicability of the proposed theory.

Keywords: Fuzzy programming · Multi-objective problem · Intuitionistic fuzzy programming · Goal programming

1 Introduction

Fuzzy programming is an extremely powerful tool for addressing a wide range of real world optimization problems. In fact, FP have appeared more visibly rather than crisp programming. Bellman and Zadeh [4] introduced the concept of fuzzy set theory in decision making problems. On the basis of [4], Tanaka et al. [20] proposed the concept of mathematical programming under fuzzy environment in first time. Zimmermann [21] gave the first formulation of fuzzy linear programming. Recently, FP has been used to solve many multi-objective problems, see [9,13,14] and references therein. Fuzzy set theory deals only with uncertainty in which sum of membership and non-membership functions is equal to one at each point of universe. When this sum is less than to one i.e. membership function and non-membership function are not complementary to each other, fuzzy set theory is not capable to handle such types of situations. To handle such cases, Atanassov [3] presented the concept of intuitionistic fuzzy set (IFS) which is one generalization of fuzzy set. IFS is quite different from fuzzy set. The reason is that IFS is characterized by two indices (membership function and non-membership function), which can be used to describe three states of fuzziness: favor, unfavor, and neutrality whereas fuzzy set is represented only by one index (membership function). In fact, IFS is capable to express fuzziness of "neither this nor that". When the information about the degree of membership

© Springer Nature Singapore Pte Ltd. 2017
K. Deep et al. (eds.), *Proceedings of Sixth International Conference
on Soft Computing for Problem Solving*, Advances in Intelligent Systems
and Computing 546, DOI 10.1007/978-981-10-3322-3_20

and the degree of non-membership of the goals of objectives in fuzzy multi-objective optimization problems are compared, then the theory of IFS is very useful. Angelov [2] applied the concept of IFS to solve the optimization problems like as transportation problem. Afterwards, a number of researchers have studied the works on IFP [1,15] and references therein, and its applications [7,17,18].

In this article the author attempts a class of IFP and developed its relationship with crisp goal programming. Both membership and non-membership functions are considered as linear in this work.

This article is organized as follows: A brief introduction of crisp goal programming is given in Sect. 2. In Sect. 3, an intuitionistic fuzzy programming is studied. In Sect. 4, a crisp goal programming from IFP is investigated. The proposed relationship between IFP and GP is illustrated through a numerical example, in Sect. 5.

2   Goal Programming

The concept of goal programming was introduced by Charnes and Copper [5] in 1961. Subsequent works on goal programming were presented by Lee [10], Charnes and Copper [6], Ignizo [8], Tamiz et al. [19] and others, see [11,16].

Now, consider the following multi-objective linear programming problem

$$\text{Optimize } G(Z) = [g_1(Z), g_2(Z), \ldots, g_m(Z)], Z = (z_1, z_2, \ldots, z_n)^T$$
subject to,
$$AZ \leq b, \quad Z \geq 0, \tag{1}$$

where A and b are $(p \times n)$ and $(p \times 1)$ matrices of real constants respectively. Let $G(\overline{Z}) = [g_1(\overline{Z}), g_2(\overline{Z}), \ldots, g_m(\overline{Z})]$, denote the vector of aspiration levels for the objectives and $g_k(Z)(k = 1, 2, \ldots, m)$ is the kth objective having its aspiration level $g_k(\overline{Z})$. Then usually following three types of goals are used.

(i) $g_k(Z) \geq g_k(\overline{Z})$, (maximization problem),
(ii) $g_k(Z) \leq g_k(\overline{Z})$, (minimization problem),
(iii) $g_k(Z) = g_k(\overline{Z})$, (equality problem).

The goals of any of the above types may be properly achieved, over achieved or under achieved. So the distance between objective function and its aspiration level is unrestricted, and is expressed by positive and negative deviational variables (h_k^+ and h_k^-) where $h_k^+ = \max(0, g_k(Z) - g_k(\overline{Z})) = \frac{1}{2}[g_k(Z) - g_k(\overline{Z}) + |g_k(Z) - g_k(\overline{Z})|]$, and $h_k^- = \max(0, g_k(\overline{Z}) - g_k(Z)) = \frac{1}{2}[g_k(\overline{Z}) - g_k(Z) + |g_k(\overline{Z}) - g_k(Z)|]$.

Introducing the deviational variables, the above types of goals take the following forms:

$$g_k(Z) - g_k(\overline{Z}) = h_k^+ - h_k^-,$$
$$g_k(\overline{Z}) - g_k(Z) = h_k^+ - h_k^-,$$
$$g_k(Z) - g_k(\overline{Z}) = h_k^+ + h_k^-,$$

respectively.

In GP, the aim of decision maker (DM) is to minimize the undesirable deviational variable. Therefore, in type (i) DM minimizes h_k^-, in type (ii) minimizes h_k^+, and in type (iii) minimizes $h_k^+ + h_k^-$.

Any of the following forms is chosen for solving a GP model.

(a) The min-max form: In this approach, DM minimizes the maximum deviational variable, let it be D. The mathematical form of this approach is as following:

$$\min D$$
subject to,
$$AZ \le b,$$
$$g_k(Z) + h_k^- - h_k^+ = g_k(\overline{Z}), \quad (k = 1, 2, \ldots, m),$$
$$D \ge f_k(h_k^-, h_k^+), \quad (k = 1, 2, \ldots, m),$$
$$h_k^- . h_k^+ = 0, \quad (k = 1, 2, \ldots, m),$$
$$Z, h_k^+, h_k^- \ge 0. \tag{2}$$

where $f_k(h_k^-, h_k^+) = h_k^-$ or h_k^+ or $h_k^+ + h_k^-$ according to the requirement of goals.

(b) The min form: The mathematical form of this approach is

$$\min \left(\sum_{k=1}^{m} f_k(h_k^-, h_k^+) \right)$$
subject to,
$$AZ \le b,$$
$$g_k(Z) + h_k^- - h_k^+ = g_k(\overline{Z}), \quad (k = 1, 2, \ldots, m),$$
$$h_k^- . h_k^+ = 0, \quad (k = 1, 2, \ldots, m),$$
$$Z, h_k^+, h_k^- \ge 0. \tag{3}$$

(c) The weighted min form: The mathematical form of this approach is

$$\min \left(\sum_{k=1}^{m} w_k f_k(h_k^-, h_k^+) \right)$$
subject to,
$$AZ \le b,$$
$$g_k(Z) + h_k^- - h_k^+ = g_k(\overline{Z}), \quad (k = 1, 2, \ldots, m),$$
$$h_k^- . h_k^+ = 0, \quad (k = 1, 2, \ldots, m),$$
$$Z, h_k^+, h_k^- \ge 0, \tag{4}$$

here $w_1, w_2, \ldots, w_m$ are called weights, which are obtained by DM.

(d) The pre-emptive priority form: In this approach, the deviational variables are ranked into a number of priority levels and are minimized in a lexicographic way. A lexicographic minimization means that the highest priority goal is considered first, then the second and so on.

The algebraic representation of a lexicographical goal program is

$$\min \left(\sum_{k \in P_i} w_k f_k(h_k^-, h_k^+), \quad i = 1, 2, \ldots, I \right)$$

subject to,

$$AZ \leq b,$$
$$g_k(Z) + h_k^- - h_k^+ = g_k(\overline{Z}), \quad (k = 1, 2, \ldots, m),$$
$$h_k^- \cdot h_k^+ = 0, \quad (k = 1, 2, \ldots, m),$$
$$Z, h_k^+, h_k^- \geq 0, \tag{5}$$

where I denotes the number of priority levels and k belongs to P_i indicates that the kth goal constraint is present in the ith priority level.

3   Intuitionistic Fuzzy Programming

In model (1), without loss of generality, the following conditions are assumed:

(H_1) Each objective is of maximization type.
(H_2) All objectives are linear, i.e. the kth objective $g_k(Z), k = 1, 2, \ldots, m$ can be expressed as:

$$g_k(Z) = c_1 z_1 + c_2 z_2 + \cdots + c_n z_n,$$

where $c_1, c_2, \ldots, c_n$ are real constants.
(H_3) Each system constraint is described in crisp sense.

Using (H_1)-(H_3), the fuzzy version of model (1) is as follows:

Find Z such that
$$G(Z) \gtrsim G(\overline{Z})$$
subject to,
$$AZ \leq b,$$
$$Z \geq 0, \tag{6}$$

where $\gtrsim$ ($\lesssim$) is fuzzification of symbol $\geq$ ($\leq$) in Zimmermann sense [21].

Let $\mu_k(Z)$ be the membership function for k^{th} objective function. Then, by using Bellman and Zadeh's approach, the crisp equivalent of model (6) is as follows:

$$\max \mu_k(Z)$$
subject to,
$$AZ \leq b,$$
$$0 \leq \mu_k(Z) \leq 1, \quad k = 1, 2, \ldots, m,$$
$$Z \geq 0. \tag{7}$$

In fuzzy optimization, Angelov [2] introduced the situation when the degree of non-acceptance is defined together with the degree of satisfaction and sum of these degrees is not exactly one. With this consideration, model (7) becomes

Find Z such that
$$\max_{Z} \left(\mu_k(Z) \right),$$
$$\min_{Z} \left(\nu_k(Z) \right),$$
subject to,
$$AZ \leq b,$$
$$\mu_k(Z) \geq \nu_k(Z), \nu_k(Z) \geq 0,$$
$$\mu_k(Z) + \nu_k(Z) \leq 1, \quad k = 1, 2, \ldots, m,$$
$$Z \geq 0, \tag{8}$$

where $\mu_k(Z)$ denotes the degree of acceptance of Z to the kth intuitionistic fuzzy objective function and $\nu_k(Z)$ denotes the degree of non-acceptance of Z from k^{th} intuitionistic fuzzy objective function.

Using max and min-operator, the intuitionistic fuzzy optimization problem (8) can be transformed to the following crisp model:

$$\max \ (\gamma - \delta)$$
subject to,
$$AZ \leq b,$$
$$\gamma \leq \mu_k(Z),$$
$$\delta \geq \nu_k(Z), \quad k = 1, 2, \ldots, m,$$
$$\gamma \geq \delta, \delta \geq 0,$$
$$\gamma + \delta \leq 1, Z \geq 0, \tag{9}$$

where γ denotes the minimal acceptable degree of objectives and δ denotes the maximal degree of rejection of objectives.

In model (9), there may be following four possibilities in selection of membership function $\mu_k(Z)$ and non-membership function $\nu_k(Z)$ of kth intuitionistic fuzzy objective:

(i) $\mu_k(Z)$ and $\nu_k(Z)$ both are linear [12];
(ii) $\mu_k(Z)$ is linear and $\nu_k(Z)$ is non-linear [2];
(iii) $\mu_k(Z)$ is non-linear and $\nu_k(Z)$ is linear [2];
(iv) $\mu_k(Z)$ and $\nu_k(Z)$ both are non-linear.

In this article, both $\mu_k(Z)$ and $\nu_k(Z)$ are choosen as linear.

Now, the linear functions $\mu_k(Z)$ and $\nu_k(Z)$ for k^{th} intuitionistic fuzzy objective function $g_k(Z)$ in sense of [21] are constructed, as follows:

$$\mu_k(Z) = \begin{cases} 1; & \text{if } g_k(Z) \geq g_k(\overline{Z}), \\ 1 - \frac{g_k(\overline{Z}) - g_k(Z)}{t_{1k}}; & \text{if } l_k \leq g_k(Z) \leq g_k(\overline{Z}), \\ 0; & \text{if } g_k(Z) \leq l_k, \end{cases} \qquad (10)$$

where $l_k = g_k(\overline{Z}) - t_{1k}$.

And

$$\nu_k(Z) = \begin{cases} 1; & \text{if } g_k(Z) \leq l_k, \\ 1 - \frac{g_k(Z) - l_k}{t_{2k}}; & \text{if } l_k \leq g_k(Z) \leq l_k + t_{2k}, \\ 0; & \text{if } g_k(Z) \geq l_k + t_{2k}, \end{cases} \qquad (11)$$

where $t_{2k} < t_{1k}$.

Then, model (9) takes the following form:

$$\max (\gamma - \delta)$$
$$\text{subject to,}$$
$$AZ \leq b,$$
$$\gamma \leq 1 - \frac{g_k(\overline{Z}) - g_k(Z)}{t_{1k}},$$
$$\delta \geq 1 - \frac{g_k(Z) - l_k}{t_{2k}}, \quad k = 1, 2, \ldots, m,$$
$$\gamma \geq \delta, \delta \geq 0,$$
$$\gamma + \delta \leq 1, Z \geq 0. \qquad (12)$$

The program (12) can be solved by using simplex method.

4   The Relationship Between IFP and GP

Mohamed [11] established a relationship between fuzzy linear program (FLP) and GP. The keyconcepts of this work are: (i) the aspiration level for each membership goal is one; (ii) the larger tolerance value indicates less important of the goal. This relationship is described as:

Theorem 1. [11] *Every FLP is equivalent to a weighted GP, where the weights are the reciprocals of the admissible violation constants.*

In this work, a relationship between intuitionistic fuzzy linear program and GP is developed. This work is motivated by [2,11]. The proposed relationship is stated in following theorem:

Theorem 2. *The intuitionistic FLP is equivalent to a weighted linear GP.*

Proof. The model (12) can be rewritten as:

$$\min (1 - \gamma) + \delta$$
$$\text{subject to,}$$
$$AZ \leq b,$$
$$1 - \gamma \geq \frac{g_k(\overline{Z}) - g_k(Z)}{t_{1k}},$$
$$\delta \geq 1 - \frac{g_k(Z) - l_k}{t_{2k}}, k = 1, 2, \ldots, m,$$
$$\delta \geq 0, 1 - \gamma \geq 0, \gamma \geq \delta,$$
$$\gamma + \delta \leq 1, Z \geq 0. \tag{13}$$

From model (13),

$$1 - \gamma \geq \max\left(0, \frac{g_k(\overline{Z}) - g_k(Z)}{t_{1k}}\right)$$

and

$$\delta \geq \max\left(0, 1 - \frac{g_k(Z) - l_k}{t_{2k}}\right)$$

With the help of definition of deviational variables, the above inequalities can convert into the following forms:

$1 - \gamma \geq h_{1k}^-$, where

$$\frac{g_k(Z)}{t_{1k}} + h_{1k}^- - h_{1k}^+ = \frac{g_k(\overline{Z})}{t_{1k}}$$
$$\text{i.e.} \quad g_k(Z) + t_{1k}h_{1k}^- - t_{1k}h_{1k}^+ = g_k(\overline{Z})$$

and $\delta \geq h_{2k}^-$, where

$$\frac{g_k(Z)}{t_{2k}} + h_{2k}^- - h_{2k}^+ = \frac{l_k + t_{2k}}{t_{2k}}$$
$$\text{i.e.} \quad g_k(Z) + t_{2k}h_{2k}^- - t_{2k}h_{2k}^+ = l_k + t_{2k}$$

with $h_{1k}^+.h_{1k}^- = 0, h_{2k}^+.h_{2k}^- = 0$ and $k = 1, 2, \ldots, m$.

Further, letting $1 - \gamma = \eta \geq 0$, then model (13) can be written as:

$$\min (\eta + \delta)$$
$$\text{subject to,}$$
$$AZ \leq b,$$
$$g_k(Z) + t_{1k}h_{1k}^- - t_{1k}h_{1k}^+ = g_k(\overline{Z}),$$
$$g_k(Z) + t_{2k}h_{2k}^- - t_{2k}h_{2k}^+ = l_k + t_{2k},$$
$$\eta \geq h_{1k}^-, \delta \geq h_{2k}^-,$$
$$\delta \leq \eta, \eta + \delta \leq 1,$$
$$h_{1k}^+.h_{1k}^- = 0, h_{2k}^+.h_{2k}^- = 0,$$
$$\text{all variables } \geq 0, k = 1, 2, \ldots, m. \tag{14}$$

By taking $t_{1k}h_{1k}^- = h_k^-$ and $t_{2k}h_{2k}^- = e_k^-$. The model (12) is equivalent to a weighted GP which is similar to model (2). In this weighted GP, the deviational variables are weighted with weights $\frac{1}{h_{1k}^-}$ for membership goals and $\frac{1}{h_{2k}^-}$ for non-membership goals. This completes the proof.

Some new IFP can be investigated which are equivalent to GP models (3), (4) and (5) in similar fashion.

5 Numerical Example

Consider the following numerical example [21]:

$$\max g_1(Z) = -z_1 + 2z_2,$$
$$\max g_2(Z) = 2z_1 + 3z_2,$$
subject to,

$$-z_1 + 3z_2 \le 21,$$
$$z_1 + 3z_2 \le 27,$$
$$4z_1 + 3z_2 \le 45,$$
$$3z_1 + z_2 \le 30,$$
$$z_1, z_2 \ge 0. \tag{15}$$

Let $g_1(\overline{Z}) = 15, g_2(\overline{Z}) = 23$ be aspiration levels for $g_1(Z)$ and $g_2(Z)$ respectively. Suppose that $t_{11} = 8, t_{12} = 10$ and $t_{21} = 5, t_{22} = 7$.
   The intuitionistic fuzzy program (12) becomes

$$\max (\gamma - \delta)$$
subject to,

$$\gamma \le -0.8750 - 0.125z_1 + 0.25z_2,$$
$$\gamma \le -1.3 + 0.2z_1 + 0.3z_2,$$
$$\delta \ge 2.4 + 0.2z_1 - 0.6z_2,$$
$$\delta \ge 2.8571 - 0.2857z_1 - 0.4286z_2,$$
$$-z_1 + 3z_2 \le 21,$$
$$z_1 + 3z_2 \le 27,$$
$$4z_1 + 3z_2 \le 45,$$
$$3z_1 + z_2 \le 30,$$
$$\gamma \ge \delta, \gamma + \delta \le 1,$$
$$z_1, z_2, \gamma, \delta \ge 0. \tag{16}$$

Using TORA software, the solution (optimal) of model (16) is $(z_1^*, z_2^*) = (0.2195, 7.0732)$, $\gamma^* = 0.8657, \delta^* = 0$ and the equivalent GP of model (16) is as follows:

$$\min (\eta + \delta)$$
subject to,

$$-z_1 + 3z_2 \leq 21,$$
$$z_1 + 3z_2 \leq 27,$$
$$4z_1 + 3z_2 \leq 45,$$
$$3z_1 + z_2 \leq 30,$$
$$-z_1 + 2z_2 + h_{11}^- - h_{11}^+ = 15,$$
$$-z_1 + 2z_2 + h_{21}^- - h_{21}^+ = 12,$$
$$2z_1 + 3z_2 + h_{12}^- - h_{12}^+ = 23,$$
$$2z_1 + 3z_2 + h_{22}^- - h_{22}^+ = 20,$$
$$\eta \geq \tfrac{h_{11}^-}{8}, \delta \geq \tfrac{h_{21}^-}{5},$$
$$\eta \geq \tfrac{h_{12}^-}{10}, \delta \geq \tfrac{h_{22}^-}{7},$$
$$\delta \leq \eta, \eta + \delta \leq 1,$$
$$h_{1k}^+ \cdot h_{1k}^- = 0, h_{2k}^+ \cdot h_{2k}^- = 0, k = 1, 2, \ldots, m,$$
$$\text{all variables} \geq 0. \tag{17}$$

The solution (optimal) of model (17) is also $(z_1^*, z_2^*) = (0.2195, 7.0732)$, $\eta^* = 0.1341$, $\delta^* = 0$.

6 Conclusion

In this article two important approaches IFP and GP are considered to solve fuzzy multi-objective optimization problems. The relationship between IFP and GP is introduced within this article. Developing of this relationship, a class of membership functions and non-membership functions is constructed in which non-membership function is not complement of membership function exactly. This article is proposed a new type of IFP form using the sense of GP. In future work the author intends to introduce a relationship between IFP, having intuitionistic fuzziness in parameters and GP.

References

1. Aggarwal, A., Khan, I.: On solving Atanassov's I-fuzzy linear programming problems: some variants of Angelov's model. Opsearch **53**(2), 375–389 (2016)
2. Angelov, P.P.: Optimization in an intuitionistic fuzzy environment. Fuzzy Sets Syst. **86**, 299–306 (1997)
3. Atanassov, K.T.: lntuitionistic fuzzy sets. Fuzzy Sets Syst. **20**, 87–96 (1986)
4. Bellman, R.E., Zadeh, L.A.: Decision-making in a fuzzy environment. Manage. Sci. **17**(4), 141–164 (1970)
5. Charnes, A., Cooper, W.W.: Management Models and Industrial Applications of Linear Programming, vol. I and II. Wiley, New York (1961)
6. Charnes, A., Cooper, W.W.: Goal programming and multiple objective optimisations. Eur. J. Oper. Res. **1**(1), 39–54 (1977)

7. Garg, H., Rani, M., Sharma, S.P., Vishwakarma, Y.: Intuitionistic fuzzy optimization technique for solving multi-objective reliability optimization problems in interval environment. Expert Syst. Appl. **41**, 3157–3167 (2014)

8. Ignizio, J.P.: On the rediscovery of fuzzy goal programming. Decis. Sci. **13**, 331–336 (1982)

9. Kumar, S.: Max-min solution approach for multi-objective matrix game with fuzzy goals. Yugoslav J. Oper. Res. **26**(1), 51–60 (2016)

10. Lee, S.M.: Goal Programming for Decision Analysis. Auerbach Publishers, Philadelphia (1972)

11. Mohamed, R.H.: The relationship between goal programming and fuzzy programming. Fuzzy Sets Syst. **89**, 215–222 (1997)

12. Nayak, P.K., Pal, M.: Intuitionistic fuzzy optimization technique for nash equilibrium solution of multi-objective bi-matrix games. J. Uncertain Syst. **5**(4), 271–285 (2011)

13. Pandey, D., Kumar, S.: Modified approach to multi-objective matrix game with vague payoffs. J. Int. Acad. Phys. Sci. **14**(2), 149–157 (2010)

14. Pandey, D., Kumar, S.: Fuzzy multi-objective fractional goal programming using tolerance. Int. J. Math. Sci. Eng. Appl. **5**(1), 175–187 (2011)

15. Rani, D., Gulati, T.R., Garg, H.: Multi-objective non-linear programming problem in intuitionistic fuzzy environment: optimistic and pessimistic view point. Expert Syst. Appl. **64**, 228–238 (2016)

16. Romero, C.: A survey of generalized goal programming (1970–1982). Eur. J. Oper. Res. **25**, 183–191 (1986)

17. Singh, S.K., Yadav, S.P.: Modeling and optimization of multi-objective non-linear programming problem in intuitionistic fuzzy environment. Appl. Math. Model. **39**, 4617–4629 (2015)

18. Singh, S.K., Yadav, S.P.: Fuzzy programming approach for solving intuitionistic fuzzy linear fractional programming problem. Int. J. Fuzzy Syst. **18**(2), 263–269 (2016)

19. Tamiz, M., Jones, D.F., El-Darzi, E.: A review of goal programming and its applications. Ann. Oper. Res. **58**, 39–53 (1995)

20. Tanaka, H., Okuda, T., Asai, K.: On fuzzy mathematical programming. J. Cybernet. **3**, 37–46 (1974)

21. Zimmermann, H.J.: Fuzzy programming and linear programming with several objective functions. Fuzzy Sets Syst. **1**, 45–55 (1978)

A Fuzzy Dual SBM Model with Fuzzy Weights: An Application to the Health Sector

Alka Arya[✉] and Shiv Prasad Yadav

Department of Mathematics, Indian Institute of Technology Roorkee,
Roorkee 247667, India
alka1dma@gmail.com, spyorfma@gmail.com

Abstract. In this paper, we focus on measuring the performance efficiencies of decision making units (DMUs) using dual slack based measure (SBM) model with fuzzy data in data envelopment analysis (DEA). In the conventional dual SBM model, the data and the weights of input and output are found as crisp quantities. However, in real world applications, input-output data and input-output weights may have vague/uncertainty due to various factors such as quality of treatment and medicines, number of medical and non-medical staffs, number of patients, etc. in health sector. To deal with such uncertainty, we can apply fuzzy set theory. In this paper, we propose a SBM model with fuzzy weights in Fuzzy DEA (FDEA) for fuzzy input and fuzzy output. This model is then reduced to a crisp LPP by using expected values of a fuzzy number (FN). Finally, we present an application of the proposed model to the health sector, consisting of two input variables as (i) sum of number of doctors and staff nurses (ii) number of pharmacists and two output variables as (i) number of inpatients (ii) number of outpatients. Both the input variables and output variables are considered as TFNs.

Keywords: Fuzzy DEA · Fuzzy dual SBM · Fuzzy weights · Hospitals efficiencies

1 Introduction

DEA is non-parametric linear programing (LP) based technique to determine the relative efficiency of homogeneous DMUs when the production process consists of multiple inputs and multiple outputs (Ramanathan 2003). There exist some mathematical programs in DEA such as: Fractional, Input minimization (Output oriented) and Output maximization (Input oriented) etc. (Cooper et al. 2007). DEA calculates maximal performance measure for each DMU relative to other DMUs. CCR model (Charnes et al. 1978) find the constant returns to scale (CRS) and BCC model (Banker et al. 1984) find the variable returns to scale (VRS), they neglects the slacks in the evaluation of efficiencies. To solve this neglection can be computed using the slack based measure (SBM) model non-radial and non-oriented DEA model (Tone 2001).

© Springer Nature Singapore Pte Ltd. 2017
K. Deep et al. (eds.), *Proceedings of Sixth International Conference on Soft Computing for Problem Solving*, Advances in Intelligent Systems and Computing 546, DOI 10.1007/978-981-10-3322-3_21

Conventional DEA deals with crisp input and crisp output data. But in real world applications, some input and/or output data possess some degree of fluctuation or imprecision or uncertainties such in health sector as quality of input resources, quality of treatment, the satisfaction level of patients, quality of medicines etc. The fluctuation can take the form of intervals, ordinal relations and fuzzy numbers etc. Therefore, to deal with such type of real life situations, we plan to extend crisp DEA to FDEA by making use of fuzzy numbers in DEA. FDEA models represent real world applications more realistically than the conventional DEA models.

The rest of the paper is organized as follows: Sect. 2 presents the literature review. Section 3 presents preliminaries required to develop the model of which include basic definitions performance efficiency, fuzzy number, triangular fuzzy number and expected values. Section 4 presents the background of primal and dual parts of the fuzzy SBM model. Section 5 presents an application to health sector to illustrate the proposed model. Section 6 concludes the finding of our work.

2   Literature Review

This section reviews some DEA based studies on health care sector around all over the world. Over the last 50 years India has built a sound health sector infrastructure (Agarwal et al. 2006). According to the literature, in the present time, the role of the health care sector has been expanding than the public health care sector in India. Determining the health care performance efficiency has an important role in developed as well as developing countries. There are some studies to determine the performance efficiencies of health care using DEA in Indian context (Mogha et al. 2014(a),(b)). The most important role in the economy of any developed as well as developing countries is health care of urban and rural areas. Sengupta (1992) was the first author to introduce the fuzzy measure, regression, entropy and fuzzy mathematical programming approach in DEA. Afsharinia et al. (2013) determined the performance efficiency of clinical units using fuzzy essence. Tsai et al. (2010) proposed the fuzzy analytic hierarchy process (FAHP) and fuzzy sensitive analysis based approach to measure the policy of Taiwan hospitals in DEA. Dotoli et al. (2015) developed a novel cross-efficiency fuzzy DEA model for evaluating different elements under uncertainty with application to the health care system. Mansourirad et al. (2010) were the first to introduce fuzzy weights in fuzzy CCR (FCCR) model and proposed a model using α-cut approach to evaluate weights for outputs in terms of TFNs. The SBM performance efficiency in DEA is extended to fuzzy forms (Jahanshahloo et al. 2004 and Saati et al. 2009). Puri and Yadav (2013) proposed a slack based measure model with fuzzy weights corresponding to fuzzy inputs and fuzzy outputs using $\alpha-$ cut approach.

3   Preliminary

This section includes some basic definitions and notions of fuzzy set theory (Zimmermann 1996), fuzzy number (FN), triangular fuzzy number (TFN), arithmetic operations on TFNs (Chen 1994) and expected values (Ghasemi 2015).

3.1   Performance Efficiency (Charnes 1978)

The performance efficiency of a DMU is defined as the ratio of the weighted sum of outputs (virtual output) to the weighted sum of inputs (virtual input). Thus, Performance efficiency $= \frac{virtual\ output}{virtual\ input}$.

DEA evaluates the relative performance efficiency of a DMU in a set of homogeneous DMUs. The relative performance efficiency of a DMU lies in the range $(0, 1]$.

3.2   Fuzzy Number (FN) (Zimmermann 1996)

An FN $\tilde{M}$ is defined as a convex normalized fuzzy set $\tilde{M}$ of the real line $\mathbb{R}$ such that

(1) there exists exactly one $x_0 \in \mathbb{R}$ with $\mu_{\tilde{M}}(x_0) = 1$. x_0 is called the mean value of $\tilde{M}$,
(2) $\mu_{\tilde{M}}$ is a piecewise continuous function, called the membership function of $\tilde{M}$.

3.3   Triangular Fuzzy Number (TFN) (Zimmermann 1996)

The TFN $\tilde{M}$ is an FN denoted by $\tilde{M} = (a, b, c)$ and is defined by the membership function $\mu_{\tilde{M}}$ given by

$$\mu_{\tilde{M}}(x) = \begin{cases} \frac{x-a}{b-a}, & a < x \leq b, \\ \frac{c-x}{c-b}, & b \leq x < c, \\ 0, & elsewhere, \end{cases}$$

for all $x \in \mathbb{R}$.

This TFN can be said to be "approximately equal to b", where b is called the modal value, and (a,c) is called support of the TFN (a,b,c).

3.4   Arithmetic Operations on TFNs (Chen 1994)

Let $\tilde{M}1 = (a_1, b_1, c_1)$ and $\tilde{M}2 = (a_2, b_2, c_2)$ be two TFNs. Then, the arithmetic operations on TFNs are given as follows:

- Addition: $\tilde{M}1 \oplus \tilde{M}2 = (a_1 + a_2, b_1 + b_2, c_1 + c_2)$.
- Subtraction: $\tilde{M}1 \ominus \tilde{M}2 = (a_1 - c_2, b_1 - b_2, c_1 - a_2)$.
- Multiplication: $\tilde{M}1 \otimes \tilde{M}2 = (min(a_1a_2, a_1c_2, c_1a_2, c_1c_2), b_1b_2, max(a_1a_2, a_1c_2, c_1a_2, c_1c_2))$

– Scalar multiplication:

$$\lambda \tilde{M}_I = \begin{cases} (\lambda a_1, \lambda b_1, \lambda c_1), & for\ \lambda \geq 0 \\ (\lambda c_1, \lambda b_1, \lambda a_1), & for\ \lambda < 0 \end{cases}$$

3.5   The Expected Values of FNs (Ghasemi 2015)

The expected interval (EI) of a TFN $\tilde{M} = (a, b, c)$ defined as follows: $EI(\tilde{M}) = [E^L(\tilde{M}), E^U(\tilde{M})]$, where
$E^L(\tilde{M}) = \frac{a+b}{2}$ and
$E^R(\tilde{M}) = \frac{b+c}{2}$.
And expected value (EV) of a TFN $\tilde{M} = (a, b, c)$ defined as follows:
$EV(\tilde{M}) = \frac{1}{2}(E^L(\tilde{M}) + E^U(\tilde{M})) = \frac{a+2b+c}{4}$.

4   Background

This paper measures the fuzzy input weights, fuzzy output weights and fuzzy efficiency of 12 community health cares of Meerut district of Uttar Pradesh (UP) State.

4.1   SBM DEA Model

Let the performance of a set of n homogeneous DMUs $(DMU_j = 1, 2, 3, ..., n)$ be determined. The performance efficiency of DMU_j is characterized by a production process of m inputs $x_{ij}(i = 1, 2, 3, ..., m)$ to produce s outputs $y_{rj}(r = 1, 2, 3, ..., s)$. Assume x_{ij_o} be the amount of the ith input used and y_{rj_o} be the amount of the rth output produced by the DMU_{j_o}. Let input data and output data be positive. The primal SBM model (Tone 2001) for DMU_{j_o} is given by the following model:

Model 1: (*Primal SBM model*)

$$\rho_{j_o} = \min \frac{1-(1/m)\sum_{i=1}^{m} s_{ij_o}^-/x_{ij_o}}{1+(1/s)\sum_{r=1}^{s} s_{rj_o}^+/y_{rj_o}} \tag{1}$$
$$subject\ to$$

$$x_{ij_o} = \sum_{j=1}^{n} x_{ij}\mu_{jj_o} + s_{ij_o}^-, \ \forall i \tag{2}$$

$$y_{rj_o} = \sum_{j=1}^{n} y_{rj}\mu_{jj_o} - s_{rj_o}^+, \ \forall r \tag{3}$$

$$\mu_{jj_o} \geq 0, \ \forall j, \ s_{ij_o}^- \geq 0, \ \forall i, \ s_{rj_o}^+ \geq 0, \ \forall r \tag{4}$$

where $s_{ij_o}^-$ and $s_{rj_o}^+$ are the slack variables in the ith input of the DMU_{j_o} and rth output of the DMU_{j_o} respectively.

Definition 1. (Tone 2001) ρ_{j_o} is called SBM efficiency (SBME) of DMU_{j_o}. DMU_{j_o} is SBM efficient if $\rho_{j_o}^* = 1$.
   This condition is equivalent to $s_{ij_o}^{-*} = 0$ and $s_{rj_o}^{+*} = 0$, i.e., no output shortfalls and no input excesses in optimal solution, otherwise inefficient.

Model 1 can be transformed into linear programming (LP) using normalization method (Charnes et al. 1978). In Model 1, multiply by a scalar $p_{j_o} > 0$ to both the numerator and denominator. The value of p_{j_o} can be adjusted in such a way that the numerator becomes 1. Thus Model 1 is reduced to the following model (Model 2):

Model 2:

$$E_{j_o} = \max p_{j_o} + \frac{1}{s}\sum_{r=1}^{s} S_{rj_o}^{+}/y_{rj_o}, \tag{5}$$
$$subject\ to$$

$$p_{j_o} - \frac{1}{m}\sum_{i=1}^{m} S_{ij_o}^{-}/x_{ij_o} = 1, \tag{6}$$

$$p_{j_o}x_{ij_o} = \sum_{j=1}^{n} x_{ij}\lambda_{jj_o} + S_{ij_o}^{-}\ \forall i, \tag{7}$$

$$p_{j_o}y_{rj_o} = \sum_{j=1}^{n} y_{rj}\lambda_{jj_o} - S_{rj_o}^{+}\ \forall r, \tag{8}$$

$$\lambda_{jj_o} \geq 0\,\forall j, S_{ij_o}^{-} \geq 0,\ \forall r, S_{rj_o}^{+} \geq 0,\ \forall i, p_{j_o} > 0, \tag{9}$$

$$where\ p_{j_o}\mu_{jj_o} = \lambda_{jj_o},\ \forall j, p_{j_o}s_{ij_o}^{-} = S_{ij_o}^{-},\ \forall i\ and\ p_{j_o}s_{rj_o}^{+} = S_{rj_o}^{+},\ \forall r$$

Let θ_{j_o}, u_{ij_o} and v_{rj_o} be the dual variables corresponding to (6), (7) and (8) respectively. Then the Dual problem LPP in Model 2 is given by:

Model 3: (*Dual SBM model*)

$$E_{j_o}^{D} = \min \theta_{j_o}, \tag{10}$$
$$subject\ to$$

$$\sum_{r=1}^{s} y_{rj}v_{rj_o} + \sum_{i=1}^{m} x_{ij}u_{ij_o} \leq 0,\ \forall j, \tag{11}$$
$$\theta_{j_o} + \sum_{i=1}^{m} x_{ij_o}u_{ij_o} + \sum_{r=1}^{s} y_{rj_o}v_{rj_o} \geq 1, \tag{12}$$
$$u_{ij_o} + \frac{\theta_{j_o}}{mx_{ij_o}} \leq 0\forall i, \tag{13}$$
$$v_{rj_o} \geq \frac{1}{sy_{rj_o}}\forall r,\ \theta_{j_o} \in \mathbb{R}, \tag{14}$$

All the variables θ_{j_o}, u_{ij_o} and v_{rj_o} are unrestricted in sign.

4.2   Proposed Fuzzy Dual SBM Model

In conventional SBM model the input-output data and input-output weights are in crisp form. But in real world application, these weights and data may have fuzzy values. Thus, in this paper, input-output data and input-output weights are taken as TFNs. Model 3 is reduced to the following model:

Model 4:

$$\tilde{E}_{j_o}^{D} = \min \tilde{\theta}_{j_o},$$
$$subject\ to$$
$$\sum_{r=1}^{s} \tilde{y}_{rj} \otimes \tilde{v}_{rj_o} + \sum_{i=1}^{m} \tilde{x}_{ij} \otimes \tilde{u}_{ij_o} \leq \tilde{0},\ \forall j,$$
$$\tilde{\theta}_{j_o} + \sum_{i=1}^{m} \tilde{x}_{ij_o} \otimes \tilde{u}_{ij_o} + \sum_{r=1}^{s} \tilde{y}_{rj_o} \otimes \tilde{v}_{rj_o} \geq \tilde{1},$$
$$m\ \tilde{u}_{ij_o} \otimes \tilde{x}_{ij_o} + \tilde{\theta}_{j_o} \leq \tilde{0}\ \forall i,\ s\ \tilde{v}_{rj_o} \otimes \tilde{y}_{rj_o} \geq \tilde{1}\ \forall r,$$

where $\tilde{x}_{ij}$ and $\tilde{y}_{rj}$ are the triangular fuzzy inputs and outputs respectively; $\tilde{u}_{ij_o}$ is the triangular fuzzy weight corresponding to the ith input and $\tilde{v}_{rj_o}$ is the triangular fuzzy weight corresponding to the rth output. By using expected values of TFN, Model 4 reduces to the following model:

Model 5:

$$EV(\tilde{E}_{j_o}^D) = \min EV(\theta_{j_o}^1, \theta_{j_o}^2, \theta_{j_o}^3),$$

$$subject\ to$$

$$EV(\textstyle\sum_{r=1}^{s} y_{rj}^1 v_{rj_o}^1, \sum_{r=1}^{s} y_{rj}^2 v_{rj_o}^2, \sum_{r=1}^{s} y_{rj}^3 v_{rj_o}^3) + EV(\sum_{i=1}^{m} x_{ij}^1 u_{ij_o}^1, \sum_{i=1}^{m} x_{ij}^2 u_{ij_o}^2,$$
$$\textstyle\sum_{i=1}^{m} x_{ij}^3 u_{ij_o}^3) \leq EV(0,0,0),\ \forall j,$$

$$EV(\theta_{j_o}^1, \theta_{j_o}^2, \theta_{j_o}^3) + EV(\textstyle\sum_{i=1}^{m} x_{ij_o}^1 u_{ij_o}^1, \sum_{i=1}^{m} x_{ij_o}^2 u_{ij_o}^2, \sum_{i=1}^{m} x_{ij_o}^3 u_{ij_o}^3)+$$
$$EV(\textstyle\sum_{r=1}^{s} y_{rj_o}^1 v_{rj_o}^1, \sum_{r=1}^{s} y_{rj_o}^2 v_{rj_o}^2, \sum_{r=1}^{s} y_{rj_o}^3 v_{rj_o}^3) \geq EV(1,1,1),$$

$$m\ EV(x_{ij_o}^1 u_{ij_o}^1, x_{ij_o}^2 u_{ij_o}^2, x_{ij_o}^3 u_{ij_o}^3) + EV(\theta_{j_o}^1, \theta_{j_o}^2, \theta_{j_o}^3) \leq EV(0,0,0),\ \forall i,$$

$$s\ EV(y_{rj_o}^1 v_{rj_o}^1, y_{rj_o}^2 v_{rj_o}^2, y_{rj_o}^3 v_{rj_o}^3) \geq EV(1,1,1),\ \forall r,$$

Using expected values in Model 5, we get Model 6.

Model 6:

$$E_{j_o}^{D1} = \min \tfrac{1}{4}(\theta_{j_o}^1 + 2\theta_{j_o}^2 + \theta_{j_o}^3),$$

$$subject\ to$$

$$\textstyle\sum_{r=1}^{s} (y_{rj}^1 v_{rj_o}^1 + 2y_{rj}^2 v_{rj_o}^2 + y_{rj}^3 v_{rj_o}^3) - \sum_{i=1}^{m} (x_{ij}^1 u_{ij_o}^1 + 2x_{ij}^2 u_{ij_o}^2 + x_{ij}^3 u_{ij_o}^3) \leq 0,\ \forall j,$$

$$(\theta_{j_o}^1 + 2\theta_{j_o}^2 + \theta_{j_o}^3) + \textstyle\sum_{i=1}^{m} (x_{ij_o}^1 u_{ij_o}^1 + 2x_{ij_o}^2 u_{ij_o}^2 + x_{ij_o}^3 u_{ij_o}^3)+$$
$$\textstyle\sum_{r=1}^{s} (y_{rj_o}^1 v_{rj_o}^1 + 2y_{rj_o}^2 v_{rj_o}^2 + y_{rj_o}^3 v_{rj_o}^3) \geq 4,$$

$$m\ (x_{ij_o}^1 u_{ij_o}^1 + 2x_{ij_o}^2 u_{ij_o}^2 + x_{ij_o}^3 u_{ij_o}^3) + (\theta_{j_o}^1 + 2\theta_{j_o}^2 + \theta_{j_o}^3) \leq 0\ \forall i,$$

$$s\ (y_{rj_o}^1 v_{rj_o}^1 + 2y_{rj_o}^2 v_{rj_o}^2 + y_{rj_o}^3 v_{rj_o}^3) \geq 4,\ \forall r,$$

$$u_{ij_o}^1 \leq u_{ij_o}^2 \leq u_{ij_o}^3\ \forall i,\ v_{rj_o}^1 \leq v_{rj_o}^2 \leq v_{rj_o}^3\ \forall r,\ \theta_{j_o}^1 \leq \theta_{j_o}^2 \leq \theta_{j_o}^3.$$

SBME of DMU_{j_o} is written as $SBME_{j_o}$ and is given by $SBME_{j_o} = (E_{j_o}^{DI})^{-1}$.

5 Application to the Health Sector

In this section, we present an application to illustrate the proposed fuzzy dual SBM model. In this paper, DMUs are CHCs in Meerut district of Uttar Pradesh, India. The performance of each CHC is determined based on two fuzzy inputs and two fuzzy outputs. The input-output data in fuzzy form are given in Table 1. For DMU_j, $j = 1, 2, 3, ..., 12$

1st Input (x_{1j}) = Sum of number of doctors and number of staff nurses
2nd Input (x_{2j}) = Number of pharmacists
1st Output (y_{1j}) = Number of inpatients
2nd Output (y_{2j}) = Number of outpatients

The fuzzy efficiencies of all CHCs are determined from Model 6, which are given in Table 2. The fuzzy efficiency scores lie between zero and 1. The fuzzy weights corresponding to fuzzy inputs and fuzzy outputs of the concerned CHCs are also determined by using Model 6, which are given in Tables 2 and 3. The

Table 1. Fuzzy input and fuzzy output data Source: Chief Medical Office, Head Office, Meerut, India.

DMUs	Fuzzy input		Fuzzy output	
	x_{1j}	x_{2j}	y_{1j}	y_{2j}
H1	(10,13,15)	(3,5,8)	(3640,3650,3655)	(134130,134137,134145)
H2	(10,12,14)	(3,5,7)	(4150,4160,4170)	(116055,116062,116068)
H3	(9,12,14)	(2,4,5)	(4360,4370,4380)	(94060,94066,94072)
H4	(6,8,11)	(1,1,3)	(485,492,500)	(24320,24329,24335)
H5	(8,10,13)	(3,4,6)	(2460,2464,2470)	(99740,99748,99760)
H6	(10,11,12)	(2,3,4)	(1360,1368,1375)	(49395,49401,49410)
H7	(9,10,12)	(1,2,6)	(1055,1062,1070)	(37765,37772,37780)
H8	(9,11,15)	(1,4,7)	(1295,1302,1310)	(82835,82841,82850)
H9	(10,12,15)	(2,5,7)	(1660,1671,1680)	(100590,100596,100605)
H10	(10,16,20)	(2,4,6)	(1010,1018,1025)	(64345,64351,64360)
H11	(9,11,14)	(3,5,8)	(1500,1504,1510)	(80050,80056,80061)
H12	(5,8,10)	(1,4,6)	(1960,1965,1972)	(58160,58167,58175)

Table 2. Efficiencies and fuzzy Input weights of 12 hospitals

DMUs	Fuzzy input weights			
	$SBME_{j_o}$	$\tilde{\theta}_{j_o}$	$\tilde{u}_{1j}$	$\tilde{u}_{2j}$
H1	1	(0,0,4)	$(0.39 \times 10^{-4}, 0.39 \times 10^{-4}, 0.39 \times 10^{-4})$	$(-0.38,-0.38,0.36)$
H2	1	(1,1,1)	$(-0.067,-0.067,0.021)$	$(-0.1,-0.1,-0.1)$
H3	1	(1,1,1)	$(-0.042,-0.042,-0.042)$	$(-0.13,-0.13,-0.13)$
H4	1	(1,1,1)	$(-0.06,-0.06,-0.06)$	$(-0.25,-0.25,-0.25)$
H5	1	(1,1,1)	$(-0.88,-0.88,-0.88)$	$(-0.4,-0.4,0.38)$
H6	1	(1,1,1)	$(-0.064,-0.064,-0.064)$	$(-0.22,-0.22,-0.22)$
H7	1	(1,1,1)	$(-0.49,-0.49,-0.49)$	$(-0.36,-0.36,-0.31)$
H8	1	(1,1,1)	$(-0.043,-0.043,-0.043)$	$(-0.48,-0.48,0.031)$
H9	1	(1,1,1)	$(-0.041,-0.041,-0.041)$	$(-0.1,-0.1,-0.1)$
H10	1	(1,1,1)	$(-0.85,0.12,0.12)$	$(-0.14,-0.14,-0.14)$
H11	0.66	(1.52,1.52,1.52)	$(-0.1,-0.1,-0.048)$	$(-0.32,-0.32,0.13)$
H12	1	(1,1,1)	$(-0.27,-0.27,0.36)$	$(-0.71,-0.71,0.73)$

Table 3. Fuzzy output weights of 12 hospitals

DMUs	Fuzzy output weights	
	$\tilde{v}_{1j}$	$\tilde{v}_{2j}$
H1	$(0.14\times10^{-3}, 0.14\times10^{-3}, 0.14\times10^{-3})$	$(0.37\times10^{-5}, 0.37\times10^{-5}, 0.37\times10^{-5})$
H2	$(0.12\times10^{-3}, 0.12\times10^{-3}, 0.12\times10^{-3})$	$(0.43\times10^{-5}, 0.43\times10^{-5}, 0.43\times10^{-5})$
H3	$(0.11\times10^{-3}, 0.11\times10^{-3}, 0.11\times10^{-3})$	$(0.53\times10^{-5}, 0.53\times10^{-5}, 0.53\times10^{-5})$
H4	$(-0.217, 0.073, 0.073)$	$(-0.091, 0.029, 0.09)$
H5	$(0.2\times10^{-3}, 0.2\times10^{-3}, 0.2\times10^{-3})$	$(-0.05, -0.05, 0.14)$
H6	$(0.36\times10^{-3}, 0.36\times10^{-3}, 0.36\times10^{-3})$	$(-0.15, -0.15, 0.46)$
H7	$(-0.07, -0.07, 0.21)$	$(-0.017, -0.017, 0.53)$
H8	$(-0.13, -0.02, 0.16)$	$(0.6\times10^{-5}, 0.6\times10^{-5}, 0.6\times10^{-5})$
H9	$(0.052, 0.052, 0.052)$	$(0.49\times10^{-5}, 0.49\times10^{-5}, 0.49\times10^{-5})$
H10	$(-0.22, 0.073, 0.073)$	$(0.87\times10^{-5}, 0.87\times10^{-5}, 0.87\times10^{-5})$
H11	$(-0.1, -0.1, 0.31)$	$(0.85\times10^{-5}, 0.85\times10^{-5}, 0.85\times10^{-5})$
H12	$(0.00025, 0.00025, 0.00025)$	$(-0.018, -0.018, .056)$

fuzzy efficiencies and weights for every CHC are obtained by executing a MAT-LAB program of Model 6. In this application H11 is SBM inefficient hospital, other hospitals are SBM efficient.

6   Conclusion

In this piece of work, we proposed a fuzzy dual SBM model (Model 4) with fuzzy weights in fuzzy DEA. Model 4 is then reduced to crisp LP SBM model (Model 6) by using expected values of FNs. Model 6 determines the fuzzy efficiencies and components of fuzzy weights corresponding to fuzzy inputs and fuzzy outputs as TFNs. Model 6 also determines the SBM efficient and SBM inefficient DMUs. These fuzzy efficiencies and fuzzy weights provided extra information to the decision maker, which is not provided by crisp dual SBM model.

Acknowledgment. The authors are thankful to the Ministry of Human Resource Development (MHRD), Govt. of India, India for financial support in pursuing this research. The authors are also thankful to Mr. Deen Bandhu, ARO, Chief Medical Office, Meerut, India for providing the valuable data of the hospitals.

References

Afsharinia, A., Bagherpour, M., Farahmand, K.: Measurement of clinical units using integrated independent component analysis-DEA model under fuzzy conditions. Int. J. Hosp. Res. **2**(3), 109–118 (2013)

Agarwal, S., Yadav, S.P., Singh, S.P.: Assessment of relative efficiency of private sector hospitals in India using DEA. J. Math. Syst. Sci. **2**, 1–22 (2006)

Banker, R.D., Charnes, A., Cooper, W.W.: Some models for the estimation of technical and scale inefficiencies in data envelopment analysis. Manag. Sci. **30**, 1078–1092 (1984)

Cooper, W.W., Seiford, L.M., Tone, K.: Data Envelopment Analysis: A Comprehensive Text with Models, Applications, References and DEA-Solver Software, 2nd edn. Springer, New York (2007)

Charnes, A., Cooper, W.W., Rhodes, E.: Measuring the efficiency of decision making units. Eur. J. Oper. Res. **2**, 429–444 (1978)

Chen, S.M.: Fuzzy system reliability analysis using fuzzy number arithmetic operations. Fuzzy Set. Syst. **66**, 31–38 (1994)

Dotoli, M., Epicoco, N., Falagario, M., Sciancalepore, F.: A cross-efficiency fuzzy Data Envelopment Analysis technique for performance evaluation of Decision Making Units under uncertainty. Comput. Ind. Eng. **79**, 103–114 (2015)

Ghasemi, M.R., Ignatius, J., Lozano, S., Emrouznejad, A., Hatami-Marbini, A.: A fuzzy expected value approach under generalized data envelopment analysis. Knowl.-Based Syst. **89**, 148–159 (2015)

Jahanshahloo, G.R., Soleimani-damaneh, M., Nasrabadi, E.: Measure of efficiency in DEA with fuzzy input-output levels: a methodology for assessing, ranking and imposing of weights restrictions. Appl. Math. Comput. **156**, 175–187 (2004)

Mansourirad, E., Rizam, M.R.A.B., Lee, L.S., Jaafar, A.: Fuzzy weights in data envelopment analysis. Int. Math. Forum **5**(38), 1871–1886 (2010)

Mogha, S.K., Yadav, S.P., Singh, S.P.: New slack model based efficiency assessment of public sector hospitals of Uttarakhand: state of India. Int. J. Syst. Assur. Eng. Manag. **5**(1), 32–42 (2014a)

Mogha, S.K., Yadav, S.P., Singh, S.P.: Estimating technical and scale efficiencies of private hospitals using a non-parametric approach: case of India. Int. J. Oper. Res. **20**(1), 21–40 (2014b)

Puri, J., Yadav, S.P.: A concept of fuzzy input mix-efficiency in fuzzy DEA and its application in banking sector. Expert Syst. Appl. **40**, 1437–1450 (2013)

Ramanathan, R.: An Introduction to Data Envelopment Analysis. Sage Publication India Pvt. Ltd., New Delhi (2003)

Saati, S., Memariani, A.: SBM model with fuzzy input-output levels in DEA. Aust. J. Basic Appl. Sci. **3**(2), 352–357 (2009)

Sengupta, J.K.: A fuzzy systems approach in data envelopment analysis. Comput. Math. Appl. **24**(9), 259–266 (1992)

Tsai, H.Y., Chang, C.W., Lin, H.L.: Fuzzy hierarchy sensitive with Delphi method to evaluate hospital organization performance. Expert Syst. Appl. **37**, 5533–5541 (2010)

Tone, K.: A slack based measure of efficiencies in data envelopment analysis. Eur. J. Oper. Res. **130**, 498–509 (2001)

Zimmermann, H.J.: Fuzzy Set Theory and Its Applications, 4th edn. Kluwer Academic Publishers, Norwell (1996)

An Approach for Purchasing a Sedan Car from Indian Car Market Under Fuzzy Environment

Mukesh Chand, Deeksha Hatwal, Shalini Singh, Varsha Mundepi,
Vidhi Raturi, Rashmi, and Shwetank Avikal[(✉)]

Department of Mechanical Engineering,
Graphic Era Hill University, Dehradun, India
Shwetank.avikal@gmail.com

Abstract. As we all know that, in this era everyone wants to buy a car that may be mid price ranged or may be a high price range car. People want that in their car should carry feature according to them. It has been seen in the past decades that the sale of automobile in India is increasing rapidly and a number of car manufactures have joined Indian car market. Hence there is a tuff competition between every car manufactures. In this case all car manufactures are trying to attract customers by providing good aesthetic and advance features in the cars. These aesthetic and features have a high impact on consumers mind. Selecting a car from Indian car market is a very large decision making problems or customers. In this work, the selection of a car from Indian car market has been treated as Multi Criteria Decision Making (MCDM) problem and a number of criterions have been selected for the study. A Fuzzy AHP (Analytic Hierarchy Process) based approach has been used for selecting a sedan car on the bases of various car criterion such as: Performance, Economy, Comfort etc. In the initial stage, some important criterion has been discussed and their weight has been calculated by FAHP. On the basis of criteria weights, the ranking of the cars has been done. The main objective behind this paper is to facilitate the consumers to have a clear idea about their preferences in purchasing a car.

Keywords: Fuzzy set · AHP (Analytic Hierarchy Process) · Car purchasing · MCDM

1 Introduction

In day to day life, generally people have to make different types of decision. Daily life is full of learning which can help people to know the diagnostic moments. Making sudden decision might be risky and holding the work can moved the opportunities. For making a right decision, people need a systematic approach for decision making. Selecting a car from Indian car market is also a decision making problem. These types of problem may carry a large number of criteria.

The Indian automobile market has become more and more competitive. The car manufacturers are working on further advancement and new technologies so that there sale may increases and hence as a result they can earn more profit. Buying a suitable car

© Springer Nature Singapore Pte Ltd. 2017
K. Deep et al. (eds.), *Proceedings of Sixth International Conference on Soft Computing for Problem Solving*, Advances in Intelligent Systems and Computing 546, DOI 10.1007/978-981-10-3322-3_22

is difficult decision making problem. Generally peoples compare a large number of attributes/features of cars such as: fuel, economy, safety, comfort etc. for decision making. Preference of each consumer may be different from one another while purchasing a car and hence purchasing a car is decision making problem that shows the preference of consumer. In the presented work, problem of selecting a sedan car from Indian car market has been considered as a multi criteria decision making problem and an approach based on Fuzzy-AHP has been used for solving the same. Analytical hierarchy process (AHP) is an approach for solving complex multi criteria decision making problems. Fuzzy AHP is the extensions of AHP method when the fuzziness of the decision maker is taken into account.

India is a growing influence of the smart technologies in the automobile sector. More research work has been done in this field. As we know that automobile has a greater impact on our day to day life. Now a day's every middle class person wants to purchase a sedan car because the purchasing power of these peoples is increasing day by day. A number of researchers have worked on same type of decision making problems and some of these have been discussed here such as:

Byun (2001) proposed AHP method to select the best product on the basis of various categories such as cost, safety, fuel economy and appearance etc.

Gungor and Isler (2005) have used AHP method for the selection of the best car among eight alternatives. Their criteria were price, acceleration, fuel consumption, safety, comfort, maintenance cost for selection of best automobile.

Terz et al. (2006) have proposed a decision support model by using the AHP methods which helps in making a right decision in automobile purchasing problem. In their study they considered performance, economy, after sale services and safety of the automobile in their model.

Zeshui (2007) has proposed the method for comparing two intuitionistic fuzzy values and develops various aggregation operator like intuitionistic fuzzy weighted averaging operator, intuitionistic fuzzy ordered weighted averaging operator and etc. and gives different properties of the operator.

Sahin and Akyer (2011) have used AHP methods and by this they selected 4*4 search and rescue vehicle. The selection criteria were fuel consumption, price, acceleration, load capacity etc.

Raut et al. (2011) have used the multi criteria decision making (MCDM) approach by using AHP method and quality function deployment fuzzy technique for preference. Their results have certified the technology, economical aspects, safety, comforts etc.

Avikal et al. (2013a) have proposed a new heuristic to assign parts to the disassembly work stations under precedence constraints. To prioritise the tasks assignment they have used Analytic hierarchy process (AHP) and PROMETHEE methods. The methodology has been explained with a case example.

Garg (2016) has proposed two methods first is a new generalized score function which is interval-valued intuitionistic fuzzy sets (IVIFSs) environment based on weighted average and second is IVIFSs method for solving MCDM problems.

Dong et al. (2015) have worked on multi criteria group decision making problem in which decision making in done on alternative criteria & are shown by triangular fuzzy number and make it to the form of incomplete reciprocal comparisons matrices.

It has been seen that a number of people have worked on the same types of problem but none of them is related to purchasing a sedan car from Indian car market. Now Fuzzy-AHP has been applied for solving the addressed problem.

2 Proposed Methodology

The problem on this work has been considered as a Multi Criteria Decision Making problem. It involves various stages:

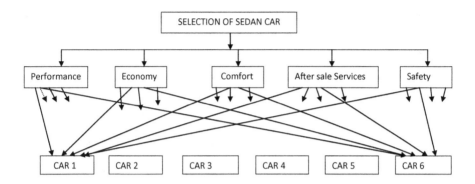

Fuzzy AHP method:

In this work, the problem has been taken as a multi criteria decision making (MCDM) problem, and fuzzy AHP has been used to evaluate the weight of each criterion. Fuzzy AHP is the extension of conventional AHP by implementing Fuzzy set theory. Fuzzy set theory allows the membership operations to work over the real number [0, 1]. Fuzzy may be defined as the membership operations and all the data about a fuzzy is describe by its membership operation. In the Fuzzy AHP triangular Fuzzy numbers have been utilized to improve the scaling scheme. The fuzzy set theory described by Avikal et al. (2013b) and Avikal and Jain (2014) has been taken as the references for the computation.

3 Computational Example

A problem of Indian car market has been taken for this study. Total 6 Sedan cars have been selected for comparison. The name of all these are:

1. Car 1 = TATA Indigo
2. Car 2 = Ford Figo Aspire
3. Car 3 = Maruti Swift Desire
4. Car 4 = Honda Amaze
5. Car 5 = Hyundai Xcent
6. Car 6 = Toyota Etios

3.1   Criteria Definition

Different five criteria (C1–C5) have been selected for this study. These criteria have been given following:

Performance (C1): Is a factor that influences car performance such as power, pickup and mileage.

Economy (C2): Basically defines initial cost of the vehicle.

Comfort/Advancement (C3): this criteria deal with the comfort available in the car and new advancement in cars.

After sale services (C4): It means how much services are available outside in market means how many dealers are there in market of that prospective brand.

Safety (C5): It defines that how safe is our car while we are driving it.

In Table 1, data of different cars has been given. The pair wise comparison matrix of criterion by using triangular Fuzzy member function has been shown in Table 2. The weights of all criterions have been calculated by the data given in Table 2 and presented in Table 3. The consistency ration of the comparison matrix is less than .1 and it means that the data is consistent. The data available in Table 2 has been normalized and presented in Table 4. Final ranking of the alternative has been done by multiplying the weight of each criterion to the respective normalized data available in Table 4. In the ranking of the alternatives, criteria 2 has negative impact because it is cost and it should be minimum and all other criterion have positive impact in final score. The final ranking of alternatives has been shown in Table 5.

Table 1. Data of different car selected for study (www.cardekho.com)

	C1 (In Lakh)	C2 (kmpl)	C3 (out of 5)	C4 (out of 5)	C5 (out of 5)
Tata Indigo	4.9	15.60	3	4	4
Ford Figo Aspire	5.2	18.20	3.5	4	3.5
Maruti Swift Desire	6.4	16.80	3.5	4.5	4
Honda Amaze	5.5	17.80	4	4.5	4
Hyundai Xcent	5.4	19.10	4	3.5	3.5
Toyota Etios	6.5	16.78	3	4	3

Table 2. Pair wise comparison matrix by fuzzy AHP

	C1	C2	C3	C4	C5
C1	1	~3	~3	~5	~3
C2	$\sim3^{-1}$	1	~1	~3	~1
C3	$\sim3^{-1}$	$\sim1^{-1}$	1	~1	$\sim3^{-1}$
C4	$\sim5^{-1}$	$\sim3^{-1}$	$\sim1^{-1}$	1	$\sim3^{-1}$
C5	$\sim3^{-1}$	$\sim1^{-1}$	~3	~3	1

Table 3. Consistency index & Consistency ratio

Criteria	Criteria weight	Consistency index & Consistency ratio
C1	0.330	
C2	0.191	CI = 0.1085
C3	0.113	CR = 0.097
C4	0.073	
C5	0.198	

Table 4. Normalizing of data

	C1	C2	C3	C4	C5
C1	0.252	0.155	0.084	0.065	0.198
C2	0.267	0.181	0.099	0.065	0.173
C3	0.330	0.167	0.099	0.073	0.198
C4	0.278	0.177	0.113	0.073	0.198
C5	0.278	0.191	0.113	0.057	0.173

Table 5. Ranking of car

Car name	Final weight	Ranking
Tata Indigo ECS	0.444	4
Ford Figo Aspire 1.2 Ti VCT Ambient	0.423	6
Maruti Swift Desire LXI	0.533	1
Honda Amaze EI VTEC	0.485	2
Hyundai Xcent 1.2 Kappa Base	0.430	5
Toyota Etios J	0.460	3

From Table 5 shows that Maruti Swift desire got the first position in car ranking, Honda amaze got second, Toyota Etios got third, Tata Indigo got forth, Hyundai Xcent got fifth and finally Ford Figo aspire got last position in car ranking.

4 Conclusion

In Indian car Market, it is very difficult to select a sedan car for middle class family. It is a multi criteria decision making problem that depends on a number of criteria. These criteria may have important impact on the decision making. In this case, it is necessary to evaluate the importance of these criteria. In the presented work, a Fuzzy AHP based approach has been proposed for evaluating the weights of all criteria selected for decision making. The ranking of sedan cars has been cone with the help of criterion's weight those have been calculated by Fuzzy-AHP. From the final ranking, it has seen that Maruti Swift desire got first rank. It got first rank as it got higher values for the entire criterion such as: Performance, economy, comfort after sale service and safety measures. All other cars got their own rank according to their final values. The application of proposed work can help the middle class peoples to purchase a sedan car from Indian car market. Some other ranking approach such as PROMOTHEE, TOPSIS and Fuzzy TOPSIS may be used for final raking in the future work to improve the results.

So application of proposed work has made ease to select a car and purchase it from Indian car market. In the future work, ranking techniques such as: PROMOTHEE, TOPSIS, ELECTRE may also be used improve the results.

References

Byun, D.: The AHP approach for selecting an automobile purchase model. Inf. Manage. **38**, 289–299 (2001)

Gungor, I., Isler, D.B.: Analitik hiyerarsi yaklasimi ile otomobil secime. ZKU Sosyal Bilimler Dergis **1**(2), 21–31 (2005)

Terz, U., Hacaloglu, S.E., Aladag, Z.: Otomobil satin alma problemi icin bir karar destek modeli. Istanbul Ticaret Universitesi Fen Bilimleri Dergis **5**(10), 43–49 (2006)

Zeshui, Xu.: Intuitionistic fuzzy aggregation operator. IEEE Trans. Fuzzy Syst. **15**(6), 1179–1187 (2007)

Sahin, Y., Akyer, H.: Ulke kaynaklannin verimli kullanimi: 4*4 arama ve kurtarma araci seciminde AHS ve Topsis yontemlerinin uygulamasi. Suleyman Demirel Universitesi Vizyoner Dergisi **3**(5), 72–87 (2011)

Raut, R.D., Bhasin, H.V., Kamble, S.S.: Multicriteria decision making for automobile purchase using an integrated analytical quality fuzzy (AQF) technique. Int. J. Serv. Oper. Manage. **10**(2), 136–167 (2011)

Avikal, S., Mishra, P.K., Jain, R.: A Fuzzy AHP and PROMETHEE method-based heuristic for disassembly line balancing problems. Int. J. Prod. Res. **52**(5), 1306–1317 (2013a)

Avikal, S., Jain, R., Mishra, P.K.: A Kano model, AHP and M-TOPSIS method based technique for disassembly line balancing under fuzzy environment. Appl. Soft Comput. **25**, 519–529 (2014)

Dong, M., Li, S., Zhang, H.: Approaches to group decision making with incomplete information based on power geometric operator and triangular fuzzy AHP. Expert Syst. Appl. **42**, 7846–7857 (2015)

Garg, H.: A new generalized improved score function of interval-valued intuitionistic fuzzy sets and applications in expert systems. Appl. Soft Comput. J. **38**, 988–999 (2016)

Avikal, S., Mishra, P.K., Jain, R.: An AHP and PROMETHEE method-based environment friendly heuristic for disassembly line balancing problems. Interdisc. Environ. Rev. **14**(1), 69–85 (2013b)

Fuzzy Subtractive Clustering for Polymer Data Mining for SAW Sensor Array Based Electronic Nose

T. Sonamani Singh, Prabha Verma, and R.D.S. Yadava$^{(\boxtimes)}$

Sensors & Signal Processing Laboratory, Department of Physics, Institute of Science, Banaras Hindu University, Varanasi 221005, India
sonamani.2065@gmail.com, pverma.bhu@gmail.com, ardius@gmail.com

Abstract. Fuzzy subtractive clustering (FSC) has been applied as data mining tool for making selection of a small set of polymers from a large set of prospective polymers having potential for being chemical interfaces for electronic nose sensor array. The basic idea behind applying FSC selection is to cluster the prospective polymers according to some measure of similarity among them in relation to their interaction with the chemicals targeted for sensing. The polymers defining the cluster centers are taken to make the selection set. The basis for defining similarity among different polymers is the partition coefficients associated with sorption of chemical analytes from vapor phase to polymer phase in thermodynamic equilibrium. The goal for selection is to identify a minimal set of polymers that provide the most diversely interaction possibilities with the target vapors. The proposed selection method has been validated by simulating responses of a polymer-coated surface acoustic wave (SAW) sensor array for detection of freshness and spoilage of milk and fish food products. The end use of the proposed selection method is suggested for developing low-cost high-performance sensor array based electronic noses for commercial and consumer applications.

Keywords: Fuzzy subtractive clustering · Data mining for sensor selection · Electronic nose · Intelligent system for food monitoring

1 Introduction

Electronic nose (E-nose) is an odor sniffing instrument that mimics mammalian smell sensing organ [1]. The development of E-nose technology is important for a variety of applications like detection of explosives and chemical weapon agents for homeland security, industrial processes monitoring, air pollutants monitoring, detection of food freshness and spoilage conditions for health safety, disease diagnostics through detection of volatile biomarkers in body odor, detection of contrabands for monitoring illicit activities etc. [1–4]. The E-noses are field-deployable practically real-time instruments having odor recognition (chemical fingerprinting) capabilities in contrast to the laboratory based analytical instruments for chemical analysis.

© Springer Nature Singapore Pte Ltd. 2017
K. Deep et al. (eds.), *Proceedings of Sixth International Conference on Soft Computing for Problem Solving*, Advances in Intelligent Systems and Computing 546, DOI 10.1007/978-981-10-3322-3_23

An E-nose instrument consists of three major parts – an array of chemical vapor sensors having broad selectivities for target chemical analytes analogous to the olfactory receptor neurons in human nose, signal read out and processing unit, and pattern recognition system. These parts are respectively analogous to the olfactory epithelium, the olfactory bulb and the olfactory cortex in brain in human nose [1, 4]. The odor recognition capability becomes more accurate and robust if the sensor array outputs carry diverse discriminatory information about chemical analytes. This capability in polymer functionalized sensor arrays comes from the set of polymer films coated on the sensing devices providing broad selectivities. Therefore, the selection of an optimal set of polymers as chemical interface materials (hence, chemical sensors) becomes crucial for developing reliable E-noses for specific applications. The sensor selection also impacts the efficiency of pattern recognition system [4].

Commonly, a large number of sensors using various prospective polymer interfaces are fabricated, and best among them are selected after rigorously evaluating their performance in some predesigned odor recognition tasks [1, 4]. This approach involves large development time and cost. A method that can throw up a short list of polymers from a large list of all prospective polymers would be of great help in reducing the development time and cost. This is where data mining tools could be helpful. The authors' group in some recent studies took up this issue and investigated whether some method could be developed for making polymer selection based on the information available about their chemical interaction abilities with target vapors. The group experimented with various data mining strategies based on some statistical and fuzzy clustering techniques by using the vapor-polymer partition coefficients data [4–10]. In these studies the polymer-coated surface acoustic wave (SAW) chemical sensors arrays were employed for defining E-nose platform. The present study is in continuation of the same approach where a 'not analyzed before' data mining method based on fuzzy subtractive clustering (FSC) is analyzed, and its performance is evaluated in comparison to earlier findings.

2 Earlier Work

In [4–10] a series of efforts were made for defining a data mining strategy for optimal selection of a small set of polymers from a large set of prospective polymers based on partition coefficient data. The partition coefficient (K) quantifies thermodynamic partitioning of a chemical analyte from vapor phase to polymer phase in equilibrium. It is defined as the ratio of analyte concentration in polymer phase (C_P) to that in vapor phase (C_V), that is, $K = C_P/C_V$. The set of partition coefficients for various vapor-polymer pairs arranged in the form of a data matrix with vapors in rows and polymers in columns is called K-matrix. Transpose of it called K^T-matrix (polymers in rows and vapors in columns) can then be treated as multivariate data with polymers denoting the data vectors. A data vector (polymer) in K^T-matrix has vapors as variables with respective K values being components of the data vector. A data vector thus quantitatively represents a polymer's interaction strength with different vapors where the polymer is a measurand and the partition coefficients its measurements. The basic idea for doing data mining here is to cluster polymers according to their similarity for vapors and select maximally dissimilar set of polymers as sensor material.

In [5, 9] the principal component analysis (PCA) and hierarchical clustering (HC) were used for making selection based on load plots and Euclidian distance, and the validation was done by analyzing discrimination of vapor analytes in body odor. In [6] a similar analysis was done for detection of explosive vapors under different ambient conditions. In [7] a heuristic method was combined with PCA and fuzzy c-means clustering (FCM) for discrimination between explosive vapors, chemical weapon agents and drugs of abuse volatiles in presence of numerous interferents represent different application scenarios. Subsequently in [8] a novel procedure for using FCM was developed for detection of freshness and spoilage markers in milk and fish headspace odor. In this method c clusters of polymers were searched repeatedly by incrementing c in each successive search by 1 $(c + 1 \leftarrow c)$ with start at $c = 2$. This was continued until a common set of polymers start reappearing in successive searches. This common set was taken to be the final set of selection. In all these studies the methods were validated by simulating SAW sensor array responses for headspace vapors and interferents, and by doing neural network classification. The FCM based selection appeared to be the best of all the procedures analyzed so far.

It appears natural then to test the efficacy of fuzzy subtractive clustering (FSC) method for this purpose. In this study we have done that for a comparative performance analysis with FCM. To avoid repeating FCM here we used the same data set as reported in [8] for milk and fish (clarified in Sect. 4). After a brief review of FSC below we present the results in Sect. 5.

3 Fuzzy Subtractive Clustering

This clustering algorithm was proposed by Chui [11] to overcome the strong dependence of the quality of solutions (cluster centers) on initial values in FCM algorithm. This method assumes that all data points are potential cluster centers, and assigns a quantitative measure for their potentials. The data point having maximum potential is chosen to be the first cluster center. After finding the first cluster center the potential of each data point is revised in such a manner that the potentials of data points closer to the selected cluster center are greatly reduced in comparison to those far away from it. This is done to avoid the selection of a nearby data point to be the next cluster center. After getting the revised potentials the data point with the maximum potential is chosen to be the second cluster center. This procedure is repeated until a termination criterion is reached. Briefly, the algorithm runs as follows.

Assuming there are N data points (total number of polymers) and d vapors (dimensionality of data space) the i-th data vector is represented as $X_i = \{K_{i1}, K_{i2}, \ldots, K_{id}\}$ with K_{ij} being the partition coefficient for j-th vapor in i-th polymer $(i = 1, 2, \ldots, N, j = 1, 2, \ldots, d)$. To every data point X_i a potential P_i is assigned according to

$$P_i = \sum_{k=1}^{N} \exp(-\alpha \|X_i - X_k\|^2) \tag{1}$$

where $\alpha = 4/r_a^2$ with r_a being a positive constant. The latter denotes a radius measure of the soft boundary of a spherical cluster with the data point at its center. The data point with the highest potential (P_1^*) is selected to be the first cluster center (X_1^*). Next, the potential reduction of each data point is done according to

$$P_i \Leftarrow P_i - P_1^* \exp(-\beta \|X_i - X_1^*\|^2), \quad i = 1, 2, \ldots, N \tag{2}$$

where $\beta = 4/r_b^2$ with $r_b > r_a$ is a positive constant that denotes the radius of neighborhood that will have measurable reduction in the potential. The data point with the highest potential is then chosen to be the second cluster center. The process is continued by monitoring the following selection or rejection criteria.

At k-th step (that is, after making k-th cluster center selection):

if $P_k^* > \bar{\varepsilon} P_1^*$ accept X_k^* to be the cluster center and continue;

else if $P_k^* < \underline{\varepsilon} P_1^*$ reject X_k^* and end the clustering process;

else let d_{min} = shortest of the distance between X_k^* and previously found cluster centers;

if $\frac{d_{min}}{r_a} + \frac{P_k^*}{P_1^*} \geq 1$ accept X_k^* as cluster center and continue;

else reject X_k^* and set the potential of X_k^* to be 0;

select the data point with the next highest potential as the new X_k^* and retest.

In the above $\underline{\varepsilon}$ and $\bar{\varepsilon}$ are two fractional threshold parameters that decide whether to reject or accept the cluster center. The polymer selection corresponds to the set of polymers associated with the final set of selected cluster centers.

4 Polymer Selection by FSC

As mentioned earlier, we consider here the same set of polymers, target vapors and K-matrices as reported in [8]. It pertains to the detection of freshness and spoilage of milk and fish products by sniffing their headspaces. A common list of 26 prospective polymers is considered for both. The volatile organics in the milk headspace is consisted of 29 chemical species, and in the fish headspace of 17 species. These include freshness and spoilage species as well as some interferents (for details see [8]). The K^T-matrices are already calculated in [8], and are given as Table 4A & 4B for the milk and Table 5 for the fish products. The FSC algorithm was implemented by using the MatLab function 'subclust' with parameters: $r_a = 0.6$, $r_b = 1.25r_a$, $\underline{\varepsilon} = 0.15$, $\bar{\varepsilon} = 0.5$ for milk data; and $r_a = 0.55$, $r_b = 1.25r_a$, $\underline{\varepsilon} = 0.204$, $\bar{\varepsilon} = 0.5$ for fish data. These parameters were empirically optimized for producing the best results. The method selected a set of six polymers. The FCM also selects a set of six polymers (taken from Table 8 in [8]). Both the results are shown in Table 1. It can be seen that except one both FCM and FSC select a different set of polymers.

Table 1. Set of polymers selected by FCM and FSC

Polymers selected by FCM					
Milk	OV25	PEI	PMCPS	SXPYR	P4 V	PLF
Fish	OV202	PEI	SXFA	SXPYR	P4 V	PBF
Polymers selected by FSC					
Milk	PMPS	PVA	FPOL	SXPHB	PMHS	PLF
Fish	PMPS	PEM	P4 V	SXPHB	PMHS	PLF

5 Validation

The SAW sensor array responses by FSC selection were calculated by using the same SAW sensor model, VOCs concentration levels and noise model as used in [8] for FCM selection. The sensor outputs in the form frequency shifts of the SAW sensor oscillators due to vapor sorption and corrupted by additive frequency noise with uniform distribution over [−30, +30] Hz were generated for 100 samples of each vapor. The concentrations were varied over ppt to ppm range similar to that Tables 1 and 2 of [8]. Data preprocessing by normalization with respect to vapor concentration and logarithmic scaling and PCA analyses were carried out as before [8], and the separability of vapor classes were examined in principal component (PC) space. Figure 1 shows the results for the milk VOCs and Fig. 2 shows similar results for the fish VOCs. From these results it is clear that FSC selection also yields qualitatively similar set of discriminating polymers as FCM in [8].

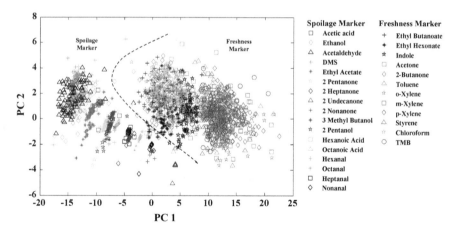

Fig. 1. FSC selection based PC1-PC2 score plots for *milk* headspace VOCs. The symbols on the right hand side are arranged according to spoilage and freshness VOCs

The visual examination of the PC score plots in Figs. 1 and 2 provides only a qualitative feel of separability between freshness versus spoilage markers. For making a comparison between FSC and FCM we need some quantitative performance measure.

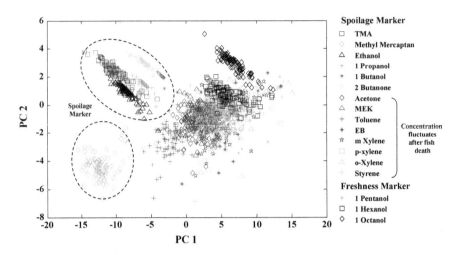

Fig. 2. FSC selection based PC1-PC2 score plots for *fish* headspace VOCs. The symbols on the right hand side are arranged according to spoilage and freshness VOCs. Certain VOCs whose concentrations fluctuate with time after fish death are indicated separately in spoilage block

In [12] a quantity measure called class-separability (denoted by J) has been defined in terms of within-class and between-class scatter matrices (denoted by S_w and S_b respectively) of principal components. The scatter matrices are basically local and global data covariances weighted by *a priori* probabilities of the classes. These are defined as

$$S_w = \sum_{i=1}^{\Omega} P_i(x - \mu_i)(x - \mu_i)^T \tag{3}$$

$$S_b = \sum_{i=1}^{\Omega} P_i(\mu_i - \mu_0)(\mu_i - \mu_0)^T \tag{4}$$

where x denotes the data matrix in the principal component space, P_i is *a priori* probability for i-th class, μ_i is the mean vector for i-th class and μ_0 is the global mean vector. Note that $P_i \cong n_i/M$ where n_i denotes the number of data points in the i-th class out of total M data points, and the summation runs over Ω vapor classes.

The local (vapor class) and global means are calculated as

$$\mu_i = \frac{1}{n_i} \sum_{i=1}^{n_i} x_i \tag{5}$$

$$\mu_0 = \sum_{i=1}^{\Omega} P_i \mu_i \tag{6}$$

The class-separability is defined as the ratio

$$J = \frac{\text{trace}\{S_w + S_b\}}{\text{trace}\{S_w\}} \tag{7}$$

These definitions suggest that trace $\{S_b\}$, trace $\{S_w\}$ and J can be taken to indicate quantitatively different aspects vapor discrimination by different polymer selections. A small trace $\{S_w\}$ value would mean high intra-class compactness, a large trace $\{S_b\}$ would mean large inter-class separation, and a high J value would indicate overall performance by a sensor array.

Table 2 below presents the values of these quantities obtained on the basis of both FSC and FCM selections. The FCM result is based on PCA data in [8]. These results are based on treating both the milk and fish data as a two-class problem for discrimination between the fresh and the spoiled products. These results show that FSC selection yields better values for all performance metrics in comparison to FCM – lower trace $\{S_w\}$ indicates more intra-cluster compactness, higher trace $\{S_b\}$ indicates more inter-class separation, higher J values means better overall performance.

Table 2. Comparison of the sensor array polymer selections by the FSC and FCM methods for sensing the milk and fish volatiles in their headspaces

	Trace $\{S_w\}$ $(\times 10^4)$	Trace $\{S_b\}$ $(\times 10^4)$	J
Milk Freshness/Spoilage (2-class)			
FCM	8.5367	4.5129	1.5286
FSC	5.1727	6.1806	2.1948
Fish Freshness/Spoilage (2-class)			
FCM	1.8093	1.7276	1.9549
FSC	1.3682	2.4703	2.8055

6 Discussion

Chemical discrimination in a sensor array based E-nose system comes from the broad interaction possibilities between the sensors coatings and the chemical analytes (both target as well as interferents). The common approach for selecting an optical set of sensors is fabrication of sensor and their evaluation and categorization based on some specific identification task. The whole procedure involves massive experimentation and analysis exercises, hence cost and time. The present data mining approach may reduce this development burden by short listing a few among many prospective polymers. The basis for selection is to segregate all prospective polymers into smaller subsets having similar analyte loading capabilities, and then select the best from each subset. Fuzzy clustering accomplishes this task by picking polymers at cluster centers in each cluster (subset). The vapor loading is determined partitioning of analytes from vapor phase into polymer phase. The polymers being similar means if subjected to interact with a set of vapor analytes there will be similar net vapor loading in each. The selection of polymer set based on this mining approach of partition coefficients therefore appears attractive. In continuation of earlier investigations we applied this approach by using two commonly used fuzzy clustering algorithms (FCM and FSC). The application of FCM has already been analyzed in [8]. In this work we applied FSC for the same problem as in [8], namely, the polymer selection for discrimination between freshness and spoilage of milk and fish products. By generating SAW sensor array based synthetic data and analyzing it as two-class (fresh versus spoiled) problem we found here that FSC performs better than FCM.

Evaluation of spoilage state of a food product however needs quantitative estimation of the spoilage markers. Usually, the vapor composition in headspace of food products is quite complex and uncertain as it arises from several factors of chemical, biological and environmental origin. Therefore, the inferences of this study alone can not be decisive on the FCM versus FSC issue. We noted that certain freshness and spoilage markers (e.g. indole as freshness marker for milk and trimethyl amine as spoilage marker for fish) are better separated with FCM selection. Therefore, further analyses are required for deciding which one is better in what respect. Possibly, both may be used in combination for doing complimentary tasks for some specific application.

7   Conclusion

In this study we analyzed the potentiality of using fuzzy subtractive clustering (FSC) as a data mining tool for making the selection of chemically selective polymers for the vapor sensor arrays in electronic nose systems. The data for mining consisted of the thermodynamic partition coefficients for different chemical analytes specific to a particular application from vapor phase into all prospective polymers. The latter were listed on the basis of available commercial and published research information. The partition coefficients were calculated on the basis of the linear solvation energy relation (LSER) by using experimentally determined solubility parameters. The FSC was applied by treating polymers as measurand and vapors as probes. The polymers are clustered according to how similar they are in regard to their affinities for various chemicals in the vapor phase. The polymers defining the cluster centres were selected for making the sensor array.

The selection procedure was validated for monitoring the freshness and spoilage of the milk and the fish food products through their headspace sniffing. The virtual sensor arrays were defined based on the FSC selected polymers by using polymer-coated SAW oscillators as chemical sensors. The analysis of the sensor array responses by principal component analysis (PCA) demonstrated that the proposed method is very effective in discriminating the spoilage volatiles against the freshness indicators. The comparison of class separability in principal component space with the previously reported fuzzy c-means (FCM) method on the same data set indicates that the FSC yields better selection than FCM at least for the applications analyzed here.

References

1. Pearce, T.C., Schiffman, S.S., Nagle, H.T., Gardner, J.W.: Handbook of Machine Olfaction. Wiley-VCH, Weinheim (2003)
2. Gardner, J.W., Yinon, J.: Electronic Noses and Sensors for the Detection of Explosives. Kluwer, Dordrecht (2004)
3. Wilson, A.D., Baietto, M.: Applications and advances in electronic nose technologies. Sensors **9**, 5099–5148 (2009)
4. Yadava, R.D.S.: Modeling, simulation, and information processing for development of a polymeric electronic nose system. In: Korotcenkov, G. (ed.) Chemical Sensors – Simulation and Modeling, vol. 3, pp. 411–502. Momentum Press, LLC, New York (2014)
5. Jha, S.K., Yadava, R.D.S.: Designing optimal surface acoustic wave electronic nose for body odor discrimination. Sens. Lett. **9**, 1612–1622 (2011)
6. Jha, S.K., Yadava, R.D.S.: Data mining approach to polymer selection for making SAW sensor array based electronic nose. Sens. Transducers J. **147**, 108–128 (2012)
7. Verma, P., Yadava, R.D.S.: A data mining procedure for polymer selection for making surface acoustic wave sensor array. Sens. Lett. **11**, 1903–1918 (2013)
8. Verma, P., Yadava, R.D.S.: Polymer selection for SAW sensor array based electronic noses by fuzzy c-means clustering of partition coefficients: model studies on detection of freshness and spoilage of milk and fish. Sens. Actuators, B **209**, 751–769 (2015)
9. Jha, S.K., Yadava, R.D.S.: statistical pattern analysis assisted selection of polymers for odor sensor array. In: 2011 IEEE International Conference Signal Processing, Communication, Computing and Networking Technologies (ICSCCN 2011), pp. 42–47 (2011)
10. Verma, P., Yadava, R.D.S.: Application of fuzzy c-means clustering for polymer data mining for making SAW electronic nose. In: Satapathy, S., Udgata, S., Biswal, B. (eds.) FICTA 2013. AISC, vol. 247, pp. 1–8. Springer, Cham (2014). doi:10.1007/978-3-319-02931-3_1
11. Chiu, S.L.: Fuzzy model identification based on cluster estimation. J. Intell. Fuzzy Systems **2**, 267–278 (1994)
12. Theodoridis, S., Koutroumbas, K.: Pattern Recognition, 3rd edn, pp. 228–230. Academic, San Diego (2006)

Clustering of Categorical Data Using Intuitionistic Fuzzy k-modes

Darshan Mehta$^{(\boxtimes)}$ and B.K. Tripathy

School of Computing Science and Engineering,
VIT University, Vellore 632014, Tamil Nadu, India
mehtadarshan@icloud.com, tripathybk@vit.ac.in

Abstract. Clustering is an important unsupervised learning algorithm that groups records according to their similarity. However, since uncertainty has become an inherent of real world datasets, crisp clustering leads to inefficiency. Hence, introduction of uncertainty based models like the fuzzy set and the intuitionistic fuzzy set is necessary to compensate for the ambiguity in data. Huang, in his fuzzy k-modes algorithm, introduced the fuzzy component in clustering categorical data by modifying the existing k-means algorithm. This correspondence describes an intuitionistic fuzzy k-modes algorithm for clustering categorical data and establishes it to be more efficient than the fuzzy k-modes algorithm. Metrics like accuracy, DB index and Dunn index are used to compare the efficiency of the two algorithms. The experimental analysis section shows that the proposed algorithm is more efficient than the existing one. Several graphical and tabular representations have been provided for easy comparison of the results.

Keywords: k-modes · Fuzzy · Intuitionistic · Clustering · Categorical data

1 Introduction

In modern world, the data gathered and information gained from it affects the organization's growth in a major way. One very common methodology for interpreting data is by classifying similar types of data into groups or clusters and separating them from the rest. This process of grouping similar data is known as clustering. The early clustering algorithms were crisp in nature, i.e., a point either belongs to a cluster or it doesn't and were mostly based on numerical data. With advancement in technology and information sciences, we started to face categorical data. Categorical values are discrete and unordered, unlike numeric data. Therefore, the clustering algorithms designed for numeric data cannot be used to cluster categorical data that exist in many real world problems. The k-modes algorithm was proposed by Huang [3] in which he modified the standard k-means process for clustering categorical data by replacing the Euclidean distance function with a simple matching dissimilarity measure and using modes to represent cluster centers, which are updated after every iteration.

The traditional clustering algorithms mentioned above cannot be applied to datasets with inherent uncertainty. So, many uncertainty based models like fuzzy set, by Zadeh in [11], rough set by Pawlak in [4] and intuitionistic fuzzy set by Atanassov in [5] were introduced. While fuzzy sets depend upon graded membership values, intuitionistic

© Springer Nature Singapore Pte Ltd. 2017
K. Deep et al. (eds.), *Proceedings of Sixth International Conference on Soft Computing for Problem Solving*, Advances in Intelligent Systems and Computing 546, DOI 10.1007/978-981-10-3322-3_24

fuzzy sets depend upon membership and non-membership values leading to hesitation values associated with every element in the domain. Several k-means algorithm are found in literature [9]. Also, a fuzzy k-modes algorithm was proposed by Huang [3] by extending the k-modes algorithm and adding the fuzzy logic to it. In this paper, we have extended this algorithm to propose an intuitionistic fuzzy k-modes algorithm. There are many ways in which the complement can be defined in an intuitionistic fuzzy set context, like the Sugeno's complement [8, 10] and the Yager's complement [6, 7]. In this article, we follow Yager's approach.

The outline of this paper is as follows: In Sect. 2, we provide the definitions and notations used in the paper. In Sect. 3, we explicate the Intuitionistic Fuzzy k-modes algorithm. In Sect. 4, examples are provided to illustrate the effectiveness of Intuitionistic Fuzzy k-modes algorithm over the results obtained by using fuzzy k-modes algorithm. Finally a concluding remark is given in Sect. 5.

2 Definitions and Notations

In this section we provide the definitions and notation used by us in the paper. Let the set of objects to be clustered be stored in a table T and be defined by a set of m attributes $A_1, A_2, \ldots, A_m$. Each attribute A_j of the table T describes a domain of values denoted by $DOM(A_j)$. A domain $DOM(A_j)$ is defined as categorical if it is finite and unordered, e.g., for any $a, b \in DOM(A_j)$, either $a = b$ or $a \neq b$.

2.1 Fuzzy k-modes

Fuzzy k-modes is an extension of k-modes which itself is an extension of the k-means algorithm. The main algorithm iterates around minimizing the cost function

$$F(W, Z) = \sum_{l=1}^{k} \sum_{i=1}^{n} w_{li}^{\alpha} d_c(Z_l, X_i) \tag{1}$$

subject to,

$$0 \leq w_{li} \leq 1, \quad 1 \leq l \leq k, \quad 1 \leq i \leq n \tag{2}$$

$$\sum_{l=i}^{k} w_{li} = 1, \quad 1 \leq i \leq n \tag{3}$$

and

$$0 \leq \sum_{i=1}^{n} w_{li} \leq n, \quad 1 \leq l \leq k \tag{4}$$

where $k(\leq n)$ is the known number of clusters, $\mathbf{X} = \{X_1, X_2, \ldots X_n\}$ is the set of n objects described by m attributes, $W = [w_{li}]$ is a $k \times n$ real valued Fuzzy partition matrix, $\alpha \in [1, \infty)$ is a weighing exponent, Z is a set of cluster centers where each $Z_i \in Z$ is represented as $[z_{l,1}, z_{l,2}, \ldots, z_{l,m}]$ for $1 \leq l \leq k$ and $d_c(Z_l, X_i)(\geq 0)$ is a simple dissimilarity measure defined as follows:

$$d_c(X, Y) = \sum_{j=1}^{m} \delta(x_j, y_j) \tag{5}$$

where

$$\delta(x_j, y_j) \begin{cases} 0, x_j = y_j \\ 1, x_j \neq y_j \end{cases} \tag{6}$$

2.2   Introduction to IFS

Fuzzy sets only generate the membership function $\mu(x)$ for $x \in X$, with non- membership value $v(x) = 1 - \mu(x)$. Whereas the Intuitionistic Fuzzy Set (IFS) proposed by Atanassov [5], takes into consideration the lack of knowledge and generates both membership $\mu(x)$ and non-membership $v(x)$. An Intuitionistic fuzzy set A in X is written as,

$$A = \{x, \mu(x), v(x) | x \in X\} \tag{7}$$

where $\mu(x) \rightarrow [0,1]$, $v(x) \rightarrow [0,1]$ are the membership and non-membership degrees of the element x in the set A with the condition $0 \leq \mu(x) + v(x) \leq 1$. At Atanassov [5] introduced a hesitation degree $\pi(x)$ which arises due to lack of knowledge in defining the membership or non-membership of an element x to the set A and is given by the following equation:

$$\pi(x) = 1 - \mu(x) - v(x); 0 \leq \pi(x) \leq 1 \tag{8}$$

The value of $v(x)$ can be calculated from Yager's Intuitionistic fuzzy complement, as:

$$v(x) = (1 - \mu(x)^\theta)^{1/\theta} \tag{9}$$

Where $\theta > 0$ is the Yager's coefficient.

2.3   Metrics

In this section we will discuss the metrics that we will use to compare our algorithm's experimental results with those of the previously existing algorithms.

Accuracy. A clustering result can be measured by the clustering accuracy r defined as:

$$r = \frac{\sum\limits_{l=1}^{k} a_l}{n} \qquad (10)$$

where, a_l is the number of instances occurring in both cluster l and its corresponding class, n is the number of instances in the dataset and k is the number of clusters.

Davies-Bouldin Index. The Davies-Bouldin criterion [12] measures the quality of the clusters generated. It is based on a ratio of intra-cluster and inter-cluster distances. The Davies-Bouldin index is defined as:

$$DB = \frac{1}{k} \sum_{i=1}^{k} \max_{j \neq 1} \{D_{i,j}\} \qquad (11)$$

where $D_{i,j}$ is the within-to-between cluster distance ratio for the ith and j^{th} clusters. In mathematical terms,

$$D_{i,j} = (\bar{d}_i + \bar{d}_j)/d_{i,j} \qquad (12)$$

$\bar{d}_k$ is the average distance between each point in the kth cluster and the centroid of the kth cluster. $d_{i,j}$ is the distance between centroids of the ith and j^{th} clusters as per Eq. (5). The optimal clustering solution has the smallest Davies-Bouldin index value.

Dunn Index. The aim of Dunn index (DI) [13] is to identify sets of clusters that are compact, i.e., ones which have a small variance between members of the cluster, and well separated from the other clusters. A higher Dunn index indicates better clustering.

The Dunn index is calculated as follows:

$$DI_k = \frac{\min_{1 \leq i < j \leq k} d_c(Z_i, Z_j)}{\max_{1 \leq l \leq k} \Delta_l} \qquad (13)$$

where $d_c(Z_i, Z_j)$ is the inter-cluster distance metric calculated as per (5), and Δ_l is the average distance of all the points in cluster l from the centroid of the cluster.

3 Algorithm

In this algorithm, we modify the fuzzy k-modes algorithm by using the Yager's Intuitionistic Fuzzy Complement (Eq. 9) instead of the standard fuzzy partition matrix.

3.1 Intuitionistic Fuzzy k-modes Algorithm

1. Choose an initial set of centroids $Z^{(1)} \in X$. Determine $W^{(1)}$ such that the cost function $F(W, Z^{(1)})$ is minimised. Set the value of t as 1.
2. Determine $Z^{(t+1)}$ such that the cost function $F(W^{(t)}, Z^{(t+1)})$ is minimized. If $F(W^{(t)}, Z^{(t+1)}) = F(W^{(t)}, Z^{(t)})$, then go to step (4); else go to step (3).
3. Determine $W^{(t+1)}$ such that the cost function $F(W^{(t+1)}, Z^{(t+1)})$ is minimized. If $F(W^{(t+1)}, Z^{(t+1)}) = F(W^{(t)}, Z^{(t+1)})$, then go to step (4); else increment the value of t by 1. And go back to step (2).
4. Calculate the cluster membership of each point in X with respect to the centroids present in the $Z^{(t+1)}$ using the fuzzy partition matrix as mentioned in the Sect. 3.4.
5. Calculate the value of various metrics present in Sect. 2.3.

3.2 Updating Fuzzy Partition Matrix

Assume $\widehat{Z}$ to be fixed and consider the problem $\min_w F(W, \widehat{Z})$ subject to the same conditions as the Eq. (1). For $\alpha > 1$, the minimizer $\widehat{W}$ is given by:

$$\widehat{W}_{li} = \begin{cases} 1, & \text{if } X_i = \widehat{Z}_l \\ 0, & \text{if } X_i = \widehat{Z}_h, h \neq l \\ 1 / \sum_{h=1}^{k} [\frac{d_c(\widehat{Z}_l, X_i)}{d_c(\widehat{Z}_h, X_i)}]^{1/(\alpha-1)}, & \text{if } X_i \neq \widehat{Z}_l \text{ and } X_i \neq \widehat{Z}_h, 1 \leq h \leq k \end{cases} \quad (14)$$

This $\widehat{W}$ is the membership function $\mu(x)$. We calculate the new W as:

$$W = 1 - v(x) \quad (15)$$

Where $v(x)$ is the calculated as per Eq. (9) using the above calculated $\widehat{W}$ as: $\mu(x)$.

3.3 Updating Cluster Centers

Assume $\widehat{W}$ to be fixed and consider the problem $\min_z F(\widehat{W}, Z)$ subject to the same conditions as the Eq. (1). Let $\mathbf{X}$ be the set of categorical objects described by categorical $A_1, A_2, \ldots, A_m$ attributes and $DOM(A_j) = \left\{ a_j^{(1)}, a_j^{(2)}, \ldots a_j^{(n_j)} \right\}$ where n_j is the number of different categories of attribute A_j for $1 \leq j \leq m$. Let the cluster centers Z_l be represented by $[z_{l,1}, z_{l,2}, \ldots, z_{l,m}]$ for $1 \leq l \leq k$. Then the cost function in Eq. (1) is minimized iff $z_{l,j} = a_j^{(r)} \in DOM(A_j)$

$$\sum_{i, x_{i,j} = a_j^{(r)}} w_{li}^a \geq \sum_{i, x_{i,j} = a_j^{(t)}} w_{li}^a, 1 \leq t \leq n_j \quad (16)$$

for $1 \leq j \leq m$.

3.4   Calculating Cluster Membership

The record X_i is assigned to the lth cluster if:

$$w_{li} = \max_{1 \leq h \leq k}\{w_{hi}\} \tag{17}$$

If the maximum is not unique, then X_i can be assigned to the first cluster achieving the maximum.

4   Experimental Results and Analysis

In this section, we provide the results obtained by implementing our algorithm on Python language and testing it with two datasets from the UCI repository, namely, Soybean [1] and Zoo [2] dataset. The Soybean dataset consists of 47 records each

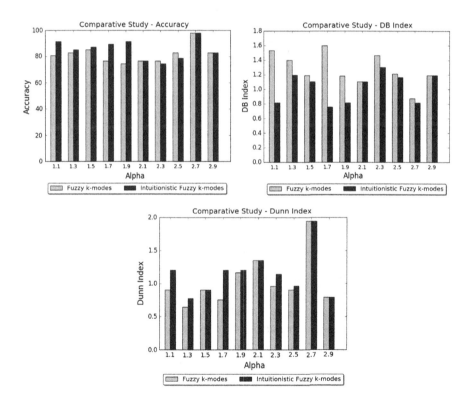

Fig. 1. Comparison of Accuracy, DB and Dunn values of different ∝ values for Soybean dataset ($\theta = 1.7$).

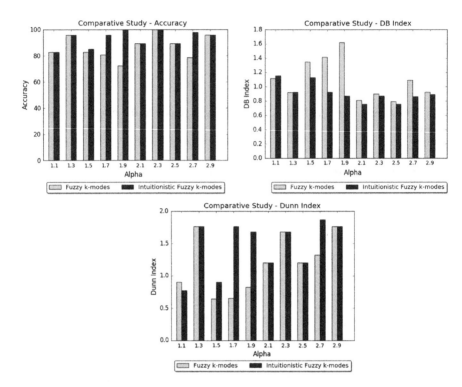

Fig. 2. Comparison of Accuracy, DB and Dunn values of different ∝ values for Soybean dataset (θ = 3.2).

described by 35 attributes. Out of these 35, we only selected 21 because the rest 14 attributes have same value for all records.

We tested our algorithm alongside the Fuzzy k-modes for different values of $\alpha \in [1.1, 3]$ at fixed intervals. We also tested it for different values of θ. We provide the results in both tabular and graphical format obtained by averaging over 100 iterations (Figs. 1, 2, 3, 4).

From the above tables, it is evident that for both Soybean and Zoo dataset, the intuitionistic fuzzy k-modes provides higher accuracy, lower DB index and higher Dunn index than the existing fuzzy k-modes algorithm. From experimentation, it is observed that the same trend is maintained for θ in the range [1.1, 8]. We also observed that for θ > 8, the behaviour of this algorithm fluctuates and no conclusions can be drawn. Since the value of θ is flexible, better results can be obtained by using the intuitionistic fuzzy k-modes algorithm (Tables 1, 2).

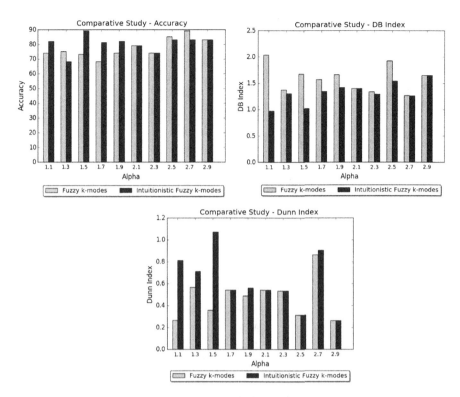

Fig. 3. Comparison of Accuracy, DB and Dunn values of different ∝ values for Zoo dataset (θ = 1.7).

Table 1. Soybean Dataset with θ = 1.7 and θ = 3.2.

Metrics	θ = 1.7		θ = 3.2	
	Fuzzy k-modes	Intuitionistic Fuzzy k-modes	Fuzzy k-modes	Intuitionistic Fuzzy k-modes
Average accuracy	0.817	0.855	0.868	0.932
Average DB index	1.276	1.027	1.095	0.914
Average Dunn index	1.033	1.148	1.196	1.461

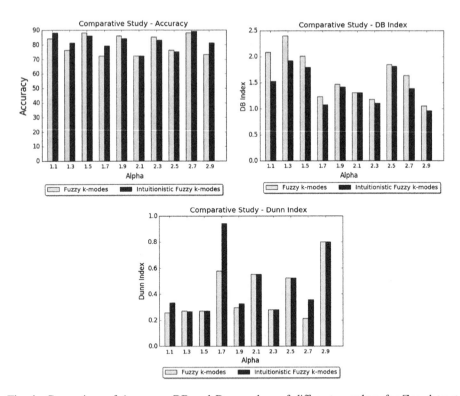

Fig. 4. Comparison of Accuracy, DB and Dunn values of different ∝ values for Zoo dataset (θ = 3.2).

Table 2. Zoo Dataset with θ = 1.7 and θ = 3.2.

Metrics	θ = 1.7		θ = 3.2	
	Fuzzy k-modes	Intuitionistic Fuzzy k-modes	Fuzzy k-modes	Intuitionistic Fuzzy k-modes
Average accuracy	0.776	0.806	0.802	0.82
Average DB Index	1.59	1.32	1.624	1.433
average Dunn index	0.474	0.626	0.405	0.465

5   Conclusion

In this paper we proposed a new categorical data clustering algorithm called the Intuitionistic Fuzzy k-modes algorithm. We could also experimentally establish by taking two datasets, the Soybean and the Zoo, with two different values of θ (=1.7 and =3.2) that for the values of α in the range [1.1, 3], the proposed algorithm is more efficient than the fuzzy k-modes algorithm for clustering categorical data.

References

1. Michalski, R.S.: Soybean (Small) Data Set. UCI Machine Learning Repository, Irvine. https://archive.ics.uci.edu/ml/datasets/Soybean+(Small)
2. Forsyth, R.: Zoo Data Set. UCI Machine Learning Repository, Irvine. https://archive.ics.uci.edu/ml/datasets/Zoo
3. Huang, Z., Ng, M.K.: A fuzzy k-modes algorithm for clustering categorical data. IEEE Trans. Fuzzy Syst. **7**(4), 446–452 (1999)
4. Pawlak, Z.: Rough sets. Int. J. Comput. Inform. Sci. **11**(5), 341–356 (1982)
5. Atanassov, K.T.: Intuitionistic fuzzy sets. Fuzzy Sets Syst. **20**(1), 87–96 (1986)
6. Senthamilarasu, S., Hemalatha, M.: A genetic algorithm based intuitionistic fuzzification technique for attribute selection. Indian J. Sci. Technol. **6**(4), 4336–4346 (2013)
7. Yager, R.R.: Some aspects of intuitionistic fuzzy sets. Fuzzy Optim. Decis. Mak. **8**(1), 67–90 (2009)
8. Sugeno, M.: Fuzzy measures and fuzzy integrals: a survey. In: Gupta, M.M., Saridis, G.N., Gaines, B.R. (eds.) Fuzzy Automata and Decision Processes, pp. 89–102 (1977)
9. Bock, H.H.: Clustering methods: a history of k-means algorithms. In: Selected Contributions in Data Analysis and Classification, pp. 161–172. Springer, Berlin (2007)
10. Chaira, T., Anand, S.: A novel intuitionistic fuzzy approach for tumor/hemorrhage detection in medical images. J. Sci. Ind. Res. **70**(6), 427–434 (2011)
11. Zadeh, L.A.: Fuzzy sets. Inf. Control **8**(3), 338–353 (1965)
12. Davies, D.L., Bouldin, D.W.: A cluster separation measure. IEEE Trans. Pattern Anal. Mach. Intell. **2**, 224–227 (1979)
13. Dunn, J.C.: A fuzzy relative of the ISODATA process and its use in detecting compact well-separated clusters. J. Cybern. **3**, 32–57 (1973)

Some Properties of Rough Sets on Intuitionistic Fuzzy Approximation Spaces and Their Application in Computer Vision

B.K. Tripathy[1(✉)] and R.R. Mohanty[2]

[1] School of Computer Science and Engineering,
VIT University, Vellore 632014, Tamilnadu, India
tripathybk@vit.ac.in
[2] Indira Memorial College, Gajapati, Chandiput 761017, Odisha, India
radharamanmohanty14@gmail.com

Abstract. The basic rough sets as introduced by Pawlak have been extended in many directions. Two such extensions are the rough sets on fuzzy approximation spaces and the rough sets by De et al. in 1999 and rough sets on intuitionistic fuzzy approximation spaces by Tripathy in 2006. There are several properties of these two types of rough sets established so far. However, in this paper we show that some of the properties established are erroneous through counter examples, establish their correct versions and illustrate their application in real life situations through an example in computer vision.

Keywords: Rough sets · Intuitionistic fuzzy approximation spaces · Computer vision · $(\alpha, \beta) - $ cut

1 Introduction

Rough set introduced by Pawlak [4] is one of the most fruitful models to handle uncertainty in data. But as this basic notion depends upon equivalence relations, which are relatively rare in real life situations several attempts have been made to make it more general. De et al. introduced rough sets on fuzzy approximation spaces, which were extended to rough sets on intuitionistic fuzzy approximation spaces by Tripathy [6, 7]. Several properties of rough sets on intuitionistic fuzzy approximation spaces are established in [9–12]. In this paper we show that some of these properties are erroneous through counter examples and establish the correct versions. Also, we show their application in the field of computer vision.

Several authors have worked on rough sets on fuzzy approximation spaces and intuitionistic fuzzy approximation spaces [1, 2, 5, 8–11]. However, we found that some of the basic properties established in De [9] and Tripathy [6, 7] are incorrect. So, the analysis done based upon these results is obviously erratic. In this article, we show the incorrectness of these results, provide the corresponding correct results and illustrate their application through an application in computer vision.

The structure of the paper henceforth is as follows. We present the definitions to be used and the notations to be followed in the paper in Sect. 2. The main results for rough

© Springer Nature Singapore Pte Ltd. 2017
K. Deep et al. (eds.), *Proceedings of Sixth International Conference on Soft Computing for Problem Solving*, Advances in Intelligent Systems and Computing 546, DOI 10.1007/978-981-10-3322-3_25

sets on intuitionistic fuzzy approximation spaces will be discussed in Sect. 3. An application of the results is to be provided in Sect. 4. In Sect. 5 we provide concluding remarks.

2 Definitions and Notations

The most successful uncertainty based model is perhaps fuzzy sets introduced by Zadeh in 1965, which was extended to define the concept of intuitionistic fuzzy sets by Atanassov [3] in 1986, where the non-membership function is not necessarily the one's complement of the membership function. We have provided the definitions of these two models in the introductory part. Another model of uncertainty is that of rough sets introduced by Pawlak [4] in 1982. Here the notion of equivalence relation R is used to approximate a set X through two other crisp sets called the lower and upper approximations of the set with respect to R denoted by $\underline{R}X$ and $\overline{R}X$ respectively. A set X is said to be R-definable if and only if $\underline{R}X = \overline{R}X$ and rough otherwise. We have defined these concepts in detail in this paper.

We next state some definitions and notations to be used in this paper.

Definition 2.1. An intuitionistic fuzzy relation on a universal set U is an intuitionistic fuzzy set defined on U x U.

Definition 2.2. $\mu_R(x,x) = 1$ and $v_R(x,x) = 0$, $\forall x \in U$.

Let R be an intuitionistic fuzzy relation over a universal set U. Then R is intuitionistic fuzzy reflexive if and only if

$$\mu_R(x, x) = 1 \text{ and } v_R(x, x) = 0, \forall x \in U. \tag{2.1}$$

R is intuitionistic fuzzy symmetric if and only if

$$\mu_R(x, y) = \mu_R(y, x) \text{ and } v_R(x, y) = v_R(y, x) \forall x, y \in U. \tag{2.2}$$

Let I = [0,1]. Then we define the set J as $J = \{(m, n)| m, n \in I \text{ and } 0 \leq m+n \leq 1\}$.

Definition 2.3. Let $(\alpha, \beta) \in J$. Then the (α, β)-cut of R is denoted by $R_{\alpha,\beta}$ and is given by

$$R_{\alpha,\beta} = \{(x, y)| \mu_R(x, y) \geq \alpha \text{ and } v_R(x, y) \leq \beta. \tag{2.3}$$

For any fuzzy proximity relation R on U and $(\alpha, \beta) \in J$, if $(x, y) \in R_{\alpha,\beta}$ then we say that x and y are (α, β)-*similar* and we denote it by $xR_{\alpha,\beta}y$.

Definition 2.4. Two elements x and y are said to be $R_{\alpha,\beta}$-*identical* denoted by $xR(\alpha, \beta)y$ if either $xR_{\alpha,\beta}y$ or there exists a sequence of elements $u_1, u_2, \ldots u_n$ in U such that $xR_{\alpha,\beta}u_1R_{\alpha,\beta}u_2\ldots u_nR_{\alpha,\beta}y$.

It may be noted that the relation $R_{(\alpha,\beta)}$ is an equivalence relation for each $(\alpha, \beta) \in J$. Here (U, R) is called an intuitionistic fuzzy approximation space. Also, it may be noted that for any $(\alpha, \beta) \in J$, $(U, R(\alpha, \beta))$ is an approximation space in the same sense as that used by Pawlak.

For any x in U we denote the equivalence class of x with respect to $R(\alpha, \beta)$ by $[x]_{R(\alpha,\beta)}$.

Definition 2.5. Let U be a universal set and R be a intuitionistic fuzzy proximity relation on U. Then for any $(\alpha, \beta) \in J$ we define the lower and upper approximations [2, 3] of a subset X in U as

$$\underline{R(\alpha, \beta)}X = \{x \in U \mid [x]_{R(\alpha,\beta)} \subseteq X\} \text{ and} \tag{2.4}$$

$$\overline{R(\alpha, \beta)}X = \{x \in U \mid [x]_{R(\alpha,\beta)} \cap X \neq \phi\}. \tag{2.5}$$

We say that X is $R(\alpha, \beta)$ discernible if and only if $\underline{R(\alpha, \beta)}X = \overline{R(\alpha, \beta)}X$. Else, X is said to be $R(\alpha, \beta)$-rough.

3   Results on Rough Sets on Intuitionistic Fuzzy Approximation Spaces

Several properties for rough sets on Intuitionistic fuzzy approximation spaces were established by Tripathy in [6, 7, 9]. We show in this section that some of these results are faulty by providing counter examples and also present corresponding rectified results and prove them. Before stating the next result we establish some results which are essential for the validity of the concepts. It is easy to see that if R and S are fuzzy proximity relations over U then $R \cup S$ and $R \cap S$ are intuitionistic fuzzy proximity relations over U. The first result incorrectly stated in [5] is that

Proposition 3.1. Let R and S be two intuitionistic fuzzy proximity relations on U. Then

$$(R \cup S)(\alpha, \beta) \subseteq R(\alpha, \beta) \cup S(\alpha, \beta) \tag{3.1}$$

$$(R \cap S)(\alpha, \beta) \supseteq R(\alpha, \beta) \cap S(\alpha, \beta) \tag{3.2}$$

Counter Example 3.1:

The following example shows the incorrectness of property (3.1).

Suppose U = $\{x_1, x_2, x_3, x_4, x_5, x_6\}$ be a universe. Two fuzzy proximity relations R and S are defined over U are given in their matrix form in Tables 1 and 2 respectively.

From the above two tables we obtain the union and intersection and is represented in the following tables. Table 3 represents $R \cup S$ and Table 4 represents $R \cap S$.

Let us consider $\alpha = 0.7$ and $\beta = 0.1$.

Therefore from Tables 1 and 2 we obtain the following:

Table 1. Fuzzy proximity relation R

R	X_1	X_2	X_3	X_4	X_5	X_6
X_1	(1,0)	(0.3, 0.5)	(0.3, 0.5)	(0.3,0.5)	(0.3, 0.5)	(0.3, 0.5)
X_2	(0.3,0.5)	(1,0)	(0.7, 0.2)	(0.7,0.2)	(0.3, 0.5)	(0.3, 0.5)
X_3	(0.3,0.5)	(0.7, 0.2)	(1,0)	(0.7,0.2)	(0.3, 0.5)	(0.3, 0.5)
X_4	(0.3,0.5)	(0.7, 0.2)	(0.7, 0.2)	(1, 0)	(0.3, 0.5)	(0.3, 0.5)
X_5	(0.3,0.5)	(0.3, 0.5)	(0.3, 0.5)	(0.3,0.5)	(1,0)	(0.7, 0.2)
X_6	(0.3,0.5)	(0.3, 0.5)	(0.3, 0.5)	(0.3,0.5)	(0.7, 0.2)	(1,0)

Table 2. Fuzzy proximity relation S

S	X_1	X_2	X_3	X_4	X_5	X_6
X_1	(1, 0)	(0.1,0.7)	(0.1,0.7)	(0.1,0.7)	(0.1,0.7)	(0.1,0.7)
X_2	(0.1,0.7)	(1, 0)	(0.1,0.7)	(0.1,0.7)	(0.1,0.7)	(0.1,0.7)
X_3	(0.1,0.7)	(0.1,0.7)	(1,0)	(0.1,0.7)	(0.8, 0.1)	(0.1,0.7)
X_4	(0.1,0.7)	(0.1,0.7)	(0.1,0.7)	(1,0)	(0.1,0.7)	(0.1,0.7)
X_5	(0.1,0.7)	(0.1.0.7)	(0.8,0.1)	(0.1,0.7)	(1,0)	(0.1.0.7)
X_6	(0.1,0.7)	(0.1.0.7)	(0.1,0.7)	(0.1,0.7)	(0.1,0.7)	(1, 0)

Table 3. Fuzzy Proximity relation $R \cup S$

$R \cup S$	X_1	X_2	X_3	X_4	X_5	X_6
X_1	(1,0)	(0.3. 0.5)	(0.3,0.5)	(0.3,0.5)	(0.3,0.5)	(0.3, 0.5)
X_2	(0.3. 0.5)	(1,0)	(0.7, 0.2)	(0.7, 0.2)	(0.3, 0.5)	(0.3, 0.5)
X_3	(0.3, 0.5)	(0.7. 0.2)	(1,0)	(0.7, 0.2)	(0.8,0.1)	(0.3,0.5)
X_4	(0.3, 0.5)	(0.7. 0.2)	(0.7, 0.2)	(1,0)	(0.3, 0.5)	(0.3, 0.5)
X_5	(0.3, 0.5)	(0.3. 0.5)	(0.8, 0.1)	(0.3, 0.5)	(1,0)	(0.7,0.2)
X_6	(0.3, 0.5)	(0.3. 0.5)	(0.3, 0.5)	(0.3, 0.5)	(0.7, 0.2)	(1,0)

Table 4. Fuzzy Proximity relation $R \cap S$

$R \cap S$	X_1	X_2	X_3	X_4	X_5	X_6
X_1	(1,0)	(0.1, 0.7)	(0.1, 0.7)	(0.1,0.7)	(0.1,0.7)	(0.1, 0.7)
X_2	(0.1, 0.7)	(1,0)	(0.1, 0.7)	(0.1,0.7)	(0.1,0.7)	(0.1, 0.7)
X_3	(0.1, 0.7)	(0.1, 0.7)	(1, 0)	(0.1,0.7)	(0.3,0.5)	(0.1, 0.7)
X_4	(0.1, 0.7)	(0.1, 0.7)	(0.1, 0.7)	(1, 0)	(0.1,0.7)	(0.1, 0.7)
X_5	(0.1, 0.7)	(0.1, 0.7)	(0.3, 0.5)	(0.1,0.7)	(1,0)	(0.1, 0.7)
X_6	(0.1, 0.7)	(0.1, 0.7)	(0.1, 0.7)	(0.1,0.7)	(0.1,0.7)	(1,0)

$$R\,(\alpha, \beta) = \{\{(2, 3), (2, 4), (3, 4), (5, 6), (3, 2), (4, 2), (4, 3), (6, 5), (1, 1), (2, 2), (3, 3),$$
$$(4, 4), (5, 5), (6, 6)\}$$

$$(3.3)$$

$$S\ (\alpha, \beta) = \{(3, 5), (5, 3), (1, 1), (2, 2), (3, 3), (4, 4), (5, 5), (6, 6)\} \quad (3.4)$$

From Tables 3 and 4 we obtain the following:

$$(R \cup S)(\alpha, \beta) = \{(2, 3), (2, 4), (3, 4), (3, 5), (5, 6), (2, 5), (2, 6), (3, 2), (4, 2), (4, 3), (5, 3),$$
$$(6, 5), (1, 1), (2, 2), (3, 3), (4, 4), (5, 5), (6, 6)\}$$
$$(3.5)$$

$$(R \cap S)(\alpha, \beta) = \{(1, 1), (2, 2), (3, 3), (4, 4), (5, 5), (6, 6)\} \quad (3.6)$$

From (3.3) and (3.4) we obtain

$$R(\alpha, \beta) \cup S(\alpha, \beta) = \{(2, 3), (2, 4), (3, 4), (5, 6), (3, 5), (5, 3), (3, 2),$$
$$(4, 2), (4, 3), (6, 5), (1, 1), (2, 2), (3, 3), (4, 4), (5, 5), (6, 6)\}$$
$$(3.7)$$

$$R(\alpha, \beta) \cap S(\alpha, \beta) = \{(1, 1), (2, 2), (3, 3), (4, 4), (5, 5), (6, 6)\} \quad (3.8)$$

It follows from (3.5) and (3.7) that $(R \cup S)(\alpha, \beta) \subseteq R(\alpha, \beta) \cup S(\alpha, \beta)$ is not true.
Counter Example 3.2:
This example shows the incorrectness of property (3.2).
Let $U = \{x_1, x_2, x_3, x_4, x_5, x_6\}$ be a universe. Fuzzy proximity relations R and S over U are given by Tables 5 and 6 respectively.
The tables for the fuzzy proximity relations $R \cup S$ and $R \cap S$ are given in Tables 7 and 8 respectively.

Table 5. Fuzzy proximity relation R

R	X_1	X_2	X_3	X_4	X_5	X_6
X_1	(1,0)	(0.3, 0.6)	(0.3, 0.6)	(0.3, 0.6)	(0.3, 0.6)	(0.3. 0.6)
X_2	(0.3, 0.6)	(1,0)	(0.7, 0.2)	(0.3, 0.6)	(0.3, 0.6)	(0.3, 0.6)
X_3	(0.3, 0.6)	(0.7, 0.2)	(1, 0)	(0.7, 0.2)	(0.3, 0.6)	(0.3. 0.6)
X_4	(0.3, 0.6)	(0.3, 0.6)	(0.7, 0.2)	(1,0)	(0.3, 0.6)	(0.3, 0.6)
X_5	(0.3, 0.6)	(0.3, 0.6)	(0.3, 0.6)	(0.3, 0.6)	(1,0)	(0.1, 0.8)
X_6	(0.3, 0.6)	(0.3, 0.6)	(0.3, 0.6)	(0.3, 0.6)	(0.1, 0.8)	(1,0)

Table 6. Fuzzy proximity relation S

S	X_1	X_2	X_3	X_4	X_5	X_6
X_1	(1,0)	(0.1,0.8)	(0.1,0.8)	(0.1, 0.8)	(0.1, 0.8)	(0.1,0.8)
X_2	(0.1,0.8)	(1. 0)	(0.1,0.8)	(0.1, 0.8)	(0.8, 0.1)	(0.1,0.8)
X_3	(0.1,0.8)	(0.1,0.8)	(1,0)	(0.1, 0.8)	(0.1, 0.8)	(0.1,0.8)
X_4	(0.1,0.8)	(0.1,0.8)	(0.1,0.8)	(1, 0)	(0.8, 0.1)	(0.1,0.8)
X_5	(0.1,0.8)	(0.8,0.1)	(0.1,0.8)	(0.8, 0.1)	(1,0)	(0.1,0.8)
X_6	(0.1,0.8)	(0.1,0.8)	(0.1,0.8)	(0.1, 0.8)	(0.1, 0.8)	(1,0)

Let us consider $\alpha = 0.7$ and $\beta = 0.1$.

Therefore from Tables 5 and 6 we obtain the following:

$$R\,(\alpha, \beta) = \{(2, 3), (3, 4), (2, 4), (3, 2), (4, 3), (4, 2), (1, 1), (2, 2), \\ (3, 3), (4, 4), (5, 5), (6, 6)\} \tag{3.9}$$

Table 7. Fuzzy proximity relation $R \cup S$

$R \cup S$	X_1	X_2	X_3	X_4	X_5	X_6
X_1	(1,0)	(0.3, 0.6)	(0.3, 0.6)	(0.3, 0.6)	(0.3, 0.6)	(0.3, 0.6)
X_2	(0.3, 0.6)	(1, 0)	(0.7, 0.2)	(0.3, 0.6)	(0.8, 0.1)	(0.3, 0.6)
X_3	(0.3. 0.6)	(0.7, 0.2)	(1, 0)	(0.7, 0.2)	(0.3, 0.6)	(0.3, 0.6)
X_4	(0.3, 0.6)	(0.3, 0.6)	(0.7, 0.2)	(1, 0)	(0.8, 0.1)	(0.3, 0.6)
X_5	(0.3, 0.6)	(0.8, 0.1)	(0.3, 0.6)	(0.8, 0.1)	(1,0)	(0.3, 0.6)
X_6	(0.3, 0.6)	(0.3, 0.6)	(0.3, 0.6)	(0.3, 0.6)	(0.3, 0.6)	(1,0)

Table 8. Fuzzy proximity relation $R \cap S$

$R \cap S$	X_1	X_2	X_3	X_4	X_5	X_6
X_1	(1,0)	(0.1,0.8)	(0.1. 0.8)	(0.1,0.8)	(0 1,0.8)	(0.1, 0.8)
X_2	(0.1,0.8)	(1,0)	(0.1. 0.8)	(0.1,0.8)	(0.3,0.6)	(0.1, 0.8)
X_3	(0.1,0.8)	(0.1,0.8)	(1,0)	(0.1,0.8)	(0.1,0.8)	(0.1, 0.8)
X_4	(0.1,0.8)	(0.1,0.8)	(0.1,0.8)	(1,0)	(0.3,0.6)	(0.1, 0.8)
X_5	(0.1,0.8)	(0.3,0.6)	(0.1,0.8)	(0.3,0.6)	(1,0)	(0.1,0.8)
X_6	(0.1,0.8)	(0.1,0.8)	(0.1,0.8)	(0.1,0.8)	(0.1,0.8)	(1,0)

$$S(\alpha, \beta) = \{(2, 5), (5, 4), (2, 4), (5, 2), (4, 5), \\ (4, 2), (1, 1), (2, 2), (3, 3), (4, 4), (5, 5), (6, 6)\} \tag{3.10}$$

and from Tables 7 and 8 we obtain the following:

$$((R \cup S)(\alpha, \beta)) = \{(2, 3), (2, 5), (3, 4), (5, 4), (2, 4), (3, 2), (5, 2), (4, 3), (4, 5), \\ (4, 2), (1, 1), (2, 2), (3, 3), (4, 4), (5, 5), (6, 6)\} \tag{3.11}$$

$$((R \cap S)(\alpha, \beta)) = \{(1, 1), (2, 2), (3, 3), (4, 4), (5, 5), (6, 6)\} \tag{3.12}$$

From (3.9) and (3.10) we obtain

$$R\,(\alpha, \beta) \cup S\,(\alpha, \beta) = \{(2, 3), (2, 5), (3, 4), (5, 4), (2, 4), (3, 2), (5, 2), (4, 3), (4, 5), \\ (4, 2), (1, 1), (2, 2), (3, 3), (4, 4), (5, 5), (6, 6)\} \tag{3.13}$$

$$R\ (\alpha, \beta) \cap S\ (\alpha, \beta) = \{(2,\ 4),\ (4,\ 2),\ (1,\ 1),\ (2,\ 2),\ (3,\ 3),\ (4,\ 4),\ (5,\ 5),\ (6,\ 6)\}$$
$$(3.14)$$

It is clear from (3.12) and (3.14) that
$(R \cap S)\,(\alpha, \beta) \supseteq R\ (\alpha, \beta) \cap S\ (\alpha, \beta)$ is not true.

Theorem 3.1. Let R and S be two fuzzy proximity relations on U. Then for any $(\alpha, \beta) \in J$,

$$(R \cup S)(\alpha, \beta) \supseteq R(\alpha, \beta) \cup S(\alpha, \beta) \tag{3.15}$$

$$(R \cap S)(\alpha, \beta) \subseteq R(\alpha, \beta) \cap S(\alpha, \beta) \tag{3.16}$$

Proof: Proof of (3.15)

$$(x, y) \in R(\alpha, \beta) \cup S(\alpha, \beta)$$
$$\Rightarrow (x, y) \in R(\alpha, \beta) \text{ or } (x, y) \in S(\alpha, \beta)$$

Now 4 cases can occur.

Case (i): $xR_{\alpha,\beta}y$ or $xS_{\alpha,\beta}y$

Then $(\mu_R(x, y) \geq \alpha$ and $v_R(x, y) \leq \beta)$ or $(\mu_S(x, y) \geq \alpha$ and $v_S(x, y) \leq \beta)$

So, $\max\{\mu_R(x, y), \mu_S(x, y)\} \geq \alpha$ and $\min\{v_R(x, y), v_S(x, y)\} \leq \beta$

$\Rightarrow \mu_{(R \cup S)}(x, y) \geq \alpha$ and $v_{(R \cup S)}(x, y) \leq \beta$

$\Rightarrow (x, y) \in (R \cup S)(\alpha, \beta)$

Case (ii): $xR_{\alpha,\beta}y$ or $\exists$ a sequence $u_1, u_2, \ldots u_n$ such that $xS_{\alpha,\beta}u_1, u_1S_{\alpha,\beta}u_2, \ldots u_nS_{\alpha,\beta}y$

If $xR_{\alpha,\beta}y$ then $\mu_R(x, y) \geq \alpha$ and $v_R(x, y) \leq \beta$.

So, $\mu_{(R \cup S)}(x, y) = \max\{\mu_R(x, y), \mu_S(x, y)\} \geq \alpha$ and

$v_{(R \cup S)}(x, y) = \min\{v_R(x, y), v_S(x, y)\} \leq \beta$

Again, Suppose $\exists$ a sequence $u_1, u_2, \ldots u_n$ such that $xS_{\alpha,\beta}u_1, u_1S_{\alpha,\beta}u_2, \ldots u_nS_{\alpha,\beta}y$.

Then $(\mu_S(x, u_1) \geq \alpha, v_S(x, u_1) \leq \beta), (\mu_S(u_1, u_2) \geq \alpha, v_S(u_1, u_2) \leq \beta), \ldots, (\mu_S(u_n, y) \geq \alpha, v_S(u_n, y) \leq \beta)$.

So, as above

$(\mu_{(R \cup S)}(x, u_1) \geq \alpha, v_{(R \cup S)}(x, u_1) \leq \beta), (\mu_{(R \cup S)}(u_1, u_2) \geq \alpha, v_{(R \cup S)}(u_1, u_2) \leq \beta), \ldots,$

$(\mu_{(R \cup S)}(u_n, y) \geq \alpha, v_{(R \cup S)}(u_n, y) \leq \beta)$.

This implies that $x(R \cup S)_{(\alpha,\beta)}u_1, u_1(R \cup S)_{(\alpha,\beta)}u_2, \ldots, u_n(R \cup S)_{(\alpha,\beta)}y$.

Thus, $x(R \cup S)(\alpha, \beta)y$.

Case (iii): $\exists$ a sequence $u_1, u_2, \ldots u_n$ such that $xR_{\alpha,\beta}u_1, u_1R_{\alpha,\beta}u_2, \ldots u_nR_{\alpha,\beta}y$ or $xS_{\alpha,\beta}y$

The proof is similar to case (ii) above.

Case (iv): $\exists$ a sequence $u_1, u_2, \ldots u_n$ such that $xR_{(\alpha,\beta)}u_1, u_1R_{(\alpha,\beta)}u_2, \ldots u_nR_{(\alpha,\beta)}y$ or

$\exists$ a sequence $v_1, v_2, \ldots v_m$ such that $xS_{(\alpha,\beta)}v_1, v_1S_{(\alpha,\beta)}v_2, \ldots v_mS_{(\alpha,\beta)}y$

In any one of the cases we apply the same argument as in case (ii) 2nd part and conclude that $x(R \cup S)(\alpha, \beta)y$.

Hence in all the cases the proof follows. The proof of (3.16) is similar.

We denote the equivalence classes generated by a fuzzy proximity relation R over U with respect to a grade $(\alpha, \beta) \in J$ by $R^*_{\alpha,\beta}$.

The following result was established in [2, 3] basing upon this notion.

Proposition 3.2. Let R and S be two fuzzy proximity relations on U and $(\alpha, \beta) \in J$. Then

$$(R \cup S)^*_{(\alpha,\beta)} \subseteq R^*_{(\alpha,\beta)} \cup S^*_{(\alpha,\beta)} \tag{3.17}$$

$$(R \cap S)^*_{(\alpha,\beta)} \supseteq R^*_{(\alpha,\beta)} \cap S^*_{(\alpha,\beta)} \tag{3.18}$$

We show below that the result is incorrect. For this we provide two counter examples where the results fail to be true.

Counter Example 3.3:

We continue with the above example where the fuzzy proximity relations R and S and their union and intersection are given in Tables 1, 2, 3 and 4. We have,

$$R^*_{(\alpha,\beta)} = \{\{1\}, \{2, 3, 4\}, \{5, 6\}\}, \ S^*_{(\alpha,\beta)} = \{\{1\}, \{2\}, \{4\}, \{6\}, \{3, 5\}\}$$

Hence,

$$(R \cup S)^*_{(\alpha,\beta)} = \{\{1\}, \{2, 3, 4, 5, 6\}\}, \ (R \cap S)^*_{(\alpha,\beta)} = \{\{1\}, \{2\}, \{3\}, \{4\}, \{5\}, \{6\}\}$$

$$R^*_{(\alpha,\beta)} \cup S^*_{(\alpha,\beta)} = \{\{1\}, \{2\}, \{4\}, \{6\}, \{2, 3, 4\}, \{3, 5\}, \{5, 6\}\}, \ R^*_{(\alpha,\beta)} \cap S^*_{(\alpha,\beta)}$$
$$= \{\{1\}\}$$

So, it is clear that (3.17) is not true.

Counter Example.3.4:

We continue with the above example where the fuzzy proximity relations R and S and their union and intersection are given in Tables 5, 6, 7 and 8. We have,

$$R^*_{(\alpha,\beta)} = \{\{1\}, \{5\}, \{6\}, \{2, 3, 4\}\}, \ S^*_{(\alpha,\beta)} = \{\{1\}, \{3\}, \{6\}, \{2, 4, 5\}\}$$

$$(R \cup S)^*_{(\alpha,\beta)} = \{\{1\}, \{6\}, \{2, 3, 4, 5\}\}, \ (R \cap S)^*_{(\alpha,\beta)}$$
$$= \{\{1\}, \{2\}, \{3\}, \{4\}, \{5\}, \{6\}\}$$

$R^*_{(\alpha,\beta)} \cap S^*_{(\alpha,\beta)} = \{\{1\}, \{6\}\}$. Hence, it is clear that (3.18) is not true.

Theorem 3.2. Let R and S be two fuzzy proximity relations on U. Then

$$(R \cup S)^*_{(\alpha,\beta)} \supseteq R^*_{(\alpha,\beta)} \cup S^*_{(\alpha,\beta)} \qquad (3.19)$$

$$(R \cap S)^*_{(\alpha,\beta)} \subseteq R^*_{(\alpha,\beta)} \cap S^*_{(\alpha,\beta)} \qquad (3.20)$$

Proof: Proof of (3.19)

Let $[x] \in R^*_{(\alpha,\beta)} \cup S^*_{(\alpha,\beta)}$. Then, $[x] \in R^*_{(\alpha,\beta)}$ or $[x] \in S^*_{(\alpha,\beta)}$.

So, for any $y \in [x]$,

$$(x,y) \in R(\alpha, \beta) \text{ or } (x,y) \in S(\alpha, \beta) \ \Leftrightarrow \ (x,y) \in (R \cup S)(\alpha, \beta) \qquad (\text{by}(3.15))$$

Hence, $[x] \in (R \cup S)^*_{(\alpha,\beta)}$.

Proof of (3.20)

Let $[x] \in (R \cap S)^*_{(\alpha,\beta)}$. Then for any $y \in [x]$, $(x,y) \in (R \cap S)(\alpha, \beta)$

$$\Rightarrow (x,y) \in R(\alpha, \beta) \cap S(\alpha, \beta) \qquad (\text{by}(3.16))$$

So, $[x] \in R^*_{(\alpha,\beta)} \cap S^*_{(\alpha,\beta)}$. This proved (3.20)

4   An Application of Rough Sets on Intuitionistic Fuzzy Approximation Spaces

Let us consider a situation where the computer needs to match two images based on the different matching properties of images.

Suppose $V = \{i_1, i_2, i_3, i_4, i_5\}$ be a set of images that needs to be matched based on the criteria; edge matching, color matching, shape and surface area matching and texture matching.

Table 9. Relation R: Matching edges

R	i_1	i_2	i_3	i_4	i_5
i_1	(1,0)	(0.30, 0.60)	(0.15, 0.80)	(0.60, 0.25)	(0.50,0.35)
i_2	(0.30, 0.60)	(1,0)	(0.40, 0.45)	(0.87,0.1)	(0.20,0.70)
i_3	(0.15, 0.80)	(0.40, 0.45)	(1,0)	(0.52,0.45)	(0.05, 0.85)
i_4	(0.60, 0.25)	(0.87,0.1)	(0.52,0.45)	(1,0)	(0.66, 0.35)
i_5	(0.50,0.35)	(0.20,0.70)	(0.05,0.85)	(0.66,0.35)	(1.0)

Table 10. Relation S: Matching colors

S	i_1	i_2	i_3	i_4	i_5
i_1	(1,0)	(0.30, 0.55)	(0.90, 0.05)	(0.60, 0.25)	(0.75, 0.20)
i_2	(0.30, 0.55)	(1,0)	(0.23, 0.76)	(0.41, 0.45)	(0.50, 0.35)
i_3	(0.90, 0.05)	(0.23, 0.76)	(1,0)	(0.60, 0.25)	(0.41, 0.45)
i_4	(0.60, 0.35)	(0.41, 0.45)	(0.60, 0.25)	(1,0)	(0.96, 0.03)
i_5	(0.75, 0.20)	(0.50, 0.35)	(0.41, 0.45)	(0.96, 0.03)	(1,0)

After the feature extraction process of the images we obtain the matching percentages of properties of two images. R, S, T, and W are the relations based on the above properties on the set V (Tables 9, 10, 11, and 12), which can be used to match images based on the required criteria. To find matching images based on two or more properties, we find the union or intersection of the relations based on the requirements.

Table 11. Relation T: Matching shape and surface area

T	i_1	i_2	i_3	i_4	i_5
i_1	(1,0)	(0.46, 0.42)	(0.30, 0.55)	(0.70, 0.25)	(0.25, 0.65)
i_2	(0.46,0.42)	(1,0)	(0.50, 0.40)	(0.35, 0.60)	(0.80, 0.15)
i_3	(0.30, 0.55)	(0.50, 0.40)	(1,0)	(0.10, 0.75)	(0.03, 0.85)
i_4	(0.70, 0.25)	(0.35, 0.60)	(0.10, 0.75)	(1,0)	(0.95, 0.02)
i_5	(0.25, 0.65)	(0.80, 0.15)	(0.03, 0.85)	(0.95, 0.02)	(1,0)

Table 12. Relation W: Matching texture

W	i_1	i_2	i_3	i_4	i_5
i_1	(1,0)	(0.09, 0.85)	(0.87,0.08)	(0.43, 0.47)	(0.67, 0.33)
i_2	(0.09, 0.85)	(1,0)	(0.15, 0.76)	(0.33, 0.58)	(0.54, 0,45)
i_3	(0.87, 0.08)	(0.15, 0.76)	(1,0)	(0.63, 0.37)	(0.70, 0.25)
i_4	(0.43, 0.47)	(0.33,0.58)	(0.63,0.37)	(1,0)	(0.30, 0.55)
i_5	(0.67, 0.33)	(0.54, 0,45)	(0.70, 0.25)	(0.30, 0.55)	(1,0)

If we fix $\alpha = 0.85$ and $\beta = 0.1$ as the percentage of matching property, we obtain the following

$$R(\alpha, \beta) = \{(2, 4), (4, 2), (1, 1), (2, 2), (3, 3), (4, 4), (5, 5)\}$$

$$R^*_{\alpha,\beta} = \{\{1\}, \{3\}, \{5\}, \{2, 4\}\}$$

$$S(\alpha, \beta) = \{(1, 3), (4, 5), (3, 1), (5, 4), (1, 1), (2, 2), (3, 3), (4, 4), (5, 5)\}$$

$$S^*_{\alpha,\beta} = \{\{2\}, \{4, 5\}, \{1, 3\}\}, T(\alpha, \beta)$$
$$= \{(4, 5), (5, 4), (1, 1), (2, 2), (3, 3), (4, 4), (5, 5)\}$$

and $T^*_{\alpha,\beta} = \{\{1\}, \{2\}, \{3\}, \{4, 5\}\}$. Again,

$$W(\alpha) = \{(1, 3), (3, 1), (1, 1), (2, 2), (3, 3), (4, 4), (5, 5)\} \text{ and}$$
$$W^*_\alpha = \{\{2\}, \{4\}, \{5\}, \{1, 3\}\}$$

Inference: From the above results we can conclude that based on the edge matching property image i_2 and i_4 are matched maximum. Similarly based on color matching image i_1 and i_3, and image i_4 and i_5 are matching. Image i_4 and i_5 are matched based on shape and surface area, and i_1 and i_3 on texture matching.

Now if we want to find the matching images based on edge matching or shape and surface area matching, we find the union of relations A_1 and A_3. The union of the two relations is as shown in Table 13.

Table 13. $R \cup T$

$R \cup T$	i_1	i_2	i_3	i_4	i_5
i_1	(1,0)	(0.46, 0.42)	(0.30, 0.55)	(0.70,0.25)	(0.50, 0.35)
i_2	(0.46, 0.42)	(1,0)	(0.50,0.40)	(0.87,0.1)	(0.80, 0.15)
i_3	(0.30, 0.55)	(0.50.0.40)	(1,0)	(0.52, 0.45)	(0.05,0.85)
i_4	(0.70,0.25)	(0.87,0.1)	(0.52, 0.45)	(1,0)	(0.95, 0.02)
i_5	(0.50, 0.35)	(0.80, 0.15)	(0.05,0.85)	(0.95, 0.02)	(1,0)

From Table 13 we find

$$(R \cup T)(\alpha) = \{(2, 4), (4, 5), (2, 5)\} \text{ and } (R \cup T)_\alpha^* = \{\{1\}, \{3\}, \{2, 4, 5\}\}$$

Inference: We can say that images i_2, i_4 and i_5 form a group of matching images based on edge matching and shape and surface area matching.

Next if we want to find the matching images based on color matching as well as texture matching, we find the intersection of relations A_2 and A_4. The intersection of the two relations is as shown in Table 14.

Table 14. $S \cap W$

$S \cap W$	i_1	i_2	i_3	i_4	i_5
i_1	(1,0)	(0.09, 0.85)	(0.87, 0.08)	(0.43, 0.47)	(0.67, 0.33)
i_2	(0.09, 0.85)	(1,0)	(0.15, 0.76)	(0.33, 0.58)	(0.50, 0.45)
i_3	(0.87, 0.08)	(0.15, 0.76)	(1,0)	(0.60,0.37)	(0.41,0.45)
i_4	(0.43, 0.47)	(0.33, 0.58)	(0.60, 0.37)	(1,0)	(0.30, 0.55)
i_5	(0.67, 0.33)	(0.50, 0.45)	(0.41, 0.45)	(0.30, 0.55)	(1,0)

From Table 14 we find

$$(S \cap W)(\alpha) = \{(1, 3), (3, 1), (1, 1), (2, 2), (3, 3), (4, 4), (5, 5)\}$$

$$(S \cap W)_\alpha^* = \{\{2\}, \{4\}, \{5\}, \{1, 3\}\}$$

Inference: We can say that images i_1 and i_3 is a set of images that are matched both in terms of color and texture similarity.

5 Conclusions

Rough sets on intuitionistic fuzzy approximation space are a generalization of basic rough sets introduced in 2006 [6, 7]. However, some of the properties established in the first paper were erroneous. In this paper we first established that the results are incorrect through counter examples and then established the correct versions of the properties. Finally, we illustrated through an application in computer vision as how the results are applicable in real life situations.

References

1. Acharjya, D.P., Tripathy, B.K.: Rough sets on fuzzy approximation space and application to distributed knowledge systems. Int. J. Artif. Intell. Soft Comput. **1**(1), 1–14 (2008)
2. Acharjya, D.P., Tripathy, B.K.: Rough sets on intuitionistic fuzzy approximation spaces and knowledge representation. Int. J. Artif. Intell. Comput. Res. **1**(1), 29–36 (2009)
3. Atanassov, K.T.: Intuitionistic fuzzy sets. Fuzzy Sets Syst. **20**, 87–96 (1986)
4. Pawlak, Z.: Rough sets. Int. J. Inf. Comput. Sci. **11**, 341–356 (1982)
5. Tripathy, B.K.: Rough sets on fuzzy approximation spaces and application to distributed knowledge systems. In: National Conference on Mathematics and its Applications, Burdwan (2004)
6. Tripathy, B.K.: Rough sets on intuitionistic fuzzy approximation spaces. In: Proceedings of the IEEE Conference on Artificial Intelligence, London, 6–8 September (2006)
7. Tripathy, B.K.: Rough sets on intuitionistic fuzzy approximation spaces. Notes Intuitionistic Fuzzy Sets **12**(1), 45–54 (2006). (Bulgaria)
8. Tripathy, B.K., Gantayat, S.S., Mohanty, D.: Properties of rough sets on fuzzy approximation spaces and knowledge representation. In: Proceeding of the National Conference on Recent Trends in Intelligent Computing, Kalyani Govt. Engineering College (W.B), 17–19 November, pp. 3–8 (2006)
9. Tripathy, B.K.: Rough sets on fuzzy approximation spaces and intuitionistic fuzzy approximation spaces. In: Abraham, A., Falcón, R., Bello, R. (eds.) Rough Set Theory: A True Landmark in Data Analysis, vol. 174, pp. 3–44. Springer, Heidelberg (2009)
10. Tripathy, B.K., Acharjya, D.P.: Association rule granulation using rough sets on intuitionistic fuzzy approximation spaces and granular computing. Ann. Comput. Sci. Ser. **9**(1), 125–144 (2011)
11. Tripathy, B.K., Acharjya, D.P.: Knowledge mining using ordering rules and rough sets on fuzzy approximation spaces. Int. J. Adv. Sci. Technol. **1**(3), 41–50 (2010)
12. Shen, Y., Wang, F.: Rough approximations of vague sets in fuzzy approximation space. Int. J. Approx. Reason. **52**, 281–296 (2011)

Interval Type-II Fuzzy Multiple Group Decision Making Based Ranking for Customer Purchase Frequency Determinants in Online Brand Community

S. Choudhury[1]([✉]), A.K. Patra[2], A.K. Parida[3], and S. Chatterjee[3]

[1] Centre for Management Studies, NERIST, Nirjuli, India
shibabrata.choudhury@gmail.com
[2] Department of Computer Science and Engineering, NERIST, Nirjuli, India
aswinipatra@gmail.com
[3] Department of Electrical Engineering, NERIST, Nirjuli, India
adikanda_2003@yahoo.co.in, sc@nerist.ac.in

Abstract. Customer's view in online brand community acts as a foundation for the co-creation of the brand, which results in increased purchase frequency. Influencing attributes as the independent variables significantly influences performance of these online platforms. Prioritization of these alternative platforms would act as a boosting factor for marketer for their inclusive growth. In this paper, an attempt has been made with the help of interval type-II fuzzy multiple group decision making process for the said purpose. A fuzzy linguistic interpretation membership has also been proposed which can further improve the computed results.

Keywords: Interval type-II fuzzy · Multi attribute group decision making · Online brand community

1 Introduction

Online brand community is bounded with a specific, non-geographical, and a set of social relation amongst admirer of the brand [1]. Online brand communities are designed to facilitate co-creation of product-brand, and other company related affairs [2]. It offers a platform beyond mere acquisition and sharing of knowledge, such as taking the lead in the community collaboration, persuading others to join the community [3]. In comparison to the traditional mechanism, online brand community is easy to implement in a cost effective manner [4]. The work presented in [5] suggests different levels of effects of different alternatives based on the level of participation in the community, which can influence both forward and backward value chain of the organization. Involvement in the community may supplement the organization's R&D through innovation, subsequently influences the brand value that became the decisive factor for frequency of visit, frequency of purchase; and so on [6–8]. In this direction, B.S. Butler in [9] has emphasized the lower form of involvement, like reading the content, is crucial to increase the sales. However, R. Rishika and et al. in their literature

© Springer Nature Singapore Pte Ltd. 2017
K. Deep et al. (eds.), *Proceedings of Sixth International Conference on Soft Computing for Problem Solving*, Advances in Intelligent Systems and Computing 546, DOI 10.1007/978-981-10-3322-3_26

in [10] revealed, high involvement creates higher scope for the company. Apart from level of involvement, company has to be alert for the negative remark, which also has equal negative pressure on the brand value of the company [11]. This sort of decision making process can never be possible without considering uncertainty into account. Fuzzy set theory is the best suitable tool to deal with uncertain (fuzziness) criteria [12]. Fuzzy set theory is based on the concept of approximation rather on the exact values. Interval type-II fuzzy [13], the advanced fuzzy set has been applied for better accuracy in result. This work emphasizes to rank the alternatives such as social networking, blog, and pay-for-click based on decision makers inputs for the attributes. The following points highlight the superiority of the proposed technique in comparison to the existing methods.

- Attributes for online brand community have been considered to choose the proposed alternatives using the interval type-II fuzzy group decision making tool.
- New interpretations of linguistic weights have also been introduced.

The paper is organized as follows: Sect. 2 reviews deterministic of purchase frequency in online brand community, Sect. 3 briefs about the basics of type-II fuzzy, and Sect. 4 illustrates arithmetic operations between type-II fuzzy sets. Section 5 discusses interval type-II fuzzy multiple group decision making for alternatives ranking in online brand community. The conclusions are discussed in Sect. 6.

2 Determination of Purchase Frequency in Online Brand Community

Figure 1 represents the proposed model of Wu, J., et al. in [14] to determine the purchase frequency in online brand community. Online brand community is no more be considered as a passive tool to realize the thought process of consumer, rather it has

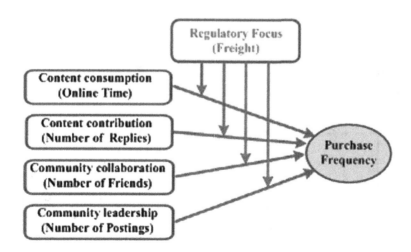

Fig. 1. Online brand community activity as proposed by Wu, J., et al. [14].

a wider canvas to involve customer in contributing, reacting, evolving new product, and many more [15]. The activities of the customers, based on their level of involvement, has been broadly classified into: (i) content consumption- customers who spends significant amount of time to study the posts of the other users for his decision making process, (ii) content contribution- these group participate actively in replying the posts, (iii) community collaboration- the number of her/his circle gives scope to interact, and (iv) community leadership- are the customers who proactively involved in the postings [16, 17]. All these activities are moderated via regulatory framework considering 'freight' as the moderating variable. Similarly, 'Purchase frequency' as explained by Wu, J., et al. (2015) has also been considered as dependent variable in the previous related work [15]. However, the proposed work considers ranking of various alternatives to facilitate purchase frequency of the online brand community.

3   Basics of Type-II Fuzzy

An interval type-II fuzzy [13] set expressed in Eq. (1) and Fig. 2.

$$\tilde{\tilde{Y}} = \left\{ (a,u), \mu_{\tilde{\tilde{Y}}}(a,u) | \forall a \in A, \ \forall u \in J_a \subseteq [0,1], \ 0 \leq \mu_{\tilde{\tilde{Y}}}(a,u) \leq 1 \right\} \tag{1}$$

Equation (2) illustrates two examples of interval type-II fuzzy sets

$$\tilde{\tilde{Y}}_1 = \left(\tilde{Y}_1^U, \tilde{Y}_1^L\right) = \left(\left(y_{11}^U, y_{12}^U, y_{13}^U, y_{14}^U; H_1\left(\tilde{Y}_1^U\right), H_2\left(\tilde{Y}_1^U\right)\right), \left(\left(y_{11}^L, y_{12}^L, y_{13}^L, y_{14}^L; H_1\left(\tilde{Y}_1^L\right), H_2\left(\tilde{Y}_1^L\right)\right)\right)\right)$$
$$\tilde{\tilde{Y}}_2 = \left(\tilde{Y}_2^U, \tilde{Y}_2^L\right) = \left(\left(y_{21}^U, y_{22}^U, y_{23}^U, y_{24}^U; H_1\left(\tilde{Y}_2^U\right), H_2\left(\tilde{Y}_2^U\right)\right), \left(\left(y_{21}^L, y_{22}^L, y_{23}^L, y_{24}^L; H_1\left(\tilde{Y}_2^L\right), H_2\left(\tilde{Y}_2^L\right)\right)\right)\right)$$

$$\tag{2}$$

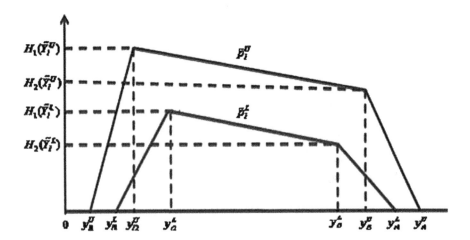

Fig. 2. Interval Type-II Trapezoidal Fuzzy sets

4 Arithmetic Operations Between Type-II Fuzzy Sets

Considering (2), following operations are illustrated [13]:

- Addition of two interval type-II fuzzy numbers:

$$
\begin{aligned}
\tilde{\tilde{Y}}_1 \oplus \tilde{\tilde{Y}}_2 &= \left(\tilde{Y}_1^U, \tilde{Y}_1^L\right) \oplus \left(\tilde{Y}_2^U, \tilde{Y}_2^L\right) \\
&= \left(\left(y_{11}^U + y_{21}^U, y_{12}^U + y_{22}^U, y_{13}^U + y_{23}^U, y_{14}^U + y_{24}^U; \min\left(H_1\left(\tilde{Y}_1^U\right), H_1\left(\tilde{Y}_2^U\right)\right), \min\left(H_2\left(\tilde{Y}_1^U\right), H_2\left(\tilde{Y}_2^U\right)\right)\right)\right), \\
&\quad \left(\left(y_{11}^L + y_{21}^L, y_{12}^L + y_{22}^L, y_{13}^L + y_{23}^L, y_{14}^L + y_{24}^L; \min\left(H_1\left(\tilde{Y}_1^L\right), H_1\left(\tilde{Y}_2^L\right)\right), \min\left(H_2\left(\tilde{Y}_1^L\right), H_2\left(\tilde{Y}_2^L\right)\right)\right)\right)
\end{aligned}
\tag{3}
$$

- Subtraction of two interval type-II fuzzy numbers:

$$
\begin{aligned}
\tilde{\tilde{Y}}_1 - \tilde{\tilde{Y}}_2 &= \left(\tilde{Y}_1^U, \tilde{Y}_1^L\right) - \left(\tilde{Y}_2^U, \tilde{Y}_2^L\right) \\
&= \left(\left(y_{11}^U - y_{24}^U, y_{12}^U - y_{23}^U, y_{13}^U - y_{22}^U, y_{14}^U - y_{21}^U; \min\left(H_1\left(\tilde{Y}_1^U\right), H_1\left(\tilde{Y}_2^U\right)\right), \min\left(H_2\left(\tilde{Y}_1^U\right), H_2\left(\tilde{Y}_2^U\right)\right)\right)\right), \\
&\quad \left(\left(y_{11}^L - y_{24}^L, y_{12}^L - y_{23}^L, y_{13}^L - y_{22}^L, y_{14}^L - y_{21}^L; \min\left(H_1\left(\tilde{Y}_1^L\right), H_1\left(\tilde{Y}_2^L\right)\right), \min\left(H_2\left(\tilde{Y}_1^L\right), H_2\left(\tilde{Y}_2^L\right)\right)\right)\right)
\end{aligned}
\tag{4}
$$

- Multiplication of two interval type-II fuzzy numbers:

$$
\begin{aligned}
\tilde{\tilde{Y}}_1 \otimes \tilde{\tilde{Y}}_2 &= \left(\tilde{Y}_1^U, \tilde{Y}_1^L\right) \otimes \left(\tilde{Y}_2^U, \tilde{Y}_2^L\right) \\
&= \left(\left(y_{11}^U \times y_{21}^U, y_{12}^U \times y_{22}^U, y_{13}^U \times y_{23}^U, y_{14}^U \times y_{24}^U; \min\left(H_1\left(\tilde{Y}_1^U\right), H_1\left(\tilde{Y}_2^U\right)\right), \min\left(H_2\left(\tilde{Y}_1^U\right), H_2\left(\tilde{Y}_2^U\right)\right)\right)\right), \\
&\quad \left(\left(y_{11}^L \times y_{21}^L, y_{12}^L \times y_{22}^L, y_{13}^L \times y_{23}^L, y_{14}^L \times y_{24}^L; \min\left(H_1\left(\tilde{Y}_1^L\right), H_1\left(\tilde{Y}_2^L\right)\right), \min\left(H_2\left(\tilde{Y}_1^L\right), H_2\left(\tilde{Y}_2^L\right)\right)\right)\right)
\end{aligned}
\tag{5}
$$

5 Interval Type-II Fuzzy Multiple Group Decision Making for Attributes Ranking of Online Brand Community

In the recent years many methods have been proposed to solve fuzzy multi attribute decision making [19–22]. The current work has adopted interval type-II multiple group decision making as proposed by Chen, S., et al. in 2010 for ranking the alternatives of online brand community.

Step-I

The current work has proposed seven linguistic term namely: "Very Low" (VL), "Low" (L), "Medium Low" (ML), "Medium" (M), "Medium High" (MH), "High" (H), "Very High" (VH). The middle point of the seven linguistic variables is "Medium" (M), is taken as the standard membership value [23]. The three higher and lower values are calculated by the Eq. (6):

$$y^U_{Higher\,(i)} = \left(\max\left(\left(y^U_{M(i)}\right)^2\right)^n, \left(1 - \left(\left(y^U_{M(i)}\right)^2\right)^n\right)\right); \; y^L_{Higher\,(i)} = \left(\max\left(\left(y^L_{M(i)}\right)^2\right)^n, \left(1 - \left(\left(y^L_{M(i)}\right)^2\right)^n\right)\right)$$

$$y^U_{Lower\,(i)} = \left(\min\left(\left(y^U_{M(i)}\right)^2\right)^n, \left(1 - \left(\left(y^U_{M(i)}\right)^2\right)^n\right)\right); \; y^L_{Lower\,(i)} = \left(\min\left(\left(y^L_{M(i)}\right)^2\right)^n, \left(1 - \left(\left(y^L_{M(i)}\right)^2\right)^n\right)\right) \quad (6)$$

$Higher = [(MH\langle n = 1\rangle), (H\langle n = 2\rangle), (VH\langle n = 3\rangle)]$

$Lower = [(ML\langle n = 1\rangle), (L\langle n = 1\rangle), (VL\langle n = 1\rangle)]$

where $i = 1, 2, 3, 4$;

Based on Eq. (6), linguistic terms for corresponding type-II fuzzy sets are shown in Table 1. The medium (M) value is considered from Chen et al. [24].

Table 1. Linguistic terms for corresponding type-II fuzzy sets

Linguistic term	Interval type − 2 fuzzy sets
Very Low (VL)	$((0, 0, 0, 0.06; 1, 1), (0, 0, 0, 0.02; 0.9, 0.9))$
Low (L)	$((0.01, 0.06, 0.06, 0.24; 1, 1), (0.03, 0.06, 0.06, 0.13; 0.9, 0.9))$
Medium Low (ML)	$((0.09, 0.25, 0.25, 0.49; 1, 1), (0.16, 0.25, 0.25, 0.36; 0.9, 0.9))$
Medium (M)	$((0.30, 0.50, 0.50, 0.70; 1, 1), (0.40, 0.50, 0.50, 0.60; 0.9, 0.9))$
Medium High (MH)	$((0.91, 0.75, 0.75, 0.51; 1, 1), (0.84, 0.75, 0.75, 0.64; 0.9, 0.9))$
High (H)	$((0.99, 0.94, 0.94, 0.76; 1, 1), (0.97, 0.94, 0.94, 0.87; 0.9, 0.9))$
Very High (VH)	$((1, 1, 1, 0.94; 1, 1), (1, 1, 1, 0.98; 0.9, 0.9))$

Step-II

For the group decision making, D_z decision matrix was constructed with z^{th} decision makers as follows:

$$D_z = \left(\tilde{\tilde{f}}s^z_{ij}\right)_{m\times n} = \begin{pmatrix} \tilde{\tilde{f}}s^z_{11} & \cdots & \tilde{\tilde{f}}s^z_{1n} \\ \vdots & \ddots & \vdots \\ \tilde{\tilde{f}}s^z_{m1} & \cdots & \tilde{\tilde{f}}s^z_{mn} \end{pmatrix} \quad (7)$$

The average decision matrix $\bar{D}_z = \left(\tilde{\tilde{f}}s_{ij}\right)_{m\times n}$

$$\tilde{\tilde{f}}s_{ij} = \left(\frac{\tilde{\tilde{f}}s^1_{ij} \oplus \tilde{\tilde{f}}s^2_{ij} \oplus \ldots \tilde{\tilde{f}}s^k_{ij}}{k}\right)$$

where $1 \le i \le m, 1 \le j \le n, 1 \le z \le k$ & k is the number of decision makers.

The work has been carried out with the help of professor in psychology (D1), marketing management (D2), and computer science engineering (D3) to rank the alternatives contributing towards purchase frequency in the online brand community. Their preference in linguistic variable has been explained in Table 1 for five attributes and corresponding alternatives (Table 2).

Weights of the attributes evaluated by decision makers are given in Table 3.

Table 2. Evaluating values of alternatives by the decision makers for different attributes

Attributes	Alternatives	Decision-makers		
		D_1	D_2	D_3
Online time	Social networking	VH	MH	VL
	Blog	M	L	VH
	Email, pay for click, etc.	H	VH	L
Freight	Social networking	H	VL	VH
	Blog	ML	H	M
	Email, pay for click, etc.	VL	H	L
Number of replies	Social networking	M	MH	H
	Blog	H	VL	ML
	Email, pay for click, etc.	ML	H	MH
Number of friends	Social networking	VH	VH	VH
	Blog	H	VH	M
	Email, pay for click, etc.	L	ML	VL
Number of postings	Social networking	H	M	L
	Blog	VH	L	M
	Email, pay for click, etc.	M	ML	H

Table 3. Attributes weight

Attributes	Decision Makers		
	D1	D2	D3
Online time	H	VH	H
Freight	MH	H	M
Number of replies	VH	M	VH
Number of friends	H	H	H
Number of postings	M	MH	H

Step-III

The weighted matrix Wt_z, with z^{th} decision makers is shown in Eq. (8)

$$Wt_z = \left(\tilde{\tilde{W}} t_i^z \right)_{1 \times m} = \left(\tilde{\tilde{W}} t_1^z \; \tilde{\tilde{W}} t_2^z \ldots \tilde{\tilde{W}} t_m^z \right) \tag{8}$$

Similarly the average weighted matrix is $\bar{W}t = \left(\tilde{\tilde{W}} t_i \right)_{1 \times m}$

Where

$$\tilde{\tilde{W}} t_i = \left(\frac{\tilde{\tilde{W}} t_i^1 \oplus \tilde{\tilde{W}} t_i^2 \oplus \ldots \tilde{\tilde{W}} t_i^k}{k} \right) \tag{9}$$

where $1 \leq i \leq m, 1 \leq z \leq k$ & k is the number of decision makers.

$\tilde{\tilde{W}}t_1$ = $((0.99,0.96,0.96,0.82;1,1)(0.98,0.96,0.96,0.91,0.9,0.9))$

$\tilde{\tilde{W}}t_2$ = $((0.73,0.73,0.73,0.66;1,1), (0.74,0.73,0.73,0.70,0.9,0.9))$

$\tilde{\tilde{W}}t_3$ = $((0.77,0.83,0.83,0.86;1,1), (0.80,0.83,0.83,0.86,0.9,0.9))$

$\tilde{\tilde{W}}t_4$ = $((0.99, 0.94, 0.94, 0.76; 1, 1), (0.97, 0.94, 0.94, 0.87, 0.9, 0.9))$

$\tilde{\tilde{W}}t_5$ = $((0.73, 0.73, 0.73, 0.66; 1, 1), (0.74, 0.73, 0.73, 0.70, 0.9, 0.9))$

Step-IV

The weighted decision matrix is shown in Eq. 10

$$WDM = \left(\sum_{i=1,f_i \in F_1}^{m} \left(\tilde{\tilde{W}}t_i \otimes \tilde{\tilde{f}}s_{ij}\right) \ominus \sum_{i=1,f_i \in F_2}^{m} \left(\tilde{\tilde{W}}t_i \otimes \tilde{\tilde{f}}s_{ij}\right)\right)$$

$$WDM = \left(\tilde{Y}_j\right)_{1 \times n} \text{ where } n = \text{number of decision makers}$$

(10)

The computed decision weight of all three decision makers are mentioned in Table 4:

Table 4. Weight decision matrix

$\tilde{Y}_1$	2.12	1.99	1.99	1.58	1	1	2.07	1.99	1.99	1.84	0.9	0.9
$\tilde{Y}_2$	1.36	1.56	1.56	1.57	1	1	1.46	1.56	1.56	1.61	0.9	0.9
$\tilde{Y}_3$	1.31	1.44	1.44	1.42	1	1	1.38	1.44	1.44	1.46	0.9	0.9

In this work F_1, known as set of *benefit* attributes, comprises of attributes 'online time', 'number of friends', 'number of replies', 'number of postings'. F_2, known as set of cost attributes consists of the only attributes 'freight'.

Step-V

As suggested in [24], the type-II fuzzy set $\tilde{\tilde{l}}_i = \left(\tilde{l}_i^U, \tilde{l}_i^L\right)$ has likelihood or preference matrix l^U.

$$l^U = \begin{pmatrix} l(\tilde{Y}_1^U \geq \tilde{Y}_1^U) \, l(\tilde{Y}_1^U \geq \tilde{Y}_2^U) & \cdots & l(\tilde{Y}_1^U \geq \tilde{Y}_n^U) \\ \vdots & \ddots & \vdots \\ l(\tilde{Y}_n^U \geq \tilde{Y}_1^U) \, l(\tilde{Y}_n^U \geq \tilde{Y}_2^U) & \cdots & l(\tilde{Y}_n^U \geq \tilde{Y}_n^U) \end{pmatrix}$$

(11)

Each element has been calculated in Eq. (12)

$$l(\tilde{Y}_1^U \geq \tilde{Y}_2^U) = \max(1 - \max(E_{21}, 0), 0)$$

$$\text{where } E_{21} = \frac{SD_{21}}{AD_{21}} = \frac{\sum_{k=1}^{4} \max(y_{2k}^U - y_{1k}^U, 0) + (y_{24}^U - y_{11}^U) + \sum_{k=1}^{2} \max\left(\left(H_k\left(\tilde{Y}_2^U\right) - H_k\left(\tilde{Y}_1^U\right)\right), 0\right)}{\sum_{k=1}^{4} \left|y_{2k}^U - y_{1k}^U\right| + (y_{14}^U - y_{11}^U) + (y_{24}^U - y_{11}^U) + \sum_{k=1}^{2} \left|H_k\left(\tilde{Y}_2^U\right) - H_k\left(\tilde{Y}_1^U\right)\right|}$$

(12)

Step-VI

The rank has been calculated for the upper value of type–II fuzzy sets $\tilde{Y}_i^U$ as shown in Eq. (13), similarly $\tilde{Y}_i^L$ can also be calculated.

$$Rank\left(\tilde{Y}_i^U\right) = \frac{1}{n(n-1)}\left(\sum_{k=1}^{n} l(\tilde{Y}_i^U \geq \tilde{Y}_k^U) + \frac{n}{2} - 1\right) \tag{13}$$

where $1 \leq i \leq n$ and $\sum_{i=1}^{n} Rank\left(\tilde{\tilde{Y}}_i^U\right) = 1$.

The final rank can be calculated as shown in Eq. (14)

$$Rank\left(\tilde{\tilde{Y}}\right) = \frac{Rank\left(\tilde{Y}_i^U\right) + Rank\left(\tilde{Y}_i^U\right)}{2} \tag{14}$$

where $1 \leq i \leq n$ and $\sum_{i=1}^{n} Rank\left(\tilde{\tilde{Y}}_i^U\right) = 1$.

Therefore, using (14), the rank values $Rank(\tilde{Y}_1)$, $Rank(\tilde{Y}_2)$, and $Rank(\tilde{Y}_3)$ are found to be 0.5833, 0.2557, and 0.1668 respectively. The above results along with the computation of the same using the existing methodology presented in [24] can be observed from Table 5.

Table 5. Comparative rank for online brand community

	Computed using proposed method	Computed using existing method [24]
$Rank(\tilde{Y}_1)$	0.5833	0.4250
$Rank(\tilde{Y}_2)$	0.2557	0.3045
$Rank(\tilde{Y}_3)$	0.1668	0.2732

The comparison of the rank computed for the online brand community using the proposed and the existing techniques are presented in Table 5. It can be observed from Table 5, $Rank(\tilde{Y}_1) \succ Rank(\tilde{Y}_2)$ for both the methods. However, the result in case of the proposed one, strongly validates the above relationship in comparison to the existing method based on the difference between the two defined alternatives. Similarly the proposed method can be a superior option in case of the relationship of $Rank(\tilde{Y}_2) \succ Rank(\tilde{Y}_3)$ and therefore incase of $Rank(\tilde{Y}_1) \succ Rank(\tilde{Y}_3)$. Since $Rank(\tilde{Y}_1) \succ Rank(\tilde{Y}_2) \succ Rank(\tilde{Y}_3)$, the preference order of the alternatives is social networking > blog > email and pay-per-click.

6   Conclusions

The prioritization for the alternatives in online brand community with the help of interval type-II fuzzy group decision has been successfully presented in this paper. As per the concurrent expert views, it has been found that "social networking" is superior medium amongst all considered alternatives for online brand community. Therefore, the proposed interval type-II fuzzy mechanism for the said purpose is more suitable in comparison to the existing methods used for the prioritization process.

References

1. McAlexander, J.H., Schouten, J.W., Koenig, H.F.: Building brand community. J. Market. **66**, 38–54 (2002)
2. Abrahams, A.S., Jiao, J., Wang, G.A., Fan, W.: Vehicle defect discovery from social media. Decis. Support Syst. **54**(1), 87–97 (2012)
3. Ahuja, M.K, Galvin, J.E.: Socialization in virtual groups. J. Manage. **29**(2), 161–185 (2003)
4. Cova, B., Pace, S.: Brand community of convenience products: new forms of customer empowerment- the case "my Nutella The community". Eur. J. Market. **40**(9/10), 1087–1105 (2006)
5. Bateman, P.J., Gray, P.H., Butler, B.S.: The impact of community commitment on participation in online communities. Inf. Syst. Res. **22**(4), 841–854 (2010)
6. Borle, S., Singh, S.S., Jain, D.C.: Customer lifetime value measurement. Manage. Sci. **54**(1), 100–112 (2007)
7. Goh, K.Y., Heng, C.S., Lin, Z.: Social media brand community and consumer behavior: quantifying the relative impact of user-and marketer-generated content. Inf. Syst. Res. **24**(1), 88–107 (2013)
8. Oestreicher-Singer, G., Zalmanson, L.: Content or community? A digital business strategy for content providers in the social age. MIS Q. **37**(2), 591–616 (2013)
9. Butler, B.S.: Membership size, communication activity, and sustainability: a resource based model of online social structures. Inf. Syst. Res. **12**(4), 346–362 (2001)
10. Rishika, R., Kumar, A., Jankiraman, R., Bezawada, R.: The effect of customers' social media participation on customer visit frequency and profitability: an empirical investigation. Inf. Syst. Res. **24**, 108–127 (2013)
11. Lin, H., Fan, W., Chau, P.Y.: Determinants of users' of social networking sites: a self regulation perspective. Inf. Manage. **51**(5), 595–603 (2014)
12. Zadeh, L.A.: Fuzzy sets. Inf. Control **8**, 338–353 (1965)
13. Mendel, J.M., John R.I., Liu, F.L.: Interval type-II fuzzy logical systems made simple. IEEE Trans. Fuzzy Syst. **14**(6), 808–821 (2006)
14. Wu, J., Huang, L., Zhao, J.L., Hua, Z.: The deeper, the better? Effect of online brand community activity on customer purchase frequency. Inf. Manage. **52**, 813–823 (2015)
15. Zhou, M., Lei, L., Wang, J., Fan, W., Wang, A.G.: Social media adoption and corporate disclosure. J. Inf. Syst. **28**(2), 23–50 (2015)
16. Sun, Y., Fang, Y., Lim, K.H.: Understanding sustained participation in transactional virtual communities. Decis. Support Syst. **53**(1), 12–22 (2012)
17. Wasko, M.M., Faraj, S.: Why should I share? Examining social capital and knowledge contribution in electronic networks of practice. MIS Q. **29**(1), 35–57 (2005)

18. Garrett, J., Gopalkrishna, S.: Customer value impact of sales contests. J. Acad. Market. Sci. **38**(6), 775–786 (2010)
19. Chang, J.R., Cheng, C.H., Kuo, C.Y.: Conceptual procedure for ranking fuzzy numbers based on adaptive two-dimensions dominance. Soft Comput. **10**(2), 94–103 (2006)
20. Chen, S.J., Chen, S.M.: A new method for handling multi criteria fuzzy decision making problems using FN-IOWA operators. Cybern. Syst. **34**(2), 109–137 (2003)
21. Fu, G.: Fuzzy optimization method for multi criteria decision making: an application to reservoir flood control operation. Expert Syst. Appl. **34**(1), 145–149 (2008)
22. Hua, Z., Gong, B., Xu, X.: A DS-AHP approach for multi-attribute decision making problem with incomplete information. Expert Syst. Appl. **34**(3), 2221–2227 (2008)
23. Chu, T.C.: Fuzzy logic control in decision making. In: International IEEE/IAS Conference on Industrial Automation and Control: Emerging Technologies, 1995, Taipei, pp. 441–451 (1995). doi:10.1109/IACET.1995.527601
24. Chen, S.M., Lee, L.W.: Fuzzy multi attributes group decision-making based on ranking values and the arithmetic operations of interval type-2 fuzzy sets. Expert Syst. Appl. **37**, 824–833 (2010)
25. Xu, Z.S.: A ranking arithmetic for fuzzy mutual complementary judgment matrices. J. Syst. Eng. **16**(4), 311–314 (2001)

User Localization in an Indoor Environment Using Fuzzy Hybrid of Particle Swarm Optimization & Gravitational Search Algorithm with Neural Networks

Jayant G. Rohra[1], Boominathan Perumal[1(✉)],
Swathi Jamjala Narayanan[1], Priya Thakur[1], and Rajen B. Bhatt[2]

[1] VIT Univerity, Vellore 632014, India
`jayant.rohra95@gmail.com`, `swathi.jns@gmail.com`,
`boominathan.p@vit.ac.in`, `pthakur93@yahoo.com`
[2] Robert Bosch Research Labs, Pittsburgh, USA
`rajen.bhatt@gmail.com`

Abstract. Detecting users in an indoor environment based on Wi-Fi signal strength has a wide domain of applications. This can be used for objectives like locating users in smart home systems, locating criminals in bounded regions, obtaining the count of users on an access point etc. The paper develops an optimized model that could be deployed in monitoring and tracking devices used for locating users based on the Wi-Fi signal strength they receive in their personal devices. Here, we procure data of signal strengths from various routers, map them to the user's location and consider this mapping as a classification problem. We train a neural network using the weights obtained by the proposed fuzzy hybrid of Particle Swarm Optimization & Gravitational Search Algorithm (FPSOGSA), an optimization strategy that results in better accuracy of the model.

Keywords: Neural networks · Optimization methods · PSO · GSA · PSOGSA · User localization · Wi-Fi signal strength · Fuzzy logic

1 Introduction

Advancements in location based services have enabled wide prospects in mobile computing. Many strategies have been adopted to provide users with custom locality based services. These strategies have shown a tremendous boom in e-commerce revenues, embedded smart systems, location based recommender systems and various other fields. Technologies like the GPS, Bluetooth and Wi-Fi could be exploited to provide such services. Bulusu et al. [1] used GPS methods for user localization, but these methods were used to achieve precision only in certain ranges and cannot be applied to indoor locations due to weak satellite signals. Bluetooth is another technology that can be used to serve this purpose, but it can only be well applied for short ranges. Thus, the user localization by using the Wi-Fi access points could be a better approach. Salazar et al. [2] introduced methods to predict the behavior of people by

© Springer Nature Singapore Pte Ltd. 2017
K. Deep et al. (eds.), *Proceedings of Sixth International Conference on Soft Computing for Problem Solving*, Advances in Intelligent Systems and Computing 546, DOI 10.1007/978-981-10-3322-3_27

monitoring their daily movements. Such location detection systems could also be used in panic situations and disasters, when people require necessary rehabilitation. Nguyen et al. [3] suggests recognition techniques for patients suffering from severe brain injuries who could be monitored by observing patterns in their movements. Pei et al. [4] proposed SVM techniques and showed better classification rates compared to other existing learning techniques. Cho [5] proposed learning methodologies to categorize the locality of indoor and outdoor regions using the location service logs of smart phones. Zou et al. [6] introduced an indoor localization mechanism based on extreme machine learning strategies and depicted its easy adaptation to versatile environments.

Zadeh [7] introduced the fuzzy set theory that has been widely adopted in many domains like real-time controllers, diagnostic systems etc. Real world data has various dimensions, much more than the classical logic of true or false. The fuzzy logic is used to correlate real life scenarios representing probabilities measuring the degree of truth in the range 0 to 1. Jang and Sun [8] proposed the interesting concept of modelling neural networks with fuzzy logic and parametrizing control. The neuro-fuzzy strategy alone would not be sufficient to attain the best throughput to the neural networks. The need for our problem lies to detect users at right locations using better learning techniques. But most of the techniques proposed lack the apt usage of optimization strategies that train the model rightly. We look into metaheuristic techniques that promise sufficiently good solutions to optimization strategies. Eberhart and Kennedy [9] introduced the Particle Swarm Optimization (PSO) strategy that considers a population of candidate solutions or particles moving around the search space and are updated to their *localBest* or *globalBest* computed using their position and velocity parameters. The standard PSO algorithm suffers from major problems like the ability to explore new search spaces. Shi and Russell [10] introduced an adaptive PSO approach that uses multiple benchmark functions to test the fuzzy system applied in various dimensions of the PSO. Liu and Abraham [11] proposed a fuzzy PSO that highlights the need to explore new search spaces by introducing a turbulence factor in the velocity component of the PSO. However, these algorithms lack the assurance of obtaining the global optimum. Mirjalili et al. [12] overcame this problem by proposing a hybrid, PSOGSA that introduces the ability of the GSA to escape the local optimum and hence improve the accuracy of the neural network. But, this algorithm lacks consistency and saturates at the lower iterations of the search, when the dimensions of the problem are increased. Nandy et al. [13] proposed a bee colony based back propagation approach to train ANN. These techniques thus improvise the fact that MLP based classifiers when trained with optimization approaches do give good performance accuracies. Kawam et al. [14] used the cuckoo swarm and PSO technique to train a MLP and hence depicted the need of using optimization strategies indeed enhances the performance of the neural network considerably. But, various such techniques adopted often lack proper convergence and guarantee that the complete population is explored.

Considering these factors, we propose the FPSOGSA that overcomes the possibilities of trapping itself in the local minima and enhances the probability of a higher convergence rate. At higher iterations, it gradually decreases the error rate rather than attaining saturation, as seen in PSOGSA. It obtains better convergence, enhances the ability of optimizing the neural network and hence reducing the mean square error of the Fuzzy Neural Network (FNN). Many such approaches have been used to train various classifiers. Artificial Neural Network (ANN) is a model easy to understand and use. More importantly it is nonlinear and non-parametric in nature. ANN is largely used to solve various classification and forecasting problems with the Back Propagation (BP) algorithm. However, the BP convergence is slow and not guaranteed. Therefore, we need to use optimization strategies to attain faster convergence and higher accuracy rates. Hence, we introduce the hybrid PSOGSA strategy as an optimization strategy here. On the other hand, the ANN is said to be a black box learning approach. It cannot deal with uncertainties. To overcome this, we introduce the fuzzy component. Fuzzy is quite good in handling uncertainties and can also interpret the relationship between the input and output by producing rules. Hence, we introduce the FPSOGSA algorithm.

2   User Localization as a Classification Problem

To predict the user's location accurately, a definite and consistent model has to be trained and deployed in a tracking or monitoring device. We measure the Wi-Fi signal strength received from various routers in a bounded location and train the neural network so that it could further predict the user's location for an unknown tuple set having signal strengths. Here, we consider a setup at an office location in Pittsburgh, USA. The office has seven Wi-Fi routers and its signal strengths received from these routers categorize the location of user in the conference room, kitchen or the indoor sports room. Sample data tabulated is shown in Table 1. WS1 corresponds to the signal strength received from the router 1, WS2 corresponds to the signal strength received from the router 2, and similarly for the other routers. The class labels corresponding to the conference room, kitchen and the indoor sport are labelled 1, 2 and 3 respectively. In our setup facility, we have considered an Android device and tabulated strengths of wireless signals captured by the device. At certain locations, the signal strengths were observed by polling the wireless signal strength at a constant time interval (every 1 s considered here). This was again repeated for other locations and suitable data was collected for one thousand and five hundred observations made at this facility for seven different routers. The model developed here, could hence be reused according to the scenario of the bounded location and the number of wireless routers in the physical facility. This data is being formulated into a pattern classification dataset by considering the seven wireless routers as the input dimensions which are used to predict the user's location in an office as one of the three dimensional categories. After having a concrete dataset ready, we now train the neural network using a metaheuristic approach that enhances the chances of classifying the right class label optimally. We discuss our approach of training the model using Fuzzy PSO GSA (FPSOGSA) in Sect. 3.

Table 1. Sample Data for user localization using wireless signal strength

WS1	WS2	WS3	WS4	WS5	WS6	WS7	Class
−64	−56	−61	−66	−71	−82	−81	1
−68	−57	−61	−65	−71	−85	−85	1
−17	−66	−61	−37	−68	−75	−77	2
−16	−70	−58	−14	−73	−71	−80	2
−52	−48	−56	−53	−62	−78	−81	3
-49	-55	-51	-49	-63	-81	-73	3

3   Evolution from the Conventional PSOGSA to FPSOGSA and Training the Neural Network with the Proposed Fuzzy-PSOGSA Algorithm

Mirjalili et al. [12] proposed the PSOGSA by introducing an exploitation capability to the standard PSO algorithm that increases the probability of finding the *globalBest* solution. The novel idea of using mass interactions among particles by including the gravitational search capability, proposed by Rashedi et al. [15] further enhanced the accuracy rates of the FNN. Later in this section, we introduce fuzzy decision parameters of the PSOGSA that decide the need for further exploration of the particle in the search space. Suitable thresholds are set to decide if the particle needs to explore further dimensions. This algorithm would hence fit the need of not missing out on the globalBest, as it gives more exploration ability to the particles.

We initially consider a space with 'N' particles that have randomly allocated positions that are referred to as the current positions (CurrPos) of the particles. The positions of each of these particles have "d" dimensions and a configuration of these positions is considered to be a candidate solution. The forces between the particles in each iteration, are calculated as,

$$F_{ij}^d(t) = G(t) \frac{M_{pi}(t) \times M_{aj}(t)}{R_{ij}(t)} \left(CurrPos_j^d(t) - CurrPos_i^d(t) \right) \qquad (1)$$

where M_{pi} and M_{aj} are passive and active gravitational masses of particles i and j respectively and R_{ij} is computed as the Euclidean distance between the two particles. The total force acting on any particle i is computed as the sum of the forces acting on every other particle in the space. The time variant gravitational constant, $G(t)$ is computed as,

$$G(t) = G_0 \times \exp\left(-k \times CurrentIteration/MaxIteration \right) \qquad (2)$$

where k is a descending co-efficient and G_0 is the initial gravitational constant value at t. The mass of each particle is related to the fitness value. It is updated at every epoch using the equation,

$$M_i(t) = \frac{CurrFit_i - best}{best - worst} \tag{3}$$

where *best* is the minimum fitness value for a minimization optimization problem and *worst* is the maximum fitness value. The acceleration of the particle is computed as follows:

$$a_i^d(t) = F_i^d(t)/M_i(t) \tag{4}$$

The weight function, W is calculated using:

$$W = W_{min} - CurrentIteration \times (W_{max} - W_{min})/MaxIteration \tag{5}$$

Here we initialize W_{min} and W_{max} as suitable minimum and maximum inertia weights.

Now, the velocity of the particle is updated by using the equation:

$$\left(Vel_i^d\right)_{t+1} = W \times \left(\exp loreVel_i^d\right)_t + rand()*a_i^d + rand()*(globalBest_j - CurrPos_i^d) \tag{6}$$

where rand is any number between the range [0,1] and the *globalBest* is the best solution obtained so far. The *exploreVel* is computed by using the fuzzy inference mechanism discussed in the section later.

Consider a neural network as shown in Fig. 1 with seven input nodes as the attributes of the user localization dataset and three output nodes as the class labels. The FPSOGSA trains the neural network by using the exploration and exploitation capabilities of the particles in the search space. As the PSO suffers saturation or slow convergence at the ending few iterations, the particles sometimes do not tend to come out of their constrained search space. This means that the mean square error (MSE) does not further decrease and hence there is very little or no change found in the accuracy of the neural network. Thus, in order to provide particles with an ability to explore new search spaces, we provide an extra velocity component, *exploreVel$_{ij}$* that is inferred from a Fuzzy Inference System (FIS). This enhances the search capability of the particles by exploring new dimensions in the search space and hence increasing the chances of obtaining a better *globalBest* solution. As discussed earlier, here we update the mass and acceleration of the particles before obtaining the explore velocity from the FIS. This is because the GSA component adds mass interactions that play a vital role in achieving the global optimum and also the fact that the acceleration component is used to update the velocity of the particle in the $(t + 1)^{th}$ iteration. The FIS takes in the Normalized Current Best Fitness Value (*NCBFV*) and the velocity of the particle (*Vel$_{ij}$*) as inputs and infers the scaling factor (*S$_f$*) and the velocity threshold control parameter (*V$_{tc}$*) as the output using the Fuzzy Rules discussed below in this section. The scaling factor, *S$_f$* is obtained as a result to prevent the particle from overshooting off its domain while getting extra exploration capability in the search space.

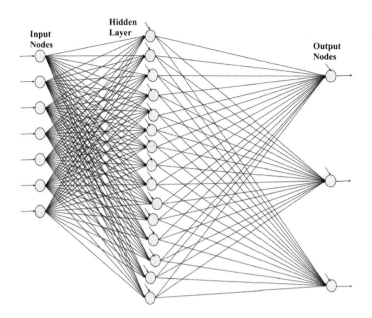

Fig. 1. Neural Network for classification of a dataset with 7 attributes and 3 class labels

The NCBFV is calculated as follows:

$$NCBFV_i = \frac{CurrFit_i - MinFit}{MaxFit - MinFit} \qquad (7)$$

where $CurrFit_i$ is the current fitness value of the particle, $MinFit$ is the least fitness value obtained by the particle till the current iteration and $MaxFit$ is the maximum fitness value obtained.

The threshold (θ) is calculated from the velocity threshold control parameter as follows:

$$\theta = e - [10(1 + V_{tc})] \qquad (8)$$

The Vel_{ij} is the latest velocity of the particle attained until the previous iteration. For the first iteration, the *exploreVelocity$_{ij}$* is considered to be the same as Vel_{ij}. For iterations after the first, the *exploreVelocity$_{ij}$* is obtained by checking for θ as follows:

$$\exp lore\, Velocity_{ij} = \left(\begin{matrix} Vel_{ij}, & Vel_{ij} \geq |\theta| \\ UDistb[-1,1] \times \max(Vel_{ij})/S_f, & Vel_{ij} \leq |\theta| \end{matrix} \right) \qquad (9)$$

where *UDistb[−1,1]* is an uniform distribution in the range [−1,1], max(*Vel$_{ij}$*) is the maximum value of the velocity obtained till now, θ is a threshold obtained from Eq. (8) and S_f is the scaling factor obtained as one of the results of the FIS. The given fuzzy inference rules are used to obtain the values of the velocity threshold control parameter

(V_{tc}) and the scaling factor (S_f), which determine the *exploreVelocity$_{ij}$* of the particle based on the threshold θ.

1. If (*NCBFV* is low) and (*Vel* is low) then (V_{tc} is high)
2. If (*NCBFV* is medium) then (V_{tc} is medium)
3. If (*NCBFV* is high) and (*Vel* is high) then (V_{tc} is low)
4. If (*NCBFV* is low) or (*Vel* is low) then (S_f is large)
5. If (*NCBFV* is medium) then (S_f is medium)
6. If (*NCBFV* is high) or (*Vel* is high) then (Sf is small)
7. If (*Vel* is high) then (V_{tc} is low) (S_f is medium)
8. If (*Vel* is low) then (V_{tc} is high) (S_f is medium)

The weight of each rule is assumed to be one. The fuzzy ranges are chosen suitably, for low/medium/high/small and large depending on the inputs parameters of the variables. Suitable triangular or Gaussian membership functions are used for fuzzification.

Finally, the position of the particle is updated to the next optimal location using:

$$(CurrPos_{ij})_{t+1} = (CurrPos_{ij})_t + (Vel_{ij})_t \tag{10}$$

3.1 Algorithm

1. Begin FPSOGSA
2. *Initialization* – Set a suitable number of iterations as *MaxIteration* to train the FNN.
    a. Initialize the dataset and normalize values in the range [-1, 1].
    b. Select a suitable number of Input, Output and Hidden nodes for the FNN depending on the dataset.
3. Obtain *Weights* to train the Neural Network (NN).
    a. Initialize randomly the weights and bias values.
    b. Choose the number of particles *(N)* and generate the initial population config-uration of particles.
    c. Compute the fitness values of each particle and store the best and worst fitness values.
       Computation– Updating and calculating the parameters of the particle in the search space.
    d. Update G using the Eq. (2) and compute the *globalBest* for each particle.
    e. Calculate the *mass*, *force* and the *acceleration* of each particle using the Eqs. (3), (1) and (4) respectively.
    f. Update the inertia weights using the Eq. (5).
       Fuzzification – to obtain the *exploreVel*
    g. Obtain and normalize the current best fitness value of the particle using the Eq. (7).

    h. Initialize the fuzzy inference system and infer the output variables, S_f and V_T using the rules defined above.

    i. Obtain the velocity threshold θ, using the Eq. (8) and compute the *exploreVel* using Eq. (9).

    j. Update the velocity of the particle and the new position using the Eqs. (6) and (10) respectively.

4. *Training* – Train the NN by passing the obtained weights
5. Obtain the mean square error of the FNN and compute the classification accuracy of the FNN.
6. Repeat the above process until *CurrentIteration = MaxIteration*.
7. End FPSOGSA.

4 Experimental Computational Results and Discussion

The inputs to the network model are the seven attributes of wireless signal strengths measured from the various routers. The outputs obtained are the class labels that classify users based on their locality. 15 hidden nodes are chosen for the neural network structure. The weights for the neural network are obtained from the optimization algorithms. The neural network shown in Fig. 1, is being trained with PSO, GSA, PSOGSA and the proposed FPSOGSA algorithms separately to obtain the initial weights required to train the neural network. These weights are further optimized over 300 iterations to obtain the best accuracy for the dataset. The classification accuracies of the neural networks after training with these algorithms for an evolution of 300 iterations are shown in Table 2. The proposed FPSOGSA boosts the performance of the neural network as it enhances the probability of exploring new search spaces and exploits the best particles so that they overcome the local minima. This is evident from the steep decrease in the Mean Square Error (MSE) values as shown in the Fig. 3. The figure also depicts the comparison in the decrease in mean square error values over three hundred iterations for the various other optimization strategies considered here.

Table 2. Classification Rates in (%)

PSO-NN	GSA-NN	PSOGSA-NN	FPSOGSA-NN	SVM	Naïve Bayes
64.66	77.53	83.28	95.16	92.68	90.47

We can clearly observe that there is very minimal error when the weights are obtained by FPSOGSA to train the neural network. Thus, the FPSOGSA is found to outperform the Particle Swarm Optimization (PSO), Gravitational Search Algorithm (GSA) and the hybrid PSOGSA (Fig. 2).

We also compare other models like the SVM and Naïve Bayes which are commonly used pattern classification approaches. However, we choose to use the neural network due to the concrete reasons explained towards the end of Sect. 1. Here, Fig. 3. shows the classification accuracies obtained using various algorithms for the dataset

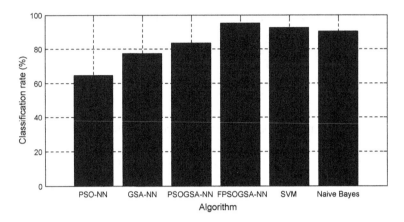

Fig. 2. Comparison of the classification accuracies of various models.

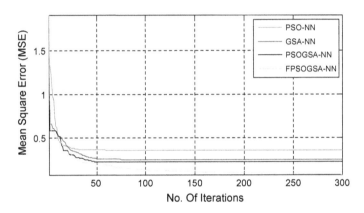

Fig. 3. Plot of MSE for user localization dataset considered for training with PSO, GSA, PSOGSA and FPSOGSA

procured. We performed tenfold cross validation and recorded the average of ten folds as the classification accuracy. From the results obtained we conclude that the FPSOGSA with neural networks give the highest classification rate.

References

1. Bulusu, N., Heidemann, J., Estrin, D.: GPS-less low-cost outdoor localization for very small devices. IEEE Personal Commun. **7**(5), 28–34 (2000)
2. Salazar, A.M., Warden, D.L., Schwab, K., Spector, J., Braverman, S., Walter, J., Ellenbogen, R.G.: Cognitive rehabilitation for traumatic brain injury a randomized trial. JAMA **283**(23), 3075–3081 (2000)

3. Nguyen, N.T., Bui, H.H., Venkatsh, S., West, G.: Recognizing and monitoring high-level behaviors in complex spatial environments. In: Proceedings of IEEE Computer Society Conference on Computer Vision and Pattern Recognition, vol. 2, II-620. IEEE (2003)
4. Pei, L., Liu, J., Guinness, R., Chen, Y., Kuusniemi, H., Chen, R.: Using LS-SVM based motion recognition for smartphone indoor wireless positioning. Sensors 12(5), 6155–6175 (2012)
5. Cho, S.B.: Exploiting machine learning Techniques for location recognition and prediction with smartphone Logs. Neurocomputing (2015)
6. Zou, H., Lu, X., Jiang, H., Xie, L.: A fast and precise indoor localization algorithm based on an online sequential extreme learning machine. Sensors 15(1), 1804–1824 (2015)
7. Zadeh, L.A.: Fuzzy sets. Inform. control 8(3), 338–353 (1965)
8. Jang, J.S., Sun, C.T.: Neuro-fuzzy modeling and control. Proc. IEEE 83(3), 378–406 (1995)
9. Eberhart, R.C., Kennedy, J.: A new optimizer using particle swarm theory. In: Proceedings of the Sixth International Symposium on Micro Machine and Human Science, vol. 1, pp. 39–43 (1995)
10. Shi, Y., Eberhart, R.C.: Fuzzy adaptive particle swarm optimization. In: Proceedings of the 2001 Congress on Evolutionary Computation, vol. 1, pp. 101–106. IEEE (2001)
11. Liu, H., Abraham, A., Zhang, W.: A fuzzy adaptive turbulent particle swarm optimisation. Int. J. Innovative Comput. Appl. 1(1), 39–47 (2007)
12. Mirjalili, S., Hashim, S.Z.M., Sardroudi, H.M.: Training feedforward neural networks using hybrid particle swarm optimization and gravitational search algorithm. Appl. Math. Comput. 218(22), 11125–11137 (2012)
13. Nandy, S., Sarkar, P.P., Das, A.: Training a feed-forward neural network with artificial bee colony based backpropagation method. arXiv preprint arXiv:1209.2548 (2012)
14. Kawam, A.A., Mansour, N.: Metaheuristic optimization algorithms for training artificial neural networks. Int. J. Comput. Inf. Technolgy 1, 156–161 (2012)
15. Rashedi, E., Nezamabadi-Pour, H., Saryazdi, S.: GSA: a gravitational search algorithm. Inform. Sci. 179(13), 2232–2248 (2009)

An Analysis of Decision Theoretic Kernalized Rough Intuitionistic Fuzzy C-Means

Ryan Serrao[1($\boxtimes$)], B.K. Tripathy[1], and A. Jayaram Reddy[2]

[1] SCOPE, VIT University, Vellore 632014, Tamil Nadu, India
{ryangodwin.serrao2013, tripathybk}@vit.ac.in
[2] SITE, VIT University, 632014 Vellore, Tamil Nadu, India
ajayaramreddy@vit.ac.in

Abstract. The decision theoretic rough set model was introduced in the 90's in order to loosen the restrictions of conventional rough approximations. Following this the conventional rough c-means was extended to the decision theoretic rough set context by Li et al. 2014. However, the Euclidean distance used as the similarity measure in this paper had the property of separability and to rectify this problem Kernel measures were used to develop the Kernel based decision theoretic rough C-means by Ryan et al. in 2016. As it is known, the hybrid models are more efficient than the individual models this approach was further extended and the Kernel based decision theoretic rough Fuzzy C-means was introduced by them recently in 2016. As a model of uncertainty intuitionistic fuzzy sets are more general than the fuzzy sets, So, we use the intuitionistic fuzzy sets instead of fuzzy sets and introduce a Kernel based Decision theoretic rough intuitionistic Fuzzy C-means in this paper. To provide variety and measure the effects, we have selected three of the most popular kernels; the Radial Basis, the Gaussian and the hyperbolic tangent kernels in our model. For the experimentation purpose we use three datasets namely the Iris, the wine and the glass data sets from the UCI repository. The efficiency measuring indices DB and D are used for evaluating the relative efficiencies of this algorithm and the other algorithms in this direction. Our results show that the proposed model provides improved results than the other two models. Some diagrams are presented to show the results visually.

Keywords: Rough sets · Clustering algorithms · Decision theoretic rough set · Similarity indices · Kernel measures

1 Introduction

The process of generating groups of similar objects (called clusters) from a given data set according to the criterion of similarity is termed as clustering and it plays an important role in data analysis. The early clustering algorithms are crisp by nature. It has been observed that uncertainty in data has become an inherent feature. This necessitated the development of uncertainty based algorithms like the fuzzy c-means

---

The original version of this chapter was revised: Incorrect author name has been corrected. The erratum to this chapter is available at DOI: 10.1007/978-981-10-3322-3_34

© Springer Nature Singapore Pte Ltd. 2017

K. Deep et al. (eds.), *Proceedings of Sixth International Conference on Soft Computing for Problem Solving*, Advances in Intelligent Systems and Computing 546, DOI 10.1007/978-981-10-3322-3_28

[2], the intuitionistic fuzzy c-means [4], the rough c-means [9] and the hybrid algorithms like the rough fuzzy c-means [10, 11] and the rough intuitionistic fuzzy c-means [3]. A probabilistic model called the decision theoretic rough set (DTRS) was introduced with the aim of relaxing the restrictions on the approximations in the conventional rough set model. The DTRS model can derive several important rough set models by using proper lost functions. Using this model a c-means algorithm was introduced recently [8] and it has been found to be more efficient than the normal rough c-mean. However, the dependence of the similarity measure in this algorithm on Euclidean distance is prone to separability in data sets. So as a remedy to this, the Euclidean distance is replaced by Kernels for measuring similarity and a Kernel based Decision Theoretic Rough C Means were proposed and analyzed [14]. However, the hybrid models are more efficient than the individual models and so the algorithm has been extended to propose and study the Kernel based Decision Theoretic Rough Fuzzy c-means algorithm, still very recently in [15]. However, it is well known that as an imprecise model, the intuitionistic fuzzy sets [4] is more general and efficient than the fuzzy sets. So, in this paper we propose a Kernel based Rough Intuitionistic Fuzzy c-means algorithm. Here, we take three kernels; namely the Gaussian, the hyper tangent and the radial basis function (RBF) for our study and compare their efficiencies. For the experimental purpose, we use three data sets of different characteristics, from the UCI repository, namely the iris data set, the wine data set and the glass data set. Measuring indices; the DB index and the D index are computed for all the three data sets and four algorithms, the Decision Theoretic Rough Intuitionistic Fuzzy C-Means (DTRIFCM), the Decision Theoretic Kernalized Rough Intuitionistic Fuzzy C-Means (DTKRIFCM) with kernels Gauss, RBF and Hyper Tangent. The results are presented in the tabular form (Table 1) and in graphical forms (Figs. 1 and 2) for visual comparison. The further structure of the paper is that we present definitions and notations in Sect. 2. In Sect. 3 we present the DTKRIFCM algorithm. The experimental set up and results form the contents of section followed by concluding remarks in Sect. 5. Finally the paper ends with a compilation of papers and other sources referred during preparation of this paper.

2   Definitions and Notations

Some definitions and notations to be used in this article are presented in this section. Pawlak [8] introduced the rough set model in 1982 where he used equivalence relations to generate two approximations called the lower and upper approximations of the set with respect to the equivalence relation. As shown by Dubois and Prade [3] the notion of rough sets and the earlier introduced notion of fuzzy sets by Zadeh [16] are complementary in nature.

2.1   Rough Sets and Uncertainty Based Models

Extending the notion of fuzzy sets, intuitionistic fuzzy sets were introduced by Atanassov as follows:

Definition 2.1.1: Given an universal set Y, an intuitionistic fuzzy set B on Y is characterised by the functions. m_B and n_B, given by, m_B, $n_B : Y \rightarrow [0,1]$. The hesitation function associated with B is denoted by π_B which is the one's complement of $(m_B + n_B)$.

We denote by U and R, a universe discourse and an equivalence relation over U respectively. For any $x \in U$ we denote its equivalence class by $[x]_R$.

Definition 2.1.2: Given any $X \subseteq U$, we associate two crisp sets $\underline{R}X$ and $\bar{R}X$ called the lower and upper approximation of X with respect to R defined as

$$\underline{R}X = \{x \in U | [x]_R \subseteq X\} \text{ and } \bar{R}X = \{x \in U | [x]_R \cap X \neq \phi\},$$

The R-boundarv $BN_R(X)$ of X is given by $BN_R(X) = \bar{R}X - \underline{R}X$. We say that X is rough with respect to R if and only if $\underline{R}X \neq \bar{R}X$, equivalentlv $BN_R(X) \neq \phi$. X is said to be R-defmable if and only if $\underline{R}X \neq \bar{R}X$, or $BN_R = \phi$.

The properties following of the properties of Rough sets are to be used in decision theoretic rough set model [14, 15]. RCM [5] does not verify all of these properties but only uses these properties to assign data points.

$$\underline{R}(X_i) \subseteq \bar{R}(X_i) \subseteq U, \, \forall X_i \subseteq U \tag{2.1}$$

$$\underline{R}(X_i) \cap \underline{R}(X_j) = \phi, \, \forall X_i, X_j \subseteq U, \, i \neq j \tag{2.2}$$

$$\underline{R}(X_i) \cap \bar{R}(X_j) = \phi, \, \forall X_i, X_j \subseteq U, \, i \neq j \tag{2.3}$$

If an object $x \in U$ is not a part of any lower approximations then it must belong to the boundary areas of two or more clusters and hence belongs to the upper approximations of these clusters.

2.2   DTRS Model

The DTRS model applies the Bayesian decision procedure for the construction of probabilistic approximations [17]. Let $\Omega = \{\omega_1, \omega_2, \ldots, \omega_s\}$ be a finite set of states and $A = \{a_1, a_2, \ldots a_m\}$ be a finite set of possible actions. Let $\lambda(a_i | \omega_j)$ denote the loss (or cost) for taking action a_i when the state is ω_j. Let $P(\omega_j | x)$ be the conditional probability of an object being in state ω_j, supposing that the object is described by x. The expected loss associated with taking the action a_i is given by

$$R(a_i | x) = \sum_{j=1}^{s} \lambda(a_i | \omega_j) P(\omega_j | x)$$

In DTRS model. $\Omega = \{A, A^c\}$ denotes the set of states indicating that indicating that an object is in A and not in A, respectively. Let $A = \{a_1, a_2, a_3\}$ be the set of actions, where a_1, a_2 and a_3 represent the three actions in classifying an object, deciding POS (A). deciding NEG(A) and deciding BND(A) respectively. Here POS $= \underline{R}(A)$, NEG $= U - \bar{R}(A)$, BND(A) $= BN_R(A)$.

The probabilities $P(A|[x])$ and $P(A^C|[x])$ are the probabilities that an object in its equivalence class belongs to A or A^C respectively. The expected loss $R(a_i|[x])$ associated with taking the corresponding actions can be expressed as:

$$R(a_i|[x]) = \lambda_{i1} P(A|[x]) + \lambda_{i2} P(A^C|[x]), \; i = 1, 2, 3.$$

The DTRCM and DTRFCM algorithms were proposed and studied in [14, 15]. Using Kernels instead of the Euclidean distance the DTKRCM and DTKRFCM algorithms were introduced and studied in [10, 11]. Making use of hybridization of DTRS model and Intuitionistic Fuzzy sets, the DTRIFCM algorithm was introduced by Sresht et al. [12, 13]. In this paper, we carry the study forward by introducing the hybrid Kernelised algorithm DTKRIFCM. For the presentation we require the following additional concepts and notations. It may be noted that some hybrid algorithms like the rough fuzzy c-means have been introduced in literature [6, 7].

2.3   Calculating Risk

The lower approximation $L(x_l)$ of $x_l \in X$ is defind as:

$$L(x_i) = \{x \in X : D(x, x_i) \leq \delta \wedge x \neq x_i\}, \text{ where} \tag{2.3.1}$$

$$\delta = \left\{ \min_{1 \leq k \leq c} D(x_l, v_k)/p \right\} \tag{2.3.2}$$

Let the conditional probability of x_l in C_i be as

$$P(C_i|x_l) = 1 / \sum_{j=1}^{c} \left[D(x_l, v_i)/D(x_l, v_j) \right]^{2/(m-1)} \tag{2.3.3}$$

Non-membership values are calculated $v_A(x)$. 't' is genarally taken as 2.

$$v_A(x) = \frac{1 - P(C_i|x_l)}{1 + t.P(C_i|x_l)}, t > 0 \tag{2.3.4}$$

We derive hesitation degree as

$$\pi_i(x_i) = 1 - P(C_i|x_l) \frac{1 - P(C_i|x_l)}{1 + t.P(C_i|x_l)}, x \in X \tag{2.3.5}$$

Modifying the fuzzy membership value as $P_{new}(C_i|x_l)$, where,

$$P_{new}(C_i|x_l) = P(C_i|x_l) + \pi_i(x_l) \tag{2.3.6}$$

T_x represents the group of clusters similar to each individual data point x.

$$T_x = \{C_i \in C : P_{new}(C_i|x) > 1/c\} \tag{2.3.7}$$

The action set is defined as $A = \{a_1, a_2, \ldots, a_c\}$. where represents allocating a data point to C_j. The loss related with taking the action a_j for x_1 when x_1 belongs to C_i is represented by $\lambda_{xl}(a_j|C_i)$ and defined as:

$$\lambda_{xl}(a_j|C_i) = \lambda_{C_i}^{a_j}(x_l) + \sum_{x \in L(x_i)} \beta(x)\lambda^{a_j}(x), \text{ where} \tag{2.3.8}$$

$$\lambda_{C_i}^{a_j}(x_l) = \begin{cases} 0, & \text{if } i = j; \\ 1, & \text{if } i \neq j. \end{cases} \tag{2.3.9}$$

$$\lambda^{a_j}(x) = \begin{cases} \{|a_j - T_x|/a_j\} = 0, \text{ if } a_j \in T_x; \\ \{|a_j - T_x|/a_j\} = 1, \text{ if } a_j \notin T_x. \end{cases} \tag{2.3.10}$$

$$\beta(x) = \exp\left[-d^2(x, x_l)/2\sigma^2\right] \tag{2.3.11}$$

The risk related with taking action a_j for x_l is represented by $R(a_j/x_i)$ and is defined as

$$R(b_j/x_l) = \sum_{i=1}^{k} \lambda_{x_i}(b_j/c_i)P_{new}(c_i/x_l) \tag{2.3.12}$$

For each data point x_l let $a_k = \arg\min_{a_i \in A}\{R(a_i/x_l)\}$. The index J_D is a measure of closeness of a_k through risk value and is given by

$$J_D = \{j|\{R(a_j|x_l)/R(a_k|x_l)\} \leq 1 + \varepsilon \wedge j \neq k\} \tag{2.3.13}$$

If $J_D = \phi$, x_l is assigned to C_k. Otherwise $\forall j \in J_D, x_l \in b_n(C_j)$.

2.3.1   DTRS

Let the classes of a classification π of U be denoted as $\pi = \{A_1, A_2, \ldots A_m\}$. The two approximations are given by:

$$\underline{apr_{(\alpha,\beta)}}(A_i) = POS_{(\alpha,\beta)}(A_i) = \{x \in U|P(A|[x]) \geq \alpha\},$$
$$\overline{apr_{(\alpha,\beta)}}(A_i) = POS_{(\alpha,\beta)}(A_i) \cup BND_{(\alpha,\beta)} = \{x \in U|P(A|[x]) \geq \beta\}$$

The approximations of a partition π in terms of the approximations of $A_i, i = 1, 2, \ldots m$ are defined as follows:

$$POS_{(\alpha,\beta)}(\pi) = \bigcup_{1 \leq i \leq m} POS_{(\alpha,\beta)}(A_i), BND_{(\alpha,\beta)}(\pi) = \bigcup_{1 \leq i \leq m} BND_{(\alpha,\beta)}(A_i), NEG_{(\alpha,\beta)}(\pi)$$

$$= U - POS_{(\alpha,\beta)}(\pi) \cup BND_{(\alpha,\beta)}(\pi)$$

The three regions defined above may not be mutually|exclusive but together they form a covering for U.

2.4   Similarity Metrics

Similarity between two data points can be calculated using several measures. Though one of the easiest and popular approaches to do this is by calculating Euclidean distance, it has several limitations. Firstly, the eventual outcomes are dependent upon the cluster centroids assigned in the beginning and secondly it is only able to segregate the data points which are linearly separable. Solution for the second limitation is provided by kernel based clustering approach which can create non-linear boundaries to segregate the data points successfully. This is ensured by converting the data present in the ordinary plane to feature plane. The feature plane is higher dimensional plane and is known as kernel space. To ensure this kind of transformation some non-linear mapping functions can be used. In this subsection we discuss on the similarity measures.

Definition 2.4.1: (Euclidean distance). Suppose $a = (a1, a2, \ldots, an)$ and $b = (b1, b2, \ldots, bn)$ are two points in the n-dimensional Euclidean space. Then the Euclidean distance d(a, b) between a and b is given by:

$$d(a, b) = \sqrt{(a_1 - b_1)^2 + (a_2 - b_2)^2 + \ldots + (a_n - b_n)^2}$$

Definition 2.4.2: (Kernel distance) [13]. Let 'a' denote a data point. Then transformation of 'a' to the feature plane which possess higher dimensionality be denoted by $\Phi(a)$. Description of inner product space is given by $K(a, b) = <\Phi(a), \Phi(a)>$. Let $a = (a1, a2, \ldots, an)$ and $b = (b1, b2, \ldots, bn)$ are two points in the n-dimensional space.

Kernel functions use in this paper are stated as follows:

(a) Radial basis kernel:

$$R(a, b) = \exp\left[-\sum_{i=1}^{n} (a_i^p - b_i^p)^q / 2\sigma^2\right]$$

Implementations of all the algorithms corresponding to radial basis kernel have been done using p = 2 and q = 2.

(b) Gaussian kernel: (RBF with p = 1 and q = 2)

$$G(a,b) = \exp\left[-\left\{\sum_{i=1}^{n} (a_i - b_i)^2\right\}/2\sigma^2\right]$$

(c) Hyper tangent kernel

$$H(a,b) = 1 - \tanh\left[-\left\{\sum_{i=1}^{n} (a_i - b_i)^2\right\}/2\sigma^2\right],$$

Where $\sigma^2 = \frac{1}{N}\sum_{i=1}^{N} \|a_i - a'\|^2$ and $a' = \frac{1}{N}\sum_{i=1}^{N} a_i$.

For all the kernels functions, N denotes the total number of existing data points and $\|x - y\|$ denotes the Euclidean distance between points x and v which pertain to Euclidean metric space. By [17].

D(a, b) denotes the complete form of kernel distance function where D(a, b) = K(a, a) + K(b, b) − 2K(a, b) and when similarity property (i.e. K(a, a) = 1) is applied, we get D(a, b) = 2(1 − K(a, b))

3   DTKRIFCM

In this section we present our proposed algorithm Decision Theoretic Kernelised Rough Intuitionistic Fuzzy C-Means.

3.1   Basic Idea

In this algorithm, we modify the Decision theoretic Rough Intuitionistic Fuzzy C means algorithm by using various different distance metrics such as the use of the Gaussian kernel, Radial Basis kernel and the hyper tangent kernel.

3.2   Algorithm Description

Input: The given data set $X = \{x_1, x_2, \ldots x_n\}$, c, w_ℓ, ε, p, σ, m, S_{max}
Output: Clustering result: $(\underline{C_1}, \overline{C_1}), \cdots, (\underline{C_c}, \overline{C_c})$ and cluster centroids

1 randomly assign the initial centroid v_i for C_i, i = 1, 2 ... c;
2 repeat
3   for i ← 1 to n do

4   for a data point $x_i \in X$, calculate

P_{new} (C j $| x_i$), j = 1, ..., c, by using (2.3.3) to (2.3.6);

5 determine x_i's neighbouring points set L (x_i) by using Eq. (2.3.1);

6 for every data point x ∈ L(x_i), determine T x by using Eq. (2.3.7);

7 calculate $R(a_j | x_i)$, j = 1, ..., c, by using Eq. (2.3.6) to Eq. (2.3.12);

8 find the action with minimal risk. a_h = argmin $a_j \in A$ { $R(a_j | x_i)$ };

Compute u_{ik} = P_{new} (C j $| x_i$) and $u_{ik} x_i$ for the cluster h and the data point x_i.

9 assign x_i to $\overline{C_h}$, i.e. $x_i \in \overline{C_h}$;

10 find the index set J_D with respect to ah by using Eq. (2.3.13);

11   if J_D = Ø then

12   assign x_i to $\underline{C_h}$, i.e. $x_i \in \underline{C_h}$;

13   else
14   assign x_i to the upper approximations of the clusters determined by J_D,

i.e. $x_i \in \overline{C_j}$ $\forall j \in J_D$;

15   end
16   end
17   calculate the new centroid for each cluster
using

(3.2.2)

$$
v_i = \begin{cases} w_{low} \dfrac{\sum_{x_k \in \underline{C_i}} u_{ik}^m x_k}{\sum_{x_k \in \underline{C_i}} u_{ik}^m} + w_{up} \dfrac{\sum_{x_k \in \overline{C_i} - \underline{C_i}} u_{ik}^m x_k}{\sum_{x_k \in \overline{C_i} - \underline{C_i}} u_{ik}^m}, & \text{if } \underline{C_i} \neq \phi \wedge (\overline{C_i} - \underline{C_i}) \neq \phi; \\[4ex] \dfrac{\sum_{x_k \in \overline{C_i} - \underline{C_i}} u_{ik}^m x_k}{\sum_{x_k \in \overline{C_i} - \underline{C_i}} u_{ik}^m}, & \text{if } \underline{C_i} = \phi \wedge (\overline{C_i} - \underline{C_i}) \neq \phi; \\[4ex] \dfrac{\sum_{x_k \in \underline{C_i}} u_{ik}^m x_k}{\sum_{x_k \in \underline{C_i}} u_{ik}^m}, & \text{Otherwise.} \end{cases}
$$

18 until the termination criterion are met;

4 Experimental Results

In this section we present the experimental results performed by us. Both the algorithms have been implemented using the python programming language in the Canopy
software interface. Three popular datasets namely Iris, Wine and Glass have been taken
from the UCI Repository [1] for experimental purposes.

First we show the DB and D index values obtained for the three datasets using
various kernels as Gauss, Radial basis and Hyper-tangent kernel. Further we represent
the clustering results of the following algorithms on the Iris dataset following with the
DB/D index variation. The DTKRIFCM performs better than the existing DTRIFCM
shown by the lower DB value [5] and higher D index [7].

Table 1. DB and D Values for different Algorithms with respect to different Datasets

	Iris Dataset		Wine Dataset		Glass Dataset	
	DB	D	DB	D	DB	D
DTRIFCM	0.3879	0.5199	1.2087	0.2820	0.9848	0.1701
DTKRIFCM (GAUSS)	0.3665	0.5313	1.0121	0.3231	0.9499	0.1896
DTKRIFCM (RBF)	0.3189	0.5796	1.1667	0.4166	0.8623	0.2179
DTKRIFCM (Hyper Tangent)	0.3545	0.5401	1.1898	0,3713	0.9378	0.1983

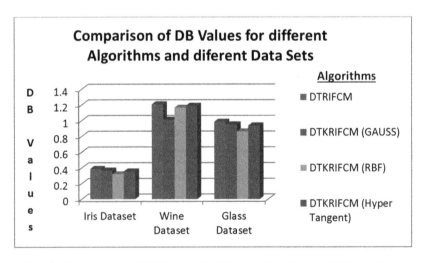

Fig. 1. Comparision of DB Values for different Algorithms and different Data

Table 1 displays the results of our experiment conducted by taking the iris dataset,
wine dataset and the glass data set to the four algorithms DTRIFCM and the three
kernelised versions of the DTKRIFCM with the three different kernels; the Gaussian
kernel, the RBF kernel and the hyper tangent kernel. We computed the DB and D
indices for all these algorithms for each of the data sets. It can be observed that the DB

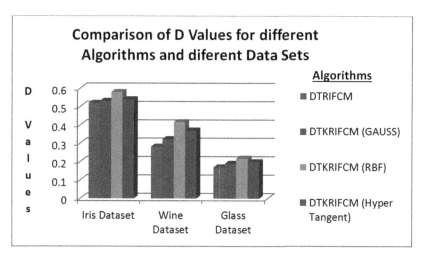

Fig. 2. Comparison of D Values for different Algorithms different Data Sets

values for all the kernelised DTKRIFCM algorithms are lower than that of DTRIFCM. However, among the kernelised versions the RBF kernel provides the best value. For the wine data set again all the kernelised versions provide lower values than the original DTRIFCM. But, here the Gaussian kernel provides the best value. For the glass data set the kernelised versions provide better DB values than the original algorithm. But, the RBF kernel produces the best value. So, we conclude that the kernelised versions of DTRIFCM are superior to the original algorithm. But among the kernels the hyper tangent kernel is the most inferior one. The RBF and Gaussian share the honours as the best for different data sets.

The D values computation shows that the Kernelised versions provide higher values in comparison to the original algorithm. Among the kernelised versions the RBF kernel provides the highest value in all cases. So, the DTKRIFCM algorithm with RBF kernel is the best for D values. We can conclude that it is the best for all data sets taking both the DB and D values into consideration as a whole.

The graphical representations are provided in Figs. 1 and 2 for visual support of the conclusions drawn above.

5 Conclusions

In this paper a modified version of the DTRIFCM is introduced namely DTKRIFCM. We have tested the new algorithm with existing popular datasets IRIS, WINE and GLASS from the UCI Repository. The two algorithms are compared based on its performance on the DB/D indexes. The DTKRIFCM performs better than the DTRIFCM. Results obtained are shown in the tabular format for evaluation and comparison. Also graphical representation of the clustered data is displayed along with the DB/D index variation for further understanding. Also, the RBF kernel outperforms all the other kernels in most cases and gives excellent clustering result.

References

1. Blake, C.L., Merz, C.J: UCI repository of machine learning databases (1998). http://www. ics.uci.edu/mlearn/mlrepository.html
2. Davis, D.L., Bouldin, D.W.: Clusters separation measure. IEEE Trans. Pattern Anal. Mach. Intell. **Pami-1**(2), 224–227 (1979)
3. Dubois, D., Prade, H.: Rough fuzzy set model. Int. J. General Syst. **46**(1), 191–208 (1990)
4. Dunn, J.C.: Fuzzy relative of ISODATA process and its use in detecting compact well-separated clusters. J. Cybern. **3**, 32–57 (1974)
5. Lingras, P., West, C.: Interval set clustering of web users with rough k-means. J. Intell. Inform. Syst. **23**(1), 5–16 (2004)
6. Maji, P., Pal, S.K.: A hybrid clustering algorithm using rough and fuzzy sets. Fundamenta Informaticae **80**(4), 475–496 (2007)
7. Mitra, S., Banka, H., Pedrycz, W.: Rough- fuzzy collaborative clustering. IEEE Trans. Syst. Man Cybern. **36**(4), 795–805 (2006)
8. Pawlak, Z.: Rough Sets: Theoretical Aspects of Reasoning About Data. Kluwer academic publishers (1991)
9. Peters, G., Crespo, F., Lingras, P., Weber, R.: Soft clustering-fuzzy and rough approaches and their extensions and derivatives. Int. J. Approximate Reasoning **54**(2), 307–322 (2013)
10. Ryan, S., Tripathy, B.K., Jayaram Reddy, J.: An analysis of decision theoretic kernalized rough c-means. In: Proceedings of the ICPCIT-2016, SRM University (2016)
11. Ryan, S., Tripathy, B.K., Jayaram Reddy, J.: An analysis of decision theoretic kernalized rough fuzzy c-means. In: ICT4SD-2016 (2016). Submitted to the International Springer Conference
12. Sresht, A., Tripathy, B.K.: A decision theoretic rough fuzzy c-means algorithm. In: Proceedings of the ICRCICN-2015 Conference (2015)
13. Agrawal, S., Tripathy, B.,K.: Decision theoretic rough intuitionistic fuzzy c-means algorithm. In: Satapathy, S.C.C., Das, S. (eds.) Proceedings of First International Conference on Information and Communication Technology for Intelligent Systems. SIST, vol. 50, pp. 71–82. Springer, Heidelberg (2016). doi:10.1007/978-3-319-30933-0_8
14. Yao, Y.: Decision-theoretic rough set models. In: Yao, J., Lingras, P., Wu, W.-Z., Szczuka, M., Cercone, Nick, J., Ślęzak, D. (eds.) RSKT 2007. LNCS (LNAI), vol. 4481, pp. 1–12. Springer, Heidelberg (2007). doi:10.1007/978-3-540-72458-2_1
15. Yao, Y.Y., Wong, S.K.M.: A decision theoretic framework for approximating concepts. Int. J. Man-Mach. Stud. **37**(6), 793–809 (1992)
16. Zadeh, L.A.: Fuzzy sets. Inform. Control **8**, 338–353 (1965)
17. Zeng, H., Ming, C.Y.: Feature selection and kernel learning for local learning based clustering. IEEE Trans. Pattern Anal. Mach. Intell. **33**(8), 1352–1547 (2011)

Analysis of Fuzzy Controller for H-bridge Flying Capacitor Multilevel Converter

P. Ponnambalam[1(✉)], K. Aroul[1], P. Prasad Reddy[2],
and K. Muralikumar[2]

[1] Faculty of School of Electrical Engineering, VIT University,
Vellore, Tamilnadu, India
p.ponnambalam@gmail.com, aroul.k@vit.ac.in
[2] School of Electrical Engineering, VIT University, Vellore, Tamilnadu, India
prasadreddyvit@gmail.com, kolamuralikumar@gmail.com

Abstract. The Normal voltage balancing property of capacitor voltage has an attractive choice of an flying capacitor converters. The study of H-bridge flying capacitor converter with 5, 7 and 9 levels are carried out and to increase the voltage levels falls the Total Harmonic Distortion. By using phase shifted carrier pulse width modulation technique (PSCPWM) the root mean square output voltage is controlled. For effective output voltage control, fuzzy controller is used. Fuzzy controllers are designed for the different voltage levels of H-bridge flying capacitor multilevel converter to control the Root Mean Square (RMS) output voltage. The Performances of the fuzzy controllers with multilevel converters of H-bridge flying capacitor for different levels are studied.

Keywords: Fuzzy logic · Fuzzy system · Flying capacitor · Phase shifted carrier pulse width modulation technique · Total harmonic distortion · Voltage balancing

1 Introduction

In Latest years it has been signified the development in multilevel inverter, more than in the medium voltage drives. Compared to the diode clamped [13] and H-bridge inverters [14–16], the H-bridge flying capacitor multilevel inverter (HBFCMLI) is the new topology [1–5]. Through the switching state selection, the flying capacitor has some distinct advantages over the absence of diodes in diode clamped and regulates flying capacitor voltage. In many industrial applications, the flying capacitors are used due to those advantages. The HBFCMLI converter, a capacitor is not an issue if voltage balancing, It has some probable changes in the voltage ratio and reduction in order to increase the voltage levels by improving the power quality [3].

The investigation of flying capacitor multilevel converter with three different voltage levels, due to the THD, the three different voltage levels are balancing of Flying Capacitor converter, through these three different voltage levels it is an important property that provides safety and effective operation [6–8]. By differentiating with single leg operation [4], high system order twice increased in H-bridge FC converter and it obtains from the single leg prototype by using mirror speculation formalism. As

© Springer Nature Singapore Pte Ltd. 2017
K. Deep et al. (eds.), *Proceedings of Sixth International Conference on Soft Computing for Problem Solving*, Advances in Intelligent Systems and Computing 546, DOI 10.1007/978-981-10-3322-3_29

behaviour of load current, modulation index, pulse proportion and load frequency, it has dynamic variation of voltage balancing. The characteristics have dealt with in this paper [9, 10].

The performance of H-bridge flying capacitor has the open loop analysing of 5, 7 and 9 level configurations, then the total harmonic distortion (THD) of those levels are compared in MATLAB simulation, the closed loop analysis of FC inverter is performed by using the fuzzy logic programs.

2 H-bridge Flying Capacitor Multilevel Inverter Analysis

In this segment the 5-level, 7-level and 9-level H-bridge Flying Capacitor Multilevel Converters are investigated through the simulation of circuits using MATLAB Simulink. Based on pulse width modulation technique for the three configurations, phase shifted carrier is used to generate the triggering pulse due to the converter circuits. This phase shifted carrier based PWM is a hybrid of phase shifted and level shifted pulse width modulation.

2.1 Five Level HBFCMLI

The 5-level H-bridge Flying capacitor (HBFC) topology is as shown in Fig. 1. The configuration of H-bridge flying capacitor multilevel inverter has positive and negative group, due to each cell, it has two complementary switches $(S_{p1} \& S_{p1}')$ for cell $P_1, S_{p2} \& S_{p2}'$ for cell $P_2, S_{p3} \& S_{p3}'$ for cell $P_3, S_{p4} \& S_{p4}'$ for P_4, S_{N1} and S_{N1}' for cell N_1, S_{N2} and S_{N2}' for cell N_2, S_{N3} and S_{N3}' for N_3, S_{N4} and S_{N4}' for N_4. For giving out the positive waveform across the load will conduct the positive group of cell, then the no. of voltage levels increases with increase in no. of cells. Due to the phase shifted carrier pulse width modulation technique, it is used to control the complementary switches such as $(S_1, S_1'), (S_2, S_2'), (S_3, S_3'), (S_4, S_4')$. By using the mirror speculation formalism, the H-bridge converter obtained from its single leg prototype. The circuit has 16 switching states. The capacitors $C_1 \& C_2$ are flying capacitors and their voltages $Vc_1 \& Vc_2$ respectively. Capacitor voltages are equal $Vc_1 = Vc_2$. The capacitor voltages which are controlled to be regulates at $-Vdc/4$, $Vdc/2$ and $Vdc/4$ correspondingly.

In the above Table 1, 5-level H-bridge flying capacitor MLI shows the 5-level output voltage for different switching states and complete switching scheme. For five level HBFCMLI is explained, in the above modes of operation. When there is an increase in the no. of switches in positive group, these should be a consecutive increase in the negative group also, by maintaining the equal no. of levels both in positive and negative. When there is increase in the output voltage levels and there will be increasing in no. of capacitors. The simulation results for output voltage of H-bridge flying capacitor at 5-level inverters is as shown in Fig. 2. From the figure, it is noticed that the 5-level output formed. The output voltage on y-axis and time period on x-axis is shown in Fig. 2. By applying Fast Fourier Transform (FFT) analysis to the five level HFCMLI as shown in Fig. 3. From the analysis, the total harmonic distortion for 5-level flying capacitor converter is 38.33% can be settled. THD value can be reduced

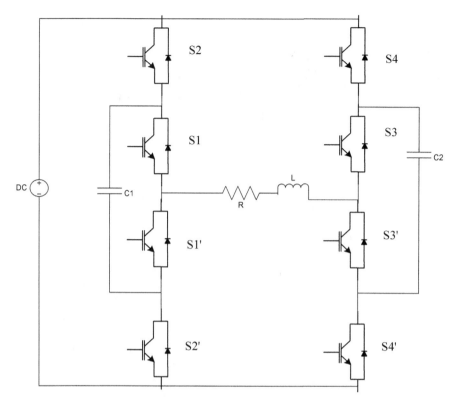

Fig. 1. Five level H-bridge flying capacitor MLI

by using proper filter circuit. The Fig. 3 it can be concluded that, harmonic order on x-axis and harmonic magnitude (% of fundamental) on y-axis. The fundamental frequency is 50 Hz for the simulation circuit.

Table 1. Switching pattern of five level HBFCMLI

Output voltage level	Switching states {(SP$_1$, SP$_2$), (SN$_1$, SN$_2$)}
$-V_{dc}$	{(0, 0), (1, 1)}
$+V_{dc}$	{(1, 1), (0, 0)}
0	{(0, 0), (1, 0)},{(1, 0) (0, 1)}{(0, 1), (1, 1)}, {(1, 0), (0, 0)}{(1, 0), (1, 0)} {(1, 1) (1, 1)},{(0, 1), (0, 1)}
$+(V/2)dc$	{(1, 1), (1, 0)}, {(0, 1), (0, 0)}
$-(V/2)dc$	{(0, 0), (0, 1)},{(1, 0) (1, 1)}

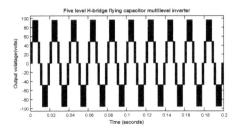
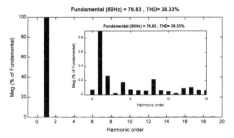

Fig. 2. Output voltage across the load for 5-level HBFCMLI

Fig. 3. FFT analysis for 5-level HBFCMLI

2.2 Seven Level HBFCMLI

For the existing 5-level H-bridge Flying Capacitor Inverter is planned to change the 7-level H-bridge Flying Capacitor by adding four more switches to the existing one and polarised capacitors are used in place of normal capacitors in the circuit with all the capacitors have taken same value. The 7-level HBFCMLI simulation result as shown in Fig. 4. From the analysis, it can be concluded that there is seven level output created. The switching sequence can be attained for creating five levels HBFCMLI by continuing the switching sequence.

The Fast Fourier Transform (FFT) analysis of 7-level flying capacitor converter is shown in Fig. 5. From the analysis it can be concluded that, the total harmonic distortion is 24.30% for seven level H-bridge flying capacitor converter. The total harmonic distortion of five level H-bridge multilevel inverter is 38.33%. The decrease in total harmonic distortion is 14.03% from five levels to seven levels. The filter circuit can be used by reducing the THD value. The seven level H-bridge Flying Capacitor Multilevel Inverter diagram shows the output voltage across the load in y-axis and the time in x-axis. The fundamental frequency is 50 Hz for seven levels HBFCMLI.

2.3 Nine Level HBFCMLI

For the modified 7-level H-bridge Flying Capacitor Inverter again planned to change the 9-level H-bridge Flying Capacitor Inverter by adding four more switches to the modified one and capacitors also same as in the 7-level HBFC with same values. The 9-level HBFCMLI simulation result as shown in Fig. 6. From the analysis it can be concluded that, there is nine level output created. The switching sequence can be achieved for creating seven levels HBFCMLI by continuing the switching sequence.

The Fast Fourier Transform (FFT) analysis of 9-level flying capacitor converter is shown in Fig. 7. From the analysis it can be concluded that, the total harmonic distortion is 17.37% for nine levels H-bridge flying capacitor converter. The total harmonic distortion (THD) for 5-level is 38.33% and for the 7-level total harmonic distortion (THD) is 24.30%. The total harmonic distortion decreases from 5-level to 7-level is 14.03% and from 7-level to 9-level the total harmonic distortion is 6.93%.

The Five level, Seven level and Nine level H-bridge Flying Capacitor Multilevel converter THD analysis as shown in Fig. 8, and it could be noticed from the figure that

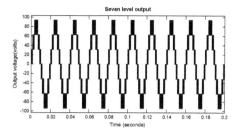

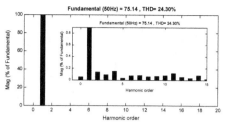

Fig. 4. Output voltage across the load for 7-level HBFCMLI

Fig. 5. FFT analysis for 7-level HBFCMLI

the reduction of THD between 5-level and 7-level is faint, while the reduction between 7-level and 9-level H-bridge flying capacitor multicell converter is less.

3 Fuzzy Controller For H-bridge Flying Capacitor Multilevel Converter

To Simulates human thinking by integrating the impression for all the physical system characteristics by using Fuzzy Logic digital control scheme. By the binary control variables into soft grades, Fuzzy Logic works with changing degrees of membership. For its low selectivity the Fuzzy Logic is used. i.e. it gives flexible response for a given input. The continuous variable systems, such as motors or positioning control the Fuzzy Control system is finest for its smooth and continuous output. To test the variables by using rules, which produce one or more outputs confessing in which rules stated [11, 12].

By achieving the appropriate output, the inputs membership degrees and response of centroid is calculated for its output of each rule is to be considered and discussing. The main characteristics of multicell converter structure, its having many switching devices. Therefore the mathematical model is complex. The Fuzzy controller technique is used to capture the uncertain information by replicating the human behavior. The Fuzzy Logic Controller (FLC) is used to control the system, when the control situation is complex, nonlinear and its mathematical model is hard.

The advantages of FLC'S are as follows

1. It does not requires the accurate mathematical model.
2. By the FLC Controller, analysis of nonlinear systems can be handling easily.
3. Uncertain inputs can be handling easily.
4. Compared to Conventional controller, the FLC is vigorous.

The Fig. 9. Shows the closed loop fuzzy controlled converter. The difference of reference voltage to output voltage gives the error signals. To calculate the fuzzy output by using these error signals. The Fuzzifier, Rule Evaluator and Defuzzifier, these three blocks are the main blocks in Fuzzy controller. The fuzzification is to be done by converting numerical variables to linguistic variables, a Rule Evaluator block is the

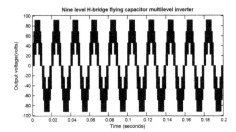

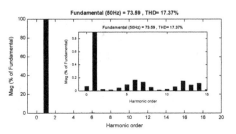

Fig. 6. Output voltage across the load for 9-level HBFCMLI

Fig. 7. FFT analysis for 9-level HBFCMLI

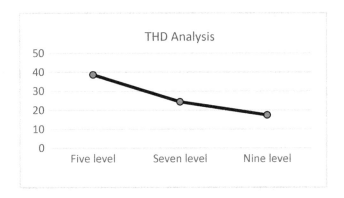

Fig. 8. THD analysis of HBFC multicell converter for different levels

decision making block and a fuzzy is the decision making block and a fuzzy set can be denoted as shown below.

$$\mu(x) \in [0, 1]$$

$$A = [x, \mu(x) \mid x \in X]$$

In the above equation X denotes collection of objects and is denoted by {x}.

The reverse of fuzzification, is called the Defuzzification. The linguistic variables are converted to crisp values by using Defuzzification. The crisp output equation is given by

$$\text{Crisp Output} = \sum A_i * X_i / \sum A_i$$

For the rule base block the linguistic control rules is used to design fuzzy controller. The fuzzification and defuzzification is required for the definition of membership function. The data base block is provides the fuzzification and defuzzification. In the block diagram, the power converter block is to be considered as phase converter, inverter or cycloconverter. The output of the FLC will be designed based on the circuit.

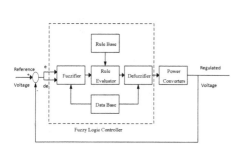

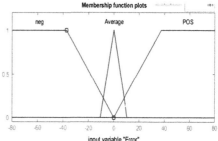

Fig. 9. Power converter FLC internal structure

Fig. 10. Fuzzy controller membership function for Input (error)

For example, by controlling the triggering angle, the output voltage can be controlled by FLC. In case of chopper it controls the duty ratio, in case of inverter and cyclo-converter it controls the Modulation index and switching sequence respectively.

3.1 Fuzzy Controller for 5- Level HBFC Multilevel Converter

For 5-level H-bridge Flying Capacitor converter is used to design the fuzzy controller with input and output membership functions. From Fig. 10 it is noticed that the fuzzy controller designed is different from that of input membership function. Figure 11 shows and explained by the designed fuzzy controller membership function output. The output variable amplitude presents in the x-axis and membership magnitude presents in the y-axis. For the fuzzy controller arrangement, the Root Mean Square (RMS) output voltage is control and the 5-level H-bridge multilevel inverter using fuzzy controller results as shown in Fig. 14. While the output voltage for load is also shown in the same figure. The Fig. 13 shows the Root mean square value (RMS) and Reference voltage.

From the Fig. 12, it can be noticed the reference voltage is initially kept at 50 v and then after 0.5 s, then it is raised to 70 v as shown in Fig. 13 it is observed that the initial part of the voltage in pulse width is less (while the reference voltage is 50 v), after 0.5 s, it is increased to raise the output voltage to 70 v for pulse width. By using fuzzy

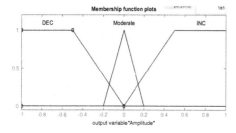
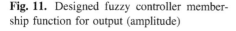

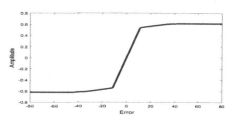

Fig. 11. Designed fuzzy controller membership function for output (amplitude)

Fig. 12. Surface view of the fuzzy controller designed

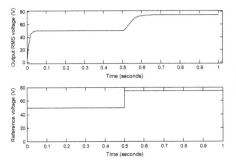
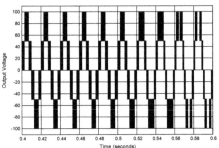

Fig. 13. Fuzzy controlled RMS Output and reference voltages for 5-level HBFC multicell converter

Fig. 14. Fuzzy controlled output voltage across load for 5-level HBFCMLI

controller, the reference voltage and RMS voltage of the five level multilevel inverter is obtained in Fig. 13. 50 v is kept at minimum reference voltage and still maintained at 0.1 s.

3.2 Fuzzy Controller for 7-Level H-Bridge Flying Capacitor Multilevel Converter

The 7-level H-bridge flying capacitor converters is similar to the five level scheme, which is designed from the fuzzy controller. The rules and input, output membership functions are same to the five level configuration, it is incorporated with its range of input member functions, and the range is −80 to +80 in the case of five level H-bridge flying capacitor, when the range is changed to −88 to +88 at seven level H-bridge flying capacitor multilevel converter. The designed fuzzy controller is simulated with seven level H-bridge Flying Capacitor multicell converter. The simulation result is obtained as shown in Fig. 16, which is the load across the output voltage, from the

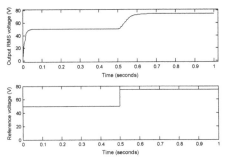
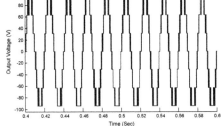

Fig. 15. 7-level HBFCMC fuzzy controlled output RMS voltage and reference voltage

Fig. 16. Fuzzy controlled output voltage across load for 7 levels HBFC multicell converter

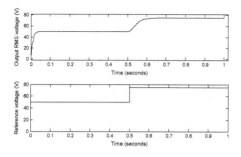 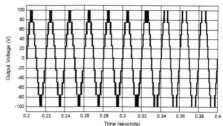

Fig. 17. Fuzzy controlled RMS output and reference voltage for 9 level H-bridge flying capacitor multicell converter

Fig. 18. Output voltage across load for fuzzy voltage for 9 level H-bridge flying capacitor controlled nine level HFCMLI

Fig. 17 it can be noticed that initially 50 v kept as reference voltage then it is changed to 80 v after 0.5 s the Root Mean Square output voltage changed along with reference voltage, which can be concluded. The reference voltage and RMS output voltages as shown in Fig. 15. During the initial period the wave form pulse width is less while pulse width increases after 0.5 s to raise the RMS output voltage to 80 v.

3.3   Fuzzy Controller for 9-Level H-bridge Flying Capacitor Multilevel Converter

The 9-level H-bridge flying capacitor converters are similar to the five level and seven level schemes, which is also designed from the fuzzy controller. Therefore there is small changes made in the input membership function, i.e. input range is kept at −90 to +90, the remaining all the input and output membership functions are similar to that of five level and seven level configurations. The designed fuzzy controller is simulated with nine level H-bridge flying capacitor multicell converter. The simulation result is obtained as shown in Fig. 18, which is the reduced output voltage across the load, from the figures it can be noticed that initially 60 v kept as reference voltage then it is changed to 80 v after 0.3 s the root mean square output voltage changed along with reference voltage, which can be concluded. To obtain the output RMS voltage of 80 v with increased pulse width, by changing the reference voltage from 60 v minimum pulse width to 80 v and the nine level H-bridge flying capacitor multilevel inverter reference voltage as shown in Fig. 16 (Table 2).

Table 2. Output parameters comparison for 5, 7 and 9-levels.

Type of inverter Configuration	Output parameters			
	Output peak voltage (Volts)	Output RMS voltage (Volts)	Voltage THD (%)	Output load current (amp)
5-level	71	50.21	38.33	0.083
7-level	74	52.33	24.30	0.064
9-level	76	53.74	17.33	0.072

4   Conclusion

The Natural voltage balancing of capacitor voltages is one of the main advantages of H-bridge flying capacitor multilevel converter. For getting better voltage balancing with capacitors in the circuit it uses phase shift carrier based PWM technique. For all the flying capacitors with different levels i.e. five level, seven level and nine level H-bridge flying capacitor converters are studied. With their performance the THD is main comparison factor, which is found that the total harmonic distortion for five levels is 38.33%, seven levels is 24.30% and nine levels is 17.37% respectively. Suitable fuzzy controllers have been designed for all the three flying capacitor multilevel converters and their performance has been verified. By using fuzzy controller technique to reduce the error to zero by taking feedback from the output of the circuit.

References

1. Meynard, T.A., Foch, H.: Multi-level conversion: high voltage choppers and voltage-source inverters. In: Proceedings of the IEEE Power Electronics Specialist Conference, pp. 397–403 (1992)
2. Meynard, T.A., Foch, H., Thomas, P., Courault, J., Jakob, R., Nahrstaedt, M.: Multicell converters: basic concepts and industry applications. IEEE Trans. Ind. Electron. **49**(5), 955–964 (2002)
3. Kou, X., Corzine, K.A., Familiant, Y.: Full binary combination schema for floating voltage source multi-level inverters. IEEE Trans. Power Electron. **17**(6), 891–897 (2002)
4. Meynard, T.A, Fadel, M., Aouda, N.: Modelling of multilevel converters. IEEE Trans. Ind. Electron **44**(3), 356–364 (1997)
5. Gateau, G., Meynar, T.A., Foch, H.: Stacked multicell converter properties and design. PESC (Vancouver) **3**, 1583–1588 (1997)
6. Song, B.M., Kim, J., Lai, J.S., Seong, K.C., Kim, H.J., Park, S.S.: A multilevel soft-switching inverter with inductor coupling. IEEE Trans. Ind. Appl. **37**(2), 628–636 (2001)
7. Sadigh A.K., Dargahi, V., Barakati, S.M.: New asymmetrical cascade multicell converter based on optimized symmetrical modules. In: IEEE International Symposium on Industrial Electronics (ISIE), pp. 408-412 (2012). ISBN 978– 1-4673-0159-6
8. Lienhardt, A.M., Gateau, G., Meynard, T.A.: Stacked multicell converter (SMC): reconstruction of flying capacitor voltages. In: 31st Annual Conference of IEEE Industrial Electronics Society (IECON) (2005)
9. McGrath, B.P., Holmes, D.G.: Analytical modeling of voltage balance dynamics for a flying capacitor multilevel converter. IEEE Trans. Power Electronics **23**(2), 543–550 (2008a)
10. McGrath, B.P., Holmes, D.G.: Natural current balancing of multicell current source converters. IEEE Trans. Power Electron. **23**(3), 1239–1246 (2008b)
11. Kevin, S.: Designing with fuzzy logic. IEEE Spect. **105**, 42–44 (1990)
12. Raviraj, V.S.C., Sen, P.C.: Comparative study of proportional – integral, sliding mode and fuzzy logic controllers for power converters. IEEE Trans. Ind. Appl. **33**(2), 518–525 (1997)
13. Nabae, A., Takahashi, I., Akagi, H.: A new neutral-point clamped PWM inverter. In: Proceedings of the Industry Applications Society Conference, pp. 761–766, September/October 1980

14. Baker, R.H.: Electric Power Converter, U.S. Patent Number 3,867,643 (1975)
15. Hammond, P.W.: A new approach to enhance power quality for medium voltage AC drives. IEEE Trans. Ind. Appl. **33**(1), 202–208 (1997)
16. Houldsworth, J.A., Grant, D.A.: The use of harmonic distortion to increase the output voltage of a three-phase PWM inverter. IEEE Trans. Ind. Appl. **20**(5), 1224–1228 (1984)

Analysis of Stacked Multicell Converter with Fuzzy Controller

P. Ponnambalam[1(✉)], M. Praveenkumar[1], Challa Babu[2], and P. Dhambi Raj[2]

[1] Faculty of School of Electrical Engineering,
VIT University, Vellore, Tamilnadu, India
p.ponnambalam@gmail.com, praveen.m@vit.ac.in
[2] School of Electrical Engineering, VIT University, Vellore, Tamilnadu, India
babu2342@gmail.com, dhambiraj@gmail.com

Abstract. In latest years, commercial companies rely upon using static power converters for developing large voltage and large current applications. For these kind applications preferably we choose multilevel inverters. The multilevel inverters are used to convert DC input power to AC output power with more number of voltage levels and with less Total Harmonic Distortion (THD). In this paper, we evolved a new technique, soft switching for unique power converter called stacked multicell converter (SMC). The fuzzy logic controller used to controls the multi levels of operation for Stacked Multicell Converter. As compared to the conventional converter, This SMC topology provides more input voltage levels, reduces the converter energy storage, and also reduces the voltage and current burden on semiconductor switching devices. This paper presents the analysis and simulation results for fuzzy logic controlled multilevel inverter of circuit topologies for open and closed loop operations.

Keywords: Stacked Multicell Converter · Soft switching controller · Phase shifted carrier Pulse Width Modulation (PWM) technique · Total Harmonic Distortion (THD)

1 Introduction

Recently all the industrial companies and research works be the more attractive towards the multilevel inverters with high power switching devices to produce large voltage and currents [1]. The inverter is one which converts DC input power into AC output of single or three phases. The voltage and current stress on the inverter is increases because the high power switching devices [2]. Now a days we are using high power semiconducting switches are Insulated Gate Bipolar Transistor (IGBT) and Metal Oxide Semi controlled Field Effect Transistor (MOSFET).

In this paper we presented a new scheme to improve dynamic performance of the multilevel inverter by using fuzzy controlled Stacked Multicell Converter. By using this scheme the output voltage levels of the converter improves and also energy stored in the flying capacitor reduces. So the switching losses of the devices reduce. A SMC works as inverter and is suits for medium and high voltage level applications.

© Springer Nature Singapore Pte Ltd. 2017
K. Deep et al. (eds.), *Proceedings of Sixth International Conference on Soft Computing for Problem Solving*, Advances in Intelligent Systems and Computing 546, DOI 10.1007/978-981-10-3322-3_30

The stacked Multicell Converter constructed by p cells and n stacks. In the SMC topology, the voltage and current stress on the semiconductor devices are reduced by distributing stress equally to all the switches. The fuzzy logic controlled SMC topology give less harmonic distortion with increasing output voltage levels [2]. So this topology most popular for the applications like Un interrupted power supplies (UPS), switched mode power supplies (SMPS) and motor drives. The new SMC topology uses the phase shifted carrier pulse width modulation technique as control technique. This technique provides desired switching pulses to the semi conducting devices by comparing the reference sinusoidal wave and phase shifted carrier wave [6]. The fuzzy logic controller monitors accurate switching pulses to the corresponding power switches.

This paper presents the analysis and simulation results for fuzzy logic control of closed loop and open loop SMC with different output voltage levels. The output levels consider for this analysis are five, seven and nine. The phase shifted carrier pulse width modulation generates triggering pulses for the power switches, to make the SMC output less harmonic [9].

2 Stacked Multicell Converter

The Stacked Multicell Converter (SMC) is behaves like Voltage Source Inverter (VSI) and it comes under the group of multilevel converter. This converter is capable to handle large voltages and gives better output control rather than the other types of converters. The voltage and current stress on the inverter is reduced, because the total stress is distributes to all semi conducting switches equally. The basic diagram of switching cell is shown in Fig. 1.

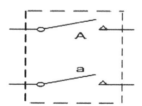

Fig. 1. Basic model of switching cell

In Fig. 1 there are 2 switches and are namely 'A' and 'a'. Both the switches never triggered at a time i.e., if **A** is in conduction means **a** must be in OFF state vice versa. The switches ON/OFF accordingly with less amount of stress on the both the switches and to obtain better output voltage. The new SMC construct p cells as columns and n stacks as rows. This SMC topology addition with flying capacitors called as flying capacitor stacked multicell converter. The SMC topology helps to reduce total energy stored by capacitors [3]. The flying capacitor is connected in such a way that to avoid damage of semi conducting devices. So the capacitor is connected among the two p cells, to share the voltage stress equally between the switches. So, the semiconducting switches protected from the damages caused by voltage stress. In the new SMC topology, there are two voltage sources which are placed top and bottom of the rows. The function of upper DC source is to increases the voltage levels in positive cycle and lower DC source is to increases voltage levels in negative half cycle. The output voltage level can increase by adding additional cell in series with the inverter circuit. The capacitors in the circuit will supplies stable voltage to the inverter circuit when the supply is shut down.

In this paper, the simulation of five level, seven level and nine level fuzzy controlled stacked multicell converters is analysed by MATLAB simulink software. The triggering signal for all the three configurations of converters are generated by phase shift carrier based pulse width modulation. These trigger pulses are given to the power semiconductor switches in the inverter circuit. This triggering configuration is hybrid.

2.1 Stacked Multicell Converter Topology for Five Level Systems

In the stacked multicell topology the output level of the converter increased by adding the p cells cascading to the previous converter system. The operational diagram of Five Level output voltage system for SMC is shown in Fig. 2.

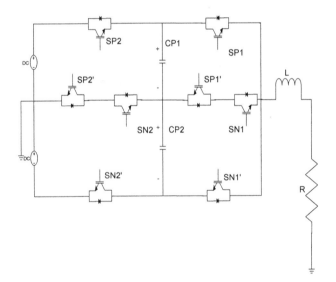

Table 1. Switching states for five level output

Output voltage level	Switching states of {SN1, SN2, SP1, SP2}
−2Vdc	{0, 0, 0, 0}
−Vdc	{0, 1, 0, 0}
	{1, 0, 0, 0}
0	{1, 1, 0, 0}
	{0, 1, 0, 1}
	{1, 0, 1, 0}
+Vdc	{1, 1, 0, 1}
	{1, 1, 1, 0}
+2Vdc	{1, 1, 1, 1}

Fig. 2. Five level inverter system

The five level inverter there are 2 columns i.e., p = 2 cells, and two rows i.e., n = 2 stacks. It consists of p × n = 4 semi conducting switching cells and flying capacitors of (p − 1) * n = 2. In Fig. 2, there are 8 switching devices, 2 voltage sources and 2 flying capacitors. The switches are divided into two groups one is positive (SP1 & SP2) group and another is negative (SP3 & SP4). In addition with the above switches there are complimentary switches and they also in positive (SP1' & SP2') group and negative (SP3' & SP4') groups [6]. The function of the flying capacitor is to distribute the available voltage so the stress on the switches reduces. The capacitor voltage is almost half of the source voltage.

The capacitor voltage is given by

$$V_C = E/2 \tag{1}$$

The value of the capacitor is find out by using formulae

$$C = (I_{rms})/(V_{ripple} * 2 * \pi * F_s) \tag{2}$$

The switching states of the five level converters are shown in Table 1. The Stacked Multilevel Converter topology for five level system turn ON and turn OFF pules generated by using Fuzzy controller. The soft switching patterns develop pulses for the SMC to get different voltage levels by changing switching modes of operations [5].

2.2 Open Loop SMC Simulation Results for Five Level Output

The number of switches in the positive group is equal to number of switches in negative group to get constant voltage level on both positive and negative cycles. The upper side DC voltage source responsible for positive cycle voltage level and lower side DC voltage source is responsible for negative cycle voltage level. The open loop Stacked Multicell Converter simulation results for five level output is shown in Fig. 3. In Fig. 3 X-axis as time period and Y-axis as output voltage.

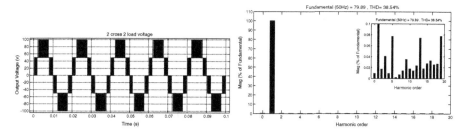

Fig. 3. SMC output voltage waveform for five level

Fig. 4. FFT analysis for five level SMC

The five level open loop SMC topology output analyzed by Fast Fourier Transform (FFT) and is show in Fig. 4. In the FFT analysis the THD for five level SMC is 38.54%. By designing the proper filter circuit Total Harmonic Distortion value of the converter is reduced. In Fig. 4 X-axis is order of the harmonic and Y-axis is THD magnitude and is simulated for power frequency of 50 Hz.

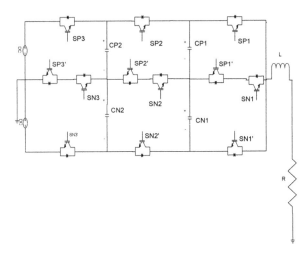

Fig. 5. Seven level inverter system

Table 2. Switching states for seven level

Voltage level	Switching states of {SN1, SN2, SN3, SP1, SP2, SP3}
−3Vdc	{0, 0, 0, 0, 0, 0}
−2Vdc	{0, 0, 1, 0, 0, 0}
	{0, 1, 0, 0, 0, 0}
	{1, 0, 0, 0, 0, 0}
−Vdc	{1, 1, 0, 0, 0, 0}
	{1, 0, 1, 0, 0, 0}
	{0, 1, 1, 0, 0, 0}
0	{1, 1, 1, 0, 0, 0}
+Vdc	{1, 1, 1, 0, 0, 1}
	{1, 1, 1, 0, 1, 0}
	{1, 1, 1, 1, 0, 0}
+2Vdc	{0, 0, 0, 0, 0, 1}
	{1, 1, 1, 1, 0, 1}
	{1, 1, 1, 0, 1, 1}
+3Vdc	{1, 1, 1, 1, 1, 1}

2.3   Stacked Multicell Converter Topology for Seven Level Output

The SMC topology for seven level output voltage topology is simple. By adding one p column in series with five level SMC topology circuit we can get seven level SMC topology. The SMC topology for seven level output configuration is shown in Fig. 5.

In seven level inverter circuit designed as columns of p = 3 cells and rows of n = 2 stacks. It consist of p × n = 6 semi conducting switching cells and flying capacitors of (p − 1) * n = 4 [5]. The SMC seven level topology consists of 12 power switches, 2 DC voltage sources and flying capacitors of 4 which are shown in Fig. 5. In this topology all the switches are high power and low frequency. The seven level SMC configurations there are three switches for both positive and negative cycles. The positive group devices represent as "SP1, SP2, and SP3" and negative group devices represents as "SN1, SN2 and SN3". This circuit also consists of complimentary devices, they are positive group complimentary devices represents as SP1′, SP2′, and SP3′ and negative group complimentary devices represented as SN1′, SN2′, SN3′. This circuit having 4 flying capacitors and they work like same as five level SMC. The number of capacitors increased because of the increase in cells and voltage level. The voltage across the flying capacitors is differs according to the placement. The first pair of capacitors voltage be the two-third of source voltage and second pair of capacitors voltage be the one-third of source voltage [4].

The capacitors voltage across first pair is

$$V_C = 2E/3 \tag{3}$$

The capacitors voltage across second pair is

$$V_C = E/3 \tag{4}$$

The switching states of the seven level converters are shown in Table 2. The SMC operates according to the switching patterns given by the fuzzy controller. The switching mode controls the output voltage levels by turn ON/OFF of switches. In this SMC seven level topology there are seven output level and each level is controlled by turning ON and turning OFF of particular switches at the given instant of time. The switching patterns for each level are different [4, 5].

2.4 SMC Open Loop Simulation Results for Seven Level Output

The SMC for seven level voltage output get by using of 12 semi conducting switches and 2 DC voltage sources as same as five level SMC topology. The simulation results for SMC for seven level is shown in Fig. 6. The upper DC source increases the level of voltage in positive half cycle and lower DC source increase the level of voltage in negative half cycle. In Fig. 6 X-axis as time period and Y-axis as output voltage (Fig. 7).

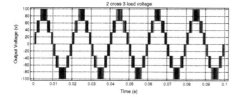
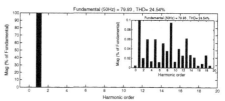

Fig. 6. SMC output voltage waveform for seven level

Fig. 7. FFT analysis for seven level SMC

2.5 Stacked Muticell Converter Topology for Nine Level Output

The SMC topology for nine level topology is getting by adding one p cell in series with seven level SMC circuit. The SMC circuit for nine level is shown in Fig. 8. The SMC topology for nine level output voltage design by using of columns of p = 4 cells and rows of n = 2 stacks. It consists of p × n = 8 semi conducting switching cells and flying capacitors of (p − 1) * n = 6. The SMC for nine level voltage circuit consists of 16 power switches, 2 DC voltage sources and fling capacitors of 6. Which are shown in Fig. 8.

The switches in nine level SMC topology also divided into two groups. The positive group switches are SP1, SP2, SP3, SP4 and negative group switches are SN1, SN2, SN3,

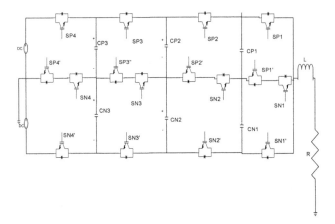

Fig. 8. Nine level inverter system

SN4. The complimentary switches for positive group is SP1', SP2', SP3', SP4' and negative group complimentary switches are SN1', SN2', SN3', SN4' [6]. The function of flying capacitors is to distribute the over voltages among the all power semi conducting switches to avoid over heating of switches. The value of THD is reduces with increasing of level in the converter circuit [7].

2.6   SMC Open Loop Simulation Results for Nine Level Output

The SMC topology for nine level output voltage circuit consists of 16 semi conducting switches. The triggering pulses for the converter circuit are generated by phase shifted carrier PWM with fuzzy controller. The simulation circuit for SMC for nine level output is shown in Fig. 9. The converter output nine level is generated by using the 2 DC sources same as in five level and seven level. In Fig. 9 X-axis as time period and Y-axis as output voltage.

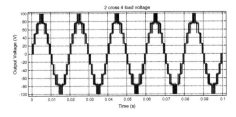

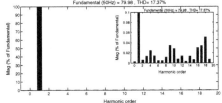

Fig. 9. SMC output voltage waveform for nine level

Fig. 10. FFT analysis for nine level SMC

The nine level open loop SMC topology output analyses by Fast Fourier Transform (FFT) and is show in Fig. 10. In the FFT analysis the THD for nine level SMC topology is 17.37%. By designing the proper filter circuit THD value of the converter is reduced. In Fig. 10 X-axis is order of the harmonic and Y-axis is THD magnitude and is simulated for power frequency of 50 Hz.

The above all FFT analysis for different levels clearly concludes that the THD value is reduces with increases number of voltage levels and is presented. The THD values of five level SMC is 38.54%, seven level SMC is 24.54% and nine level SMC is 17.37%. By reducing the THD value the purity of the sinusoidal wave is increases. The analysis for all voltage levels are represented in a graph shown in Fig. 11.

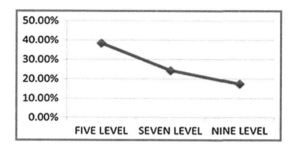

Fig. 11. Complete FFT analysis

3 Soft Switching Control of Stacked Multicell Converter

The control strategy for converters is changes day by day to improve the converter efficiency and power quality. The closed loop analysis of power electronic converters is done with P, PI and PID controller. Now, the researches going on the soft switching for power converters. One of the best soft switching controls is fuzzy. For closed loop analysis the fuzzy logic controller will give the control signals to the converters. The fuzzy logic was invented by Lotfi Zadeh in 1965. This logic is based on "Degree of Truth". By using this logic the results were more accurate rather than Boolean expressions such as '0' or '1'. Unlike the other controls the fuzzy uses range-to-point and range-to-range controls. It is just like as person feelings and interference procedure. Fuzzy develops digital signals as same as human thinking by simulating all the physical systems and are integrated. Fuzzy get output soft grades by converting from binary input variables with respect to the changes in the membership function. One of the major advantage of Fuzzy controller is low selectivity i.e., for any given input it produces adaptive responses. Fuzzy system is best suited for controlling of motors with continuous variable systems like positing systems, because output of the motor is smooth and continuous. The fuzzy system generates one or more responses for a testing variable according to the rules. The output of the system depends on input response of the rule and centroid of the response.

The multilevel inverter consists of many switching devices in the circuit. The more number of switches be the main advantage of the multilevel inverters. To analyze that many switches are mathematically complex. The Fuzzy controller produces outputs for uncertain information like human brain. The Fuzzy Logic Controller (FLC) used for nonlinear systems, control complex systems and system which having hard to find mathematical model [6].

The output of the fuzzy controller is obtained only after fuzzification of each input with membership function. This type of control is mostly suited for mechanical objects with software or hardware. Fuzzy logic is the combination of many valued logics and reasoning that approximate like as human. In the Boolean logic the variables are with the logic 0 or 1 but in the fuzzy logic the truth value varies in between 0 to 1. Let take an example using fuzzy logic controller for the temperature controller. The output of the controller range will be cool, medium or heat. The basic structure of Fuzzy Logic Controller using in closed loop operation is shown in Fig. 12.

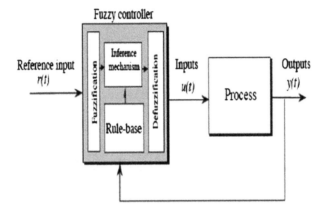

Fig. 12. Fuzzy logic controller basic model

3.1 Fuzzy Working

The fuzzy sets consists of variables which having relations with physical system variables. All the fuzzy elements have degree of membership. Classical sets obeys bivalent condition i.e., the element in the set belongs to that set or not. In this manner all the elements in the classical set are mapped with two capacity components of 1 or 0. According to the fuzzy set theory, classical bivalent sets are called as crisp sets. In crisp set information is "imprecise or imperfect". Fuzzy uses the 'If-Then' rule to analyze the inputs. In fuzzification process all the inputs are process with the membership function and all the input variables are mapped with the different relations in the knowledge base. The rule base is a decision maker to choose the precise decision suitable for the case. This can be explained with simple example on temperature samples; the temperature of the system classifies into 3 categories (i) Low (0–20 °F), (ii) Medium (20 °F–60 °F) and (iii) High (60 °F–100 °F). In the classical set any value of the temperature is the one subset of any of the above classification. If temperature is 30 °F the classical set clearly fit for Medium. But in the fuzzy set, one value of temperature is divided into two or three possible subsets at the same instant of time. For the same value of 30 °F is fix in with LOW to a specific degree say 0.4°, but it also fix in with

Fig. 13. Classical
set for samples   **Fig. 14.** Amplitude membership   **Fig. 15.** Error membership

MEDIUM to around 0.7°. For another value of temperature say 50 °F fuzzy consider in all the three cases, the value of degree 0.2 with LOW or HIGH and have a maximum value of degree 1 for MEDIUM [8]. The classical set for temperature samples is shown in Fig. 13.

The MATLAB software consists of fuzzy tool bar to simulate fuzzy outputs for physical systems. In FIS editor, there are many membership functions are preloaded. The membership block shows the duty ratio of the function and error of the system. The amplitude duty ratio function circuit is shown in Fig. 14, and input error membership function is show in Fig. 15.

3.2 Five Level SMC Topology Using Fuzzy Controller

The circuit using for this simulation is same as five level open loop SMC circuit. The closed loop operation of SMC is analyse by using fuzzy controller. The fuzzy logic is used to obtain the voltage RMS value to find error value. The V_{RMS} is compared with reference value of voltage V_{REF} to get error signal. The generated error signal fed to the fuzzy controller and this compared with open loop SMC for five level output topology. The output voltage generated by using fuzzy controller for five level SMC is shown in Fig. 16. The RMS voltage (V_{RMS}) and reference voltage (V_{REF}) is shown in Fig. 17. In figures time is taken as X-axis and voltage taken as Y-axis.

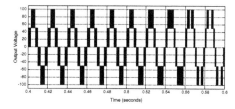

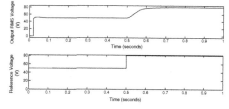

Fig. 16. Fuzzy controller output voltage for
five level SMC

Fig. 17. Five level SMC V_{RMS} and V_{REF}

By considering the Fig. 17, it is clear that the RMS output voltage increase with increase in reference voltage. From 0 to 0.5 s the reference voltage is constant at 50 V and RMS output voltage is 50 V. At the instant of 0.5 s the reference voltage magnitude increased to 80 V. The RMS output voltage is also increased to 80 V.

3.3 Seven Level SMC Topology Using Fuzzy Controller

The seven level SMC topology operation using fuzzy control gives RMS voltage as output and is shown in Fig. 18. The closed loop operation of seven level SMC is same as five level fuzzy controlled SMC. The only difference is range of membership function for the input variables. For five level fuzzy controlled SMC input range of the membership is [−80, +80]. For seven level fuzzy controlled SMC input range of membership is [−88, +88].

The RMS output voltage and reference voltage wave forms for fuzzy controlled seven level SMC is shown in Fig. 19. The V_{REF} voltage maintained constant at 50 V up to time reaches to 0.5 s. Whenever time reaches to 0.5 s the reference voltage increased to 80 V and the RMS voltage also rises to 80 V.

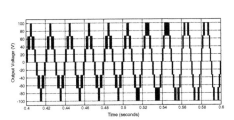

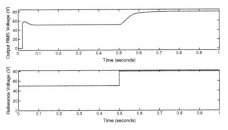

Fig. 18. Fuzzy controller output voltage for seven level SMC

Fig. 19. Seven level SMC V_{RMS} and V_{REF}

3.4 Nine Level SMC Topology Using Fuzzy Controller

The fuzzy controlled nine level SMC topology gives output voltage V_{RMS} shown in Fig. 20. The closed loop operation of seven level configuration is same for nine level SMC also. But only difference is input range of membership function. For seven level

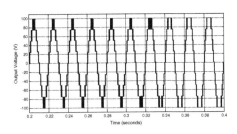

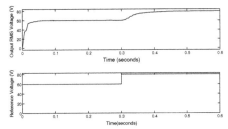

Fig. 20. Fuzzy controller output voltage for seven level SMC

Fig. 21. Nine level SMC V_{RMS} and V_{REF}

SMC input range of membership is [−88, +88]. For nine level SMC configuration input range of membership is [−89, +89].

The RMS output voltage and reference voltage wave forms for fuzzy controlled nine level SMC is shown in Fig. 21. The V_{REF} voltage maintained constant 60 V up to time reaches to 0.5 s. Whenever time reaches to 0.5 s the reference voltage increased to 80 V and the RMS voltage also rises to 80 V

4   Conclusion

This paper analysed the operation of open loop and closed loop configurations of Stacked Multicell Converter (SMC). This topology best suited for the applications using low power high voltage semi conducting switches. The proposed topology reduces the Total Harmonic Distortion (THD) by increasing the number of voltage levels. The closed loop analysis of SMC is explains with soft switching control called Fuzzy Logic Controller (FLC). The MATLAB Simulink software is used to develop Phase shifted carrier Pulse Width Modulation (PWM) to generate pulses for the power semiconducting switches in the SMC. The simulation results for Five level, Seven level and Nine level SMC with open loop and closed loop operation has been presented. The design and operation of closed loop control SMC is achieved by using FLC. The fuzzy controlled SMC output voltage satisfied the circuit requirements. The advantage of this work is to reduce THD by increasing the number of output levels by SMC. The proposed FLC stacked multicell converter for closed loop operation reduces the harmonics in the supply. In future by changing the control strategies for SMC converter improves the system performance.

References

1. Meynard, T.A., Fosch, H., Thomas, P.: Multicell converters: basic concepts and industry applications. IEEE Trans. Ind. Electron. **49**, 955–964 (2002)
2. Rodriguez, J., Franquelo, L.G., Kouro, S., Leon, J.I., Portillo, R.C., Prats, M.A.M., Perez, M. A.: Multilevel converters: an enabling technology for high-power applications. Proc. IEEE **97** (11), 1786–1817 (2009)
3. Ornov, N., Ruderman, A., du Toit Mouton, H.: Simple time domain analysis of natural balancing in flying capacitor stacked multicell converters. In: Proceedings of IEEE, pp. 1779–1785, May 2014. ISBN: 978-1-4799-4032-5/14
4. Nava, J.M.F., Sanchez, P.B.: Stacked multicell converter controlled by DSP. In: Proceedings of the 14th International Conference on Electronics, Communications and Computers (CONIELECOMP 2004), pp. 69–75 (2004)
5. Gateau, G., Meynard, T.A., Foch, H.: Stacked multicell converter properties and design. In: PESC 20001 (Vancouver), vol. 3, pp. 1583–1588, 17–22 June 2001
6. Anil, G., Sasisankar, A., Akbar, A., Basheer, F.: Fuzzy control of multicell converter. IOSR-JEEE **8**(4), 54–63 (2013)

7. Dargahi, V.: Detailed and comprehensive mathematical modeling of flying capacitor stacked multicell multilevel converters. Int. J. Comput. Math. Electr. Electron. Eng. (COMPEL) **33** (1/2), 483–526 (2014)

8. Bai, Y., Wang, D.: Fundamentals of fuzzy logic control – fuzzy sets, fuzzy rules and defuzzifications, pp. 17–36, January 2007. doi:10.1007/978-1-84628-469-4_2

9. Meynard, T.A., Foch, H., Turpin, C., Richardeau, F., Delmas, L., Gateau, G.: Multicell converters: derived topologies. IEEE Trans. Ind. Electron. **49**, 978–987 (2002)

Implementation of Fuzzy Logic on FORTRAN Coded Free Convection Around Vertical Tube

Jashanpreet Singh[1(✉)], Chanpreet Singh[2], and Satish Kumar[1]

[1] Mechanical Engineering Department,
Thapar University Patiala, Patiala 147004, India
jashanpreet.singh@thapar.edu
[2] Department of Mechanical Engineering,
Punjabi University Patiala, Patiala 147002, India

Abstract. In present study, fuzzy logic is used to predict the free convection over a heated vertical cylindrical tube. Tube has diameter 38 mm and length 500 mm. Numerical simulation involves use of implicit finite difference scheme to solve the fluidics equations for vertical tube. Tube is imposed to a fine structural grid and appropriate boundary conditions. Properties of two fluids namely air and water vapour is used for numerical simulation. Grashof number is varied from 2.04×10^6 to 2.62×10^8 and 2.69×10^6 to 3.36×10^8 for air and water vapour respectively. A computer code in FORTRAN programming language is used to draw velocity and temperature profiles. Fuzzy Interface System Mamdani is used to evaluate output membership from different fuzzy sets. Fuzzy predicted results are found in good agreement with experiments. Rayleigh number was found 1.45×10^7 to 3.63×10^8 and 2.66×10^7 to 6.67×10^8 for air and water vapour. Fuzzy logic results confirm $\pm 4\%$ agreement with experimental results.

Keywords: Free convection · Vertical cylinder · Fuzzy logic · Implicit scheme · Finite difference method

1 Introduction

Natural convection from a vertical cylinder is important in field of heat transfer. Wide range of engineering applications is heat exchanger, refrigerator condensers, electronic components etc. [1, 2]. Free convection from horizontal and vertical cylinder has been studied numerically and experimentally by many of researchers. In earlier days, Blottner [3] performed a mathematical analysis using finite difference method to find out the solution of boundary layer equations. Doulas et al. [4] investigated a parabolic equation by combining the finite element and finite difference method through a method of characteristics. Result showed smaller truncation error while those compared with standard methods for a convection-dominated problem. Saha and Hossain [5] performed a numerical study to analyze the free convection laminar flow. The effect of buoyancy was established due to mass and thermal diffusion in a stable thermally stratified medium adjacent to vertical surface. The governing equation solved through an implicit finite difference method and local non-similarity method. Results showed

© Springer Nature Singapore Pte Ltd. 2017
K. Deep et al. (eds.), *Proceedings of Sixth International Conference on Soft Computing for Problem Solving*, Advances in Intelligent Systems and Computing 546, DOI 10.1007/978-981-10-3322-3_31

various aspects of complex interaction of the two buoyant mechanisms. Fahiminia et al. [6] performed a computational analysis on laminar natural convection using an implicit scheme finite volume approach through CFD simulations. They solved the flow domain in presence of density gradients under a gravitational field on vertical surfaces. Results showed that convective heat transfer rate from fin arrays depends on fin height, fin length, fin spacing and base-ambient temperature difference. Zhanlav and Ulziibayar [7] developed a numerical model for solving obtained finite-difference scheme for the Helmholtz equation. Shiferaw and Mittal [8] performed a numerical study on 3-D Poisson's equation by second-order finite differences in cylindrical coordinates system with the Dirichlet's boundary conditions using Hockney's method. They observed the model was helpful in saving the number of computation, computational time and accuracy level. Ahmad and Bilal [9] performed a numerical study by solving the Blasius equation through neural network algorithms. They approximated the governing equations by finite difference method and simulated through Sequential Quadratic Programming algorithm and hybrid AST-INP techniques. Petrova [10] performed mathematical analysis of integrability of the Euler and Navier-Stokes equations. Results showed that these equations define solution on the tangent non-integrable manifold and integrable structures.

Fuzzy logic is used over a wide range to solve different type of problems. Zadeh [11] developed a method to design the system for conducting the experiments. A fuzzy logic is an approach including fuzzy system established by fuzzy sets namely membership function and rule table [11, 12]. Yousefi et al. [13] had performed a numerical as well experimental study the natural convection from heated horizontal cylinder placed a vertical channel. Karami et al. [14] had performed a simulation study the free convection heat transfer over an isothermal horizontal cylinder in a vertical channel using fuzzy logic. The aim of present work is to investigate the effect of various parameters on heat transfer over a vertical cylinder. The governing equations are solved using Finite difference method with implicit scheme and simulated in FORTRAN 95 program code. Fuzzy logic is implemented in MATLAB R2012 7.14.0.739 software package. Fuzzy model Mamdani is used as Fuzzy interface system [15]. The aim of using fuzzy approach in present study is to evaluate the effect of membership functions i.e. parameters on output.

2   Physical Model of Problem

A two-dimensional model of coordinate system (x, y) in region around a vertical tube is shown in Fig. 1(a). Symbol Q represents the heat flow and velocity is denoted by u. The far region exists at $y \to \infty$. The tube diameter (d) and length (L) are considered as 38 and 500 mm respectively. Three basic fluidic equations are employed to physical model. The continuity, momentum and energy equations are given as Eqs. 1, 2 and 3 respectively.

$$\frac{\partial v}{\partial y} + \frac{\partial u}{\partial x} = 0 \tag{1}$$

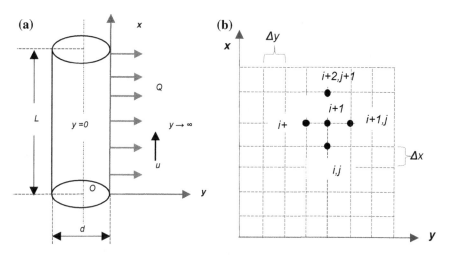

Fig. 1. (a) Physical model of vertical cylindrical tube (b) Grid of flow geometry

$$u\frac{\partial u}{\partial x} = g\beta(T - T_a) + v\left(\frac{\partial^2 u}{\partial y^2}\right) - v\frac{\partial u}{\partial y} \tag{2}$$

$$u\frac{\partial T}{\partial x} = \alpha\frac{\partial^2 T}{\partial y^2} - v\frac{\partial T}{\partial y} \tag{3}$$

The governing equations are solved by using implicit finite difference method which follows Taylor series expansion [16]. The combined algorithmic approach is used to solve the governing equations for a heated vertical tube. Combined approach utilizes the both Forward difference and Central difference algorithms. The partial derivatives of x are solved by forward-difference and derivatives of y are solved by central-difference scheme. The truncated equations are follows.

Continuity equation is truncated as:

$$v_{i+1,j+1} = v_{i+1,j-1} - \frac{2\Delta y}{\Delta x}\left[u_{i+1,j} - u_{i,j+1}\right] \tag{4}$$

Similarly, truncated form of Momentum equation is given as:

$$u_{i+1,j} = \frac{u_{i,j} - \frac{v_{i,j}}{u_{i,j}} \cdot \frac{\Delta x}{2\Delta y}\left[u_{i+1,j+1} - u_{i+1,j-1}\right] + \frac{\beta.g.\Delta x\left[T_{i,j} - T_a\right]}{u_{i,j}} + \frac{v.\Delta x}{u_{i,j}.\Delta y^2}\left[u_{i+1,j+1} + u_{i+1,j-1}\right]}{\left[1 + \frac{2v\Delta x}{u_{i,j}.\Delta y^2}\right]} \tag{5}$$

Similarly, the Energy equation is given as follows:

$$T_{i+1,j} = \frac{\frac{\alpha \Delta x}{2u_{i,j}\Delta y^2}\left[T_{i+1,j+1}+T_{i+1,j-1}\right] - \frac{V_{i,j}\Delta x}{u_{i,j}2\Delta y}\cdot\left[T_{i+1,j+1}-T_{i+1,j-1}\right] + T_{i,j}}{\left[1 + \frac{2\alpha\Delta x}{2\Delta y^2 u_{i,j}}\right]} \tag{6}$$

Discretized governing equations such as Eqs. 4–6 are subjected to boundary values condition given as:

$$y = 0, \quad T = Ts, \quad u = 0, \ v = 0 \tag{7}$$

$$y \rightarrow \infty, \quad T = Ta, \quad u \rightarrow 0, \ v \neq 0 \tag{8}$$

In Eq. 7, the condition $y = 0$ interprets the region from which the boundary layer starts i.e. no slip boundary condition. In Eq. 8, the condition at infinity interprets the region just outside the boundary layer i.e. free slip boundary condition. Equations 4–6 are programmed in FORTRAN 95 language for simulation purpose. Grid subjected to this program is shown in Fig. 1(b). Grid independency test has been carried out on $m \times n$ grid at 1001×51, 1001×101 and 2001×101 points. The results comes out from this grid are approximately same but 1001×101 is selected as a balanced grid. The points along the surface of cylindrical tube in x-axis are denoted as i points. Space perpendicular to tube at right angle i.e. along y-axis denotes by j points. Where i points varies from 1 to $n + 1$ and j points varies from 1 to $m + 1$. LINUX-based

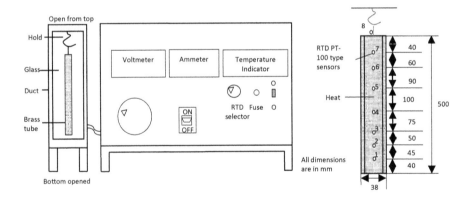

Fig. 2. Schematic diagram of (**a**) experimental set up of vertical tube (**b**) heater unit

Table 1. Properties of air and water vapour at $T_f = 310$ K [16]

Property	Units	Air	Water vapour
Kinematic viscosity (α)	m^{2}s^{-1}	23×10^{-6}	14×10^{-6}
Thermal diffusivity (v)	m^{2}s^{-1}	16×10^{-6}	14×10^{-6}
Thermal conductivity (k)	W/m K	0.0268	0.01845
Prandtl number (Pr)	–	0.712	0.9978

Ubuntu 2.6.321 operating system is used for numeric calculations. Typical CPU time needed for a single run was the order of 10–15 s.

3 Experiments

Experiments are performed in order to validate the numerical simulation. Experiments are conducted on laboratory scale equipment (manufactured by ARE educational equipment Pvt. Ltd., Maharashtra). Schematic diagram of setup is shown in Fig. 2 (*a*) which consists of hollow brass tube having diameter 38 mm and length 500 mm. Figure 2(*b*) represents the heater assembly which consist of RTD-100 type sensors to measure the surface temperature of tube at different locations. Heater input of 6.76 W is established at steady-state achieved after 4 h. The surface temperature (T_s) was maintained as 310 K. Ambient (T_a) and film temperature (T_f) were found as 294 and 302 K respectively. Volumetric coefficient of thermal expansion (β) is 3.33×10^{-3} K^{-1} at the given film temperature. Table 1 shows the properties of air and water vapour at T_f.

4 Results and Discussion

4.1 Variation of Parameters Along Length of Vertical Tube

Figure 3 represents the velocity profiles for air and water vapour which evaluate the hydraulic boundary layer thickness at various locations along length of tube. Convectional currents involves during the natural convection which initially gave rise to the velocity. Velocity gradient drops slowly under the influence of outer fluid and falls to zero when travels in *y*-axis.

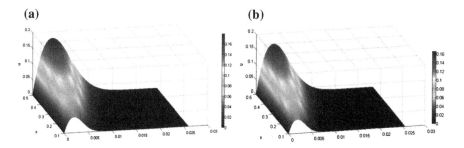

(a)   **(b)**

Fig. 3. Velocity profiles for **(a)** air and **(b)** water vapour

Figure 4 represents the temperature profiles for air and water vapour which evaluate the thermal boundary layer thickness at various locations along length of tube. Temperature gradient attains highest value at heated surface of tube and rapidly drops as moves in *y*-axis. Thermal boundary layer δ_T for air and water vapour is shown in Fig. 5. Characteristics lines are drawn for δ_T with respect to *x* which evaluates that air

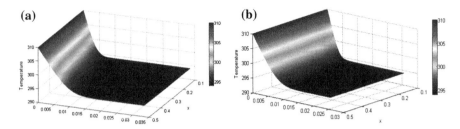

Fig. 4. Temperature profiles for **(a)** air and **(b)** water vapour

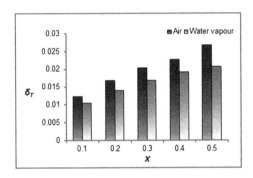

Fig. 5. Variation of δ_T along x

having $Pr = 0.7122$ show higher values as compare to water vapour having $Pr = 0.9978$. Higher the boundary layer thickness results in high heat transfer.

4.2   Dependency of Rayleigh Number, Ra

In literature Raleigh number can be determined from Grashof and Prandtl number given as:

$$Ra_x = Gr_x . Pr \qquad (9)$$

It is clear from the linear relation that Ra is higher for air than that of water vapour. Rayleigh number varies from 10^6 to 10^9. The minimum value of Ra and Gr, for air; is 1.46×10^6 and 2.04×10^6 & for water vapour; is 2.68×10^6 and 2.69×10^6; respectively. The maximum value of Ra and Gr, for Air; is 2.4×10^8 and 3.36×10^8 & water vapour is 2.62×10^8 and 2.61×10^8; respectively.

Figure 6(a) shows the dependency of boundary layer thickness on Rayleigh number. Characteristic curve of boundary layer thickness for air is away from water vapour. From Figure, it can be clearly seen that as the value of Prandlt number increases the larger boundary layer is formed. Figure 6(b) shows the dependency of Rayleigh number

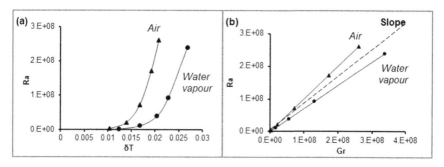

Fig. 6. Variation of Ra with respect **(a)** δ_T, and **(b)** Gr

on Grashof number. The curve for Air passes nearby and slightly below from slope whereas the curve for water vapour passes much above from the slope. It is observed that as the Pr increases the linearity between the Ra and Pr disturbs.

4.3 Dependency of Nusselt Number, Nu

In this section, the Nusselt number is calculated from numerical results of thermal boundary layer. Nusselt number has been calculated from numerical heat transfer coefficient. Heat transfer coefficient h can be calculated from slope of temperature gradient dT/dy along length follows:

$$h_x = -k\frac{\left(\frac{dT}{dy}\right)_x}{T_s - T_a} \tag{10}$$

$$Nu_x = \frac{h_x x}{k} \tag{11}$$

Figure 7 shows the results of Nusselt number against length of tube. It is observed that results of Nu shows the higher value characteristics curve at Pr of 0.9978 than 0.7122.

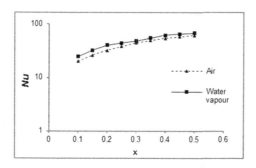

Fig. 7. Variation of $Log\ (Nu)$ along x

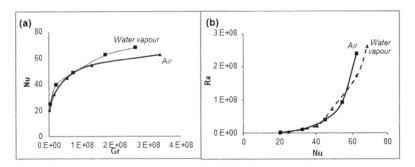

Fig. 8. Variation of *Nu* with respect to (a) *Gr* (b) *Ra*

Characteristics line for air passes above the line of water vapour. It indicates that the *Nu* increases with increase in high *Pr*.

Figure 8(a) shows the characteristic curves in-between *Nu* and *Gr* for air and water vapour at *Pr* of 0.9978 and 0.7122 respectively. Curve for air is smooth as compare to that of Water vapour. It is because of higher value of *Pr*, the higher value of *Pr* i.e. above 1.0 the turbulence starts. For *Pr* which is close to 1.0 have turbulence behavior. This is the reason behind noise in curve for air at *Pr* = 0.7122. Figure 8(b) show the characteristic curves in-between *Nu* and *Ra* at *Pr* = 0.9978 and 0.7122. Curve for air is again found smooth as compare to water vapour. Turbulence is observed when *Pr* approaches to 1.

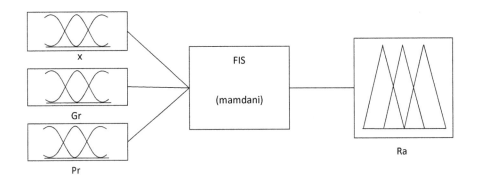

Fig. 9. Fuzzy interface system

5   Defuzzification of Free Convection

The aim of present study is to consider the effect of location, Grashof and Prandtl numbers on Rayleigh number from vertical tube. In order to implement fuzzy logic, input and output levels are determined. The value of vertical location in five levels is ranging from 0.1 to 0.5 m, Grashof number in five levels ranging from 2.04×10^6 to

(a)   **(b)**

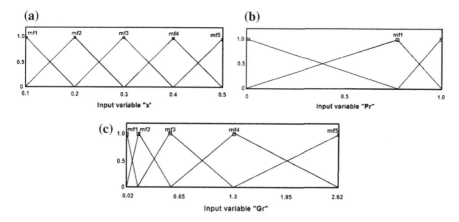

(c)

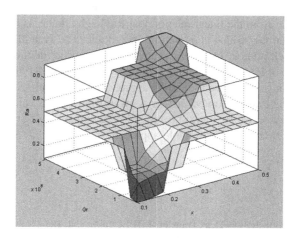

Fig. 10. Membership functions of **(a)** vertical location **(b)** Prandtl number **(c)** Grashof number

2.62×10^8, Prandtl number at single level 0.712 are chosen. Rayleigh number is chosen as output variable. Fuzzy interface system (FIS) Mamdani is used as shown in Fig. 9. Triangular member functions are used for input and output variables [17]. Figure 10 represents input membership functions used in FIS. Twelve rules are applied to by accounting various combinations of input variables. Output comes out in form of Raleigh number.

Figure 11 indicates that as the location along vertical cylinder increases the Rayleigh number as increases. It was observed that Grashof number increases with increase in vertical location which directly related with Rayleigh number as seen in Eq. 9. The Value of Rayleigh number was found in deviation of ±4% with experimental data as mentioned in Fig. 12. Therefore, fuzzy logic results show good

Fig. 11. Evolution of Gr and Ra at various location of cylinder along length

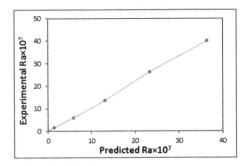

Fig. 12. Comparison of Experimental vs. predicted Rayleigh number

agreement with experiments. So, Fuzzy logic approach can be reliable approach to predict Rayleigh number.

6 Conclusion

In present work, the numerical modeling of fluidic equations has been done and subjected to the implicit scheme of finite difference method. For this purpose, a combined algorithmic approach had been selected for discretization of governing equations and these are subjected to adequate boundary conditions. Thermal boundary layer thickness is obtained. Thermal boundary layer thickness comes out numerically has been plotted with respect to Ra and Nu. It can be concluded that the thermal heating in order of lateral distance is continuous and increases with transverse distance. However, numerical thermal boundary layer thickness shows $\pm 10\%$ deviation with analytical boundary layer thickness. Thermal boundary layer shows higher value for high value of Pr. It can be concluded that for high value of Pr the thermal boundary layer will form larger. Rayleigh number varies from 10^6 to 10^9 so laminar behavior of convection was observed. It is clear from the linear relation that Rayleigh number is higher for high for value of Prandtl number which is proved in fuzzy logic. The Prandtl number increases the value of Nusselt number increases. Water vapour showed the turbulence behavior. For a higher value of Prandtl number the turbulence behavior observed. Prandtl number becomes dominating in convection as Pr approaches to 1. Curve with $Pr = 0.9978$ contains noise. The Prandtl number for air is away from critical value so it shows the ideal laminar behavior. Fuzzy logic showed $\pm 4\%$ agreement with experimental results which proves the fuzzy logic as reliable approach.

References

1. Som, S.K.: Introduction to Heat Transfer. PHI Learning Pvt. Ltd., New Delhi (2008)
2. Sparrow, E.M., Gregg, J.L.: Laminar free convection heat transfer from the outer surface of a vertical circular cylinder. Trans. ASME **78**, 1823–1829 (1956)

3. Blottner, F.G.: Finite difference methods of solution of the boundary-layer equations. Am. Inst. Aeronaut. Astronaut. J. **82**, 193–205 (1970)
4. Douglas Jr., J., Russell, T.F.: Numerical methods for convection-dominated diffusion problems based on combining the method of characteristics with finite element or finite difference procedures. SIAM J. Numer. Anal. **19**(5), 871–885 (1982)
5. Saha, S.C., Hossain, M.A.: Natural convection flow with combined buoyancy effects due to thermal and mass diffusion in a thermally stratified media. Nonlinear Anal. Model. Control **9** (1), 89–102 (2004)
6. Fahiminia, M., Naserian, M., Goshayeshi, H., Majidian, D.: Investigation of natural convection heat transfer coefficient on extended vertical base plates. Energy Power Eng. **3**, 174–180 (2011)
7. Zhanlav, T., Ulziibayar, V.: The best finite-difference scheme for the Helmholtz equation. Am. J. Comput. Math. **2**, 207–212 (2012)
8. Shiferaw, A., Mittal, R.C.: Fast finite difference solutions of the three dimensional Poisson's equation in cylindrical coordinates. Am. J. Comput. Math. **3**, 356–361 (2013)
9. Ahmad, I., Bilal, M.: Numerical solution of Blasius equation through neural network algorithms. Am. J. Comput. Math. **4**, 223–232 (2014)
10. Petrova, L.: The peculiarity of numerical solving the Euler and Navier-Stokes equations. Am. J. Comput. Math. **4**, 304–310 (2014)
11. Zadeh, L.A.: Fuzzy sets. J. Inf. Control **8**, 338–353 (1965)
12. Acilar, A.M., Arslan, A.: Optimization of multiple input–output fuzzy membership functions using clonal selection algorithm. Expert Syst. Appl. **38**(3), 1374–1381 (2011)
13. Yousefi, T., Harsini, I. Ashjaee, M.: Free convection from an isothermal horizontal cylinder in a vertical channel. In: International Conference on Heat Transfer in Components and Systems for Sustainable Energy Technologies (Heat SET), Chambery (2007)
14. Karami, A., Yousefi, T., Rezaei, E., Amiri, A.: Modeling of the free convection heat transfer from an isothermal horizontal cylinder in a vertical channel via the fuzzy logic. Int. J. Multiphys. **6**(1), 7–16 (2012)
15. Belarbi, K., Titel, F., Bourebia, W., Benmahammed, K.: Design of Mamdani fuzzy logic controllers with rule base minimisation using genetic algorithm. J. Eng. Appl. Artif. Intell. **18**, 875–880 (2005)
16. Murlidhar, K., Biswas, G.: Advanced Engineering Fluid Mechanics. Narosa Publishing House, New Delhi (1996)
17. Pedrycz, W.: Why triangular membership functions? J. Fuzzy Sets Syst. **64**, 21–30 (1994)

Availability Analysis of the Butter Oil Processing Plant Using Intuitionistic Fuzzy Differential Equations

Neha Singhal[✉] and S.P. Sharma

Department of Mathematics, Indian Institute of Technology Roorkee,
Roorkee 247667, India
nehasinghal.iitr@gmail.com, sspprfma@iitr.ac.in

Abstract. The main objective of this manuscript is to discuss the availability analysis of the industrial plant. Conventionally availability studies assume that probability in Markov models are accurate. However in reality, data is either insufficient or contain uncertainty which violates this assumption. Keeping this in view the availability of Butter oil processing plant is evaluated after developing the intuitionistic fuzzy differential equations for the system by using its Markov model. (α, β)-Cut method has been used to evaluate intuitionistic fuzzy availability of the system.

Keywords: Availability · Intuitionistic fuzzy differential equations · Markov model · Runge-Kutta fourth order method

1 Introduction

The majority of industrial systems are repairable and consist of various subsystems and each subsystem is composed of many components. The probability that the system performs in expected manner depends directly on the performance of each of its components. With the growing complexity of systems, the study of their reliability and availability becomes more important. The understanding of this role requires an attempt to characterize, study and examine the systems behavior by shrinking the likelihood of the failures and thus increasing their operational availability and designed life.

For measuring the performance of the system, many techniques such as event tree, fault tree analysis (FTA), petri nets (PNs), reliability block diagrams (RBDs) and Markovian approach etc. are available in the literature [1,3].

These independent models analyze reliability of a repairable system using different approaches and have been applied in different scenario. Out of these, Markov analysis is considered to be the most widespread technique being used nowadays.

© Springer Nature Singapore Pte Ltd. 2017
K. Deep et al. (eds.), *Proceedings of Sixth International Conference on Soft Computing for Problem Solving*, Advances in Intelligent Systems and Computing 546, DOI 10.1007/978-981-10-3322-3_32

In various engineering problems, the binary assumption in reliability theory is not acceptable. In 1965, L.A. Zadeh [4] discussed the basic concepts of fuzzy set theory. But in real life, there are many situations concerned with the degree of hesitation. Then in 1983 Atanassov [5,6] introduced the concept of intuitionistic fuzzy set (IFS) theory as the generalization of fuzzy set theory. Many authors [7,8] worked in theoretical as well as in practical applications of intuitionistic fuzzy set theory.

Thus binary state assumption in reliability theory is replaced by intuitionistic fuzzy state assumption. Kumar et al. [9] discussed fuzzy reliability analysis of dual-fuel steam turbine propulsion system in LNG carriers considering data uncertainty. Garg [10] proposed an approach for analyzing the reliability of industrial system using fuzzy Kolmogorov differential equations. Knezevic and Odoom [2] discussed reliability modeling of repairable systems using petrinets and fuzzy lambda-tau methodology. For the evaluation of reliability, the system is mathematically modeled in terms of the differential equations from its transition diagram having uncertainties in the involved parameters. Fuzzy differential equations have also been studied [11–13]. Garg [14] discussed an approach for solving fuzzy differential equations using Runge-Kutta and Biogeography-based optimization. Many authors [15,16] discussed intuitionistic fuzzy differential equations.

In this present paper intuitionistic fuzzy differential equations have been derived with the help of the existing crisp Markov model of Butter oil processing plant. The rest of the paper has been divided into five sections. In Sect. 2 some basic definitions are given. Section 3 describes the methodology for the availability analysis of the system by solving the intuitionistic fuzzy differential equations. In Sect. 4 case study of butter oil processing plant with the given assumptions has been discussed. The final results by considering the case of transient is discussed in Sect. 5.

2 Preliminaries

In (1983) Atanassov [5] proposed the concept of IFS as the extension of the notion of fuzzy set theory. In IFS theory, the characteristic function of the element x in the universe X is expressed in terms of their membership (called acceptance) as well as non membership (called rejection) value such that their sum always belongs to unit interval $[0, 1]$. A brief description of main concepts and definitions [6,7] regarding to intuitionistic fuzzy set is given below.

2.1 Intuitionistic Fuzzy Set (IFS)

Definition: Let X be a universe of discourse. Then the IFS $\tilde{A}$ in X is given by $\tilde{A} = \{< x, \mu_{\tilde{A}}(x), \nu_{\tilde{A}}(x) >| \ x \in X\}$, where the functions

$$\mu_{\tilde{A}} : X \to [0, 1] \text{ and } \nu_{\tilde{A}} : X \to [0, 1]$$

are subjected to the condition $0 \leq \mu_{\tilde{A}}(x) + \nu_{\tilde{A}}(x) \leq 1 \; \forall x \in X$. The values $\mu_{\tilde{A}}(x)$ and $\nu_{\tilde{A}}(x)$ represent the membership and non membership degree of element x to a set $\tilde{A}$ respectively.

2.2 (α, β)-Cut

Definition: An $(\alpha, \beta)-$cut of IFS $\tilde{A}$ denoted as $\tilde{A}[\alpha, \beta]$ [16] is defined by $\tilde{A}[\alpha, \beta] = \tilde{A}^{\alpha} \cap \tilde{A}_{\beta}$, where $\tilde{A}^{\alpha} = \{x \in X \mid \mu_{\tilde{A}}(x) \geq \alpha\}$ and $\tilde{A}_{\beta} = \{x \in X \mid \nu_{\tilde{A}}(x) \leq \beta\}$ for $\alpha \in (0, 1]$ and $\beta \in [0, 1)$ such that $\alpha + \beta \leq 1$.

In this paper, we separately define $\tilde{A}^{0}$ as the closure of the union of all $\tilde{A}^{\alpha}$'s for $\alpha \in (0, 1]$. Similarly, $\tilde{A}_{1}$ as the closure of the union of all $\tilde{A}_{\beta}$'s for $\beta \in [0, 1)$.

2.3 Convex Intuitionistic Fuzzy Set

Definition: An IFS $\tilde{A}$ in universe U is convex iff its membership function $\mu_{\tilde{A}}(x)$ is fuzzy convex while non membership function $\nu_{\tilde{A}}(x)$ is fuzzy concave. i.e.

$$\mu_{\tilde{A}}(\lambda x_1 + (1 - \lambda x_2)) \geq min(\mu_{\tilde{A}}(x_1), \mu_{\tilde{A}}(x_2))$$

and

$$\nu_{\tilde{A}}(\lambda x_1 + (1 - \lambda x_2)) \leq max(\nu_{\tilde{A}}(x_1), \nu_{\tilde{A}}(x_2))$$

$\forall x_1, x_2 \in U, \; 0 \leq \lambda \leq 1$

2.4 Normal Intuitionistic Fuzzy Set

Definition: An IFS $\tilde{A}$ in X is normal if there exists at least one point $x_0 \in X$ such that $\mu_{\tilde{A}}(x_0) = 1$.

2.5 Intuitionistic Fuzzy Number (IFN)

Definition: An intuitionistic fuzzy subset $\tilde{A} = \{< x, \mu_{\tilde{A}}(x), \nu_{\tilde{A}}(x) >\mid x \in R\}$ of the real line R is called Intuitionistic Fuzzy Number (IFN) if

1. $\tilde{A}$ is normal and convex IFS,
2. $\mu_{\tilde{A}}(x)$ is upper semi-continuous and $\nu_{\tilde{A}}(x)$ is lower semi-continuous,
3. $\tilde{A} = \{x \in X, \nu_{\tilde{A}}(x) < 1\}$ is bounded.

2.6 Triangular Intuitionistic Fuzzy Number (TIFN)

A TIFN $\tilde{A}$ with parameters $a_1' \leq a_1 \leq a_2 \leq a_3 \leq a_3'$ is a subset of IFS in R, denoted as $\tilde{A} = \langle (a_1, a_2, a_3); (a_1', a_2, a_3') \rangle$ with membership and non-membership functions defined respectively by

$$\mu_{\tilde{A}}(x) = \begin{cases} \frac{x - a_1}{a_2 - a_1}, & a_1 \leq x \leq a_2 \\ \frac{a_3 - x}{a_3 - a_2}, & a_2 \leq x \leq a_3 \\ 0, & \text{otherwise} \end{cases} \quad \text{and} \quad \nu_{\tilde{A}}(x) = \begin{cases} \frac{a_2 - x}{a_2 - a_1'}, & a_1' \leq x \leq a_2 \\ \frac{x - a_2}{a_3' - a_2}, & a_2 \leq x \leq a_3' \\ 1, & \text{otherwise} \end{cases}$$

3   Methodology

For the evaluation of reliability, the system is mathematically modeled in terms of the differential equations from its transition diagram having uncertainties using markov model. These equations are then converted into intuitionistic fuzzy differential equations for handling the uncertainties.

Strategy followed through this approach is described here.

Step 1. Collection of the data from various resources: Information in the form of system components' failure rates $(\lambda's)$ and repair rates $(\mu's)$ from the database, historical records, literature and expert opinion etc.

Step 2. Conversion of crisp numbers into vague numbers: As the collected data is either out of date or collected under different environmental conditions, leads to the problem of uncertainty in the failure and repair rates. So crisp numbers in the extracted data are converted into intuitionistic fuzzy numbers having known spreads as suggested by system reliability analyst.

Step 3. In this step, a the system of intuitionistic fuzzy differential equations has been derived with the help of the markov model of the system. A linear first order intuitionistic fuzzy differential equation can be written as:

$$\frac{d\tilde{y}(t)}{dt} = \tilde{a}\tilde{y}(t) + g(t), \text{ with initial conditions } \tilde{y}(0) = \tilde{\gamma}, \tag{1}$$

where $\tilde{\gamma}$ and $\tilde{a}$ are triangular intuitionistic fuzzy numbers with $g(t)$ as continuous function on the interval I.

Let $\tilde{y}(t)$ be the intuitionistic fuzzy subset of real numbers for $t \in I$. Let $\tilde{y}(t)[\alpha, \beta] = \tilde{y}(t)^\alpha \cap \tilde{y}(t)_\beta$ where $\tilde{y}(t)^\alpha$ and $\tilde{y}(t)_\beta$ are closed and bounded intervals for all t. Let

$$\tilde{y}(t)^\alpha = [\tilde{y}(t)^\alpha_{(L)}, \tilde{y}(t)^\alpha_{(R)}]$$
$$\tilde{y}(t)_\beta = [\tilde{y}(t)_{\beta(L)}, \tilde{y}(t)_{\beta(R)}],$$

where $\tilde{y}(t)^\alpha_{(L)}$ and $\tilde{y}(t)^\alpha_{(R)}$ are functions of t and α. $\tilde{y}(t)_{\beta(L)}$ and $\tilde{y}(t)_{\beta(R)}$ are functions of t and β.

Assume that all $\tilde{y}^\alpha_{(L)}, \tilde{y}^\alpha_{(R)}, \tilde{y}_{\beta(L)}$ and $\tilde{y}_{\beta(R)}$ are continuously differentiable functions on t for all α and β. Now substitute the (α, β)-cuts of $\tilde{y}(t)$ into Eq. (1). Then using the concepts of interval arithmetic, system of differential equations (1) reduces to following differential equations:

$$\tilde{y}'(t)^\alpha_{(L)} = b + g(t), \tag{2}$$

where $b = \min(\tilde{a}_{(L)}^{\alpha} \tilde{y}_{(L)}^{\alpha}, \tilde{a}_{(L)}^{\alpha} \tilde{y}_{(R)}^{\alpha}, \tilde{a}_{(R)}^{\alpha} \tilde{y}_{(L)}^{\alpha}, \tilde{a}_{(R)}^{\alpha} \tilde{y}_{(R)}^{\alpha})$

$$\tilde{y}'(t)_{(R)}^{\alpha} = c + g(t), \tag{3}$$

where $c = \max(\tilde{a}_{(L)}^{\alpha} \tilde{y}_{(L)}^{\alpha}, \tilde{a}_{(L)}^{\alpha} \tilde{y}_{(R)}^{\alpha}, \tilde{a}_{(R)}^{\alpha} \tilde{y}_{(L)}^{\alpha}, \tilde{a}_{(R)}^{\alpha} \tilde{y}_{(R)}^{\alpha})$

$$\tilde{y}'(t)_{\beta(L)} = b' + g(t), \tag{4}$$

where $b' = \min(\tilde{a}_{\beta(L)} \tilde{y}_{\beta(L)}, \tilde{a}_{\beta(L)} \tilde{y}_{\beta(R)}, \tilde{a}_{\beta(R)} \tilde{y}_{\beta(L)}, \tilde{a}_{\beta(R)} \tilde{y}_{\beta(R)})$

$$\tilde{y}'(t)_{\beta(R)} = c' + g(t), \tag{5}$$

where $c' = \max(\tilde{a}_{\beta(L)} \tilde{y}_{\beta(L)}, \tilde{a}_{\beta(L)} \tilde{y}_{\beta(R)}, \tilde{a}_{\beta(R)} \tilde{y}_{\beta(L)}, \tilde{a}_{\beta(R)} \tilde{y}_{\beta(R)})$

with the initial conditions $\tilde{y}(0)_{(L)}^{\alpha} = \tilde{\gamma}_{(L)}^{\alpha}, \tilde{y}(0)_{(R)}^{\alpha} = \tilde{\gamma}_{(R)}^{\alpha}, \tilde{y}(0)_{\beta(L)} = \tilde{\gamma}_{\beta(L)}$ and $\tilde{y}(0)_{\beta(R)} = \tilde{\gamma}_{\beta(R)}$.

Then these converted ordinary differential equations (2)–(5) for each $\alpha, \beta \in [0,1]$ can be solved by Runge-Kutta fourth order method. Clearly $\tilde{y}(t)$ is an intuitionistic fuzzy solution for all t if the obtained values of $\tilde{y}(t)_{(L)}^{\alpha}$, $\tilde{y}(t)_{(R)}^{\alpha}$, $\tilde{y}(t)_{\beta(L)}$ and $\tilde{y}(t)_{\beta(R)}$ define the (α, β)-cuts

$$([\tilde{y}(t)_{(L)}^{\alpha}, \tilde{y}(t)_{(R)}^{\alpha}], [\tilde{y}(t)_{\beta(L)}, \tilde{y}(t)_{\beta(R)}])$$

of triangular intuitionistic fuzzy numbers.

The basic assumptions used in this methodology are :

- Failure rates and repair rates are independent of each other and their unit is per day.
- There is no simultaneous failure of the systems.
- Subsystem B fails through reduced state only.
- Repaired components function like new components and switch-over devices used for standby systems are perfect.

4   Case Study

Gupta et al. [17] used Markov model with crisp parameters to calculate crisp reliability. In the present paper intuitionistic fuzzy parameters are used in place of crisp parameters and the above described methodology is applied to Butter Oil processing plant. Markov model of Butter Oil Processing plant is shown in Fig. 1

4.1   System Description

A butter-oil manufacturing plant is a complex engineering system comprising of various subsystems situated mostly in northern part of India. The main subsystems of plant are briefly described below [17]:

- Separator (A): In this subsystem, motor, bearings and gearbox are connected in series. Failure of this subsystem causes the complete failure of the system.
- Pasteuriser (B): Herein two units of pasteurisers are arranged in parallel configuration with one operative and other in cold standby. Complete failure of pasteuriser occurs when both the components fail.
- Continuous Butter Making (C): The CBM consists of gearbox, motor and bearings in series. Failure of this subsystem causes the complete failure of the system.
- Melting Vats (D): This system consists of monoblock, motors, pumps and bearings in series. Failure of this subsystem causes the complete failure of the system.
- Butter-Oil Clarifier (E): The unit consists of gearbox and motor in series. Failure of this subsystem will cause the complete failure of the system.
- Packaging (F): This subsystem consists of printed circuit board and pneumatic cylinder in series. Failure of this subsystem causes the complete failure of the system.

The failure and repair rates corresponding to each subsystem of the system are given as:

Failure rate $(\lambda) = [0.008\ 0.0054\ 0.0027\ 0.0009\ 0.0027\ 0.0055\ 0.01111]$

Repair rate $(\mu) = [0.41\ 0.40\ 0.70\ 0.30\ 0.65\ 6.00]$

As the data collected for evaluation of reliability contains uncertainty. So, to account for uncertainties and vagueness in data, the obtained crisp data are converted into intuitionistic fuzzy numbers as suggested by decision makers/system analyst. An input data for intuitionistic fuzzy failure rate (λ_i) and intuitionistic fuzzy repair rate (μ_i) for ith component of the system is in the form of triangular intuitionistic fuzzy numbers with 15 % in both the directions with membership and 20 % in both the directions with non-membership functions.

4.2 Notations:

- B_1 indicates that subsystem B is working in reduced state.
- λ_i, i = 1, 2, ..., 7 represent the failure rates of the subsystems A, C, D, E, F, B_1 and B respectively.
- μ_i, i = 1, 2, ..., 6 represent the repair rates of the subsystems A, C, D, E, F and B respectively.
- The symbols a, b, c, d, e and f represent the failed state of the subsystems A, B, C, D, E and F respectively.
- $P_j(t)$, j = 1, 2, ..., 13 represent the probability that the system is in jth state at time t.

4.3 Mathematical Formulation

Using the concepts of probability and markov modeling, following intuitionistic fuzzy differential equations corresponding to the transition diagram (Fig. 1) are formulated as:

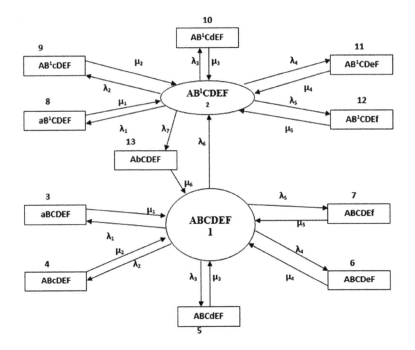

Fig. 1. Transition diagram of Butter oil processing plant

$$\frac{d\tilde{P}_1(t)}{dt} \oplus \tilde{\delta}_1 \tilde{P}_1(t) = \sum_{j=1}^{5} \tilde{\mu}_j \tilde{P}_{j+2}(t) \oplus \tilde{\mu}_6 \tilde{P}_{13}(t) \tag{6}$$

$$\frac{d\tilde{P}_2(t)}{dt} \oplus \tilde{\delta}_2 \tilde{P}_2(t) = \sum_{j=1}^{5} \tilde{\mu}_j \tilde{P}_{j+7}(t) \oplus \tilde{\lambda}_6 \tilde{P}_1(t) \tag{7}$$

$$\frac{d\tilde{P}_{i+2}(t)}{dt} \oplus \tilde{\mu}_i \tilde{P}_{i+2}(t) = \tilde{\lambda}_i \tilde{P}_1(t), \qquad i = 1, 2, ...5 \tag{8}$$

$$\frac{d\tilde{P}_{i+7}(t)}{dt} \oplus \tilde{\mu}_i \tilde{P}_{i+7}(t) = \tilde{\lambda}_i \tilde{P}_2(t), \qquad i = 1, 2, ...5 \tag{9}$$

$$\frac{d\tilde{P}_{13}(t)}{dt} \oplus \tilde{\mu}_6 \tilde{P}_{13}(t) = \tilde{\lambda}_7 \tilde{P}_2(t) \tag{10}$$

with $\tilde{\delta}_1 = \sum_{j=1}^{6} \tilde{\lambda}_j$ and $\tilde{\delta}_2 = \sum_{j=1}^{5} \tilde{\lambda}_j \oplus \tilde{\lambda}_7$

with the initial conditions:
$\tilde{P}_1(0) = \langle(0.94, 0.96, 0.98); (0.935, 0.96, 0.985)\rangle$
$\tilde{P}_2(0) = \langle(0.004, 0.005, 0.006); (0.0035, 0.005, 0.0065)\rangle$ and
$\tilde{P}_j(0) = \langle(0, 0, 0); (0, 0, 0)\rangle$ for $j = 3$ to 13.

The availability function $\tilde{A}v(t)$ of the system in terms of $\tilde{P}_1(t)$ and $\tilde{P}_2(t)$ can be obtained by

$$\tilde{A}v(t) = \tilde{P}_1(t) \oplus \tilde{P}_2(t)$$

5 Results

Intuitionistic fuzzy system availability is evaluated by the set of first order intuitionistic fuzzy differential equations at different $(\alpha, \beta)-$ cuts and mission time $t = 365$ days. Solution obtained from the differential equations (6)–(10) is summarized in Table 1 for $\alpha, \beta = 0, 0.2, 0.4, 0.8, 1.0$. From the analysis, it has been observed that results computed by proposed approach are better than the existing results. Based on these probabilities, the corresponding $(\alpha, \beta)-$ cut of the overall system availability, for the mission time $t = 365$ days by proposed approach lies in the interval $[0.9036677, 0.9438700]$. Similar effect on the overall system availability at different level of uncertainties is computed and summarized in Table 2. From these results it has been concluded that

- The results computed by the traditional method (crisp) [17] do not give the exact idea about the behavior of the system. As these methods deal with the precise data and cannot deal with the data containing uncertainties.
- Results provided by the proposed method deal with the various degrees of membership and non-membership functions. For instance, the system availability corresponding to $\alpha = 0.7$ and $\beta = 0.1$ lies in $[0.9212317, 0.9263048]$.
- On the other hand, the results are computed by the proposed approach by handling the uncertainties in the data in the form of intuitionistic triangular fuzzy numbers. From this, corresponding to different presumption levels, the system availability has been computed for $t = 365$ days.

The complete results of system availability are summarized in Table 2. With the help of (α, β)-cut, approximated value of membership and non-membership function of intuitionistic fuzzy availability at $t = 365$ days is defined here.

$$\mu_{\tilde{A}v}(x) = \begin{cases} \frac{x - 0.9036677}{0.0201005}, & 0.9036677 \leq x \leq 0.9237682 \\ 1, & x = 0.9237682 \\ \frac{0.9438700 - x}{0.0201018}, & 0.9237682 \leq x \leq 0.9438700 \\ 0, & \text{otherwise} \end{cases}$$

$$\nu_{\tilde{A}v}(x) = \begin{cases} \frac{0.9237682 - x}{0.0253642}, & 0.8984040 \leq x \leq 0.9237682 \\ 0, & x = 0.9237682 \\ \frac{x - 0.9237682}{0.0253666}, & 0.9237682 \leq x \leq 0.9491348 \\ 1, & \text{otherwise} \end{cases}$$

System availability at $t = 365$ days in term of intuitionistic fuzzy set is shown in Fig. 2.

Table 1. Solution of Intuitionistic fuzzy Kolmogorov's differential equation at t=365 days

j	$\bar{P}_j^\alpha$ for $\alpha = 0$		$\bar{P}_j^\alpha$ for $\alpha = 0.2$		$\bar{P}_j^\alpha$ for $\alpha = 0.4$		$\bar{P}_j^\alpha$ for $\alpha = 0.6$		$\bar{P}_j^\alpha$ for $\alpha = 0.8$		$\bar{P}_j^\alpha$ for $\alpha = 1$	
	$\bar{P}_{j(L)}^\alpha$	$\bar{P}_{j(R)}^\alpha$	$\bar{P}_{j(L)}^\alpha$	$\bar{P}_{j(R)}^\alpha$	$\bar{P}_{j(L)}^\alpha$	$\bar{P}_{j(R)}^\alpha$	$\bar{P}_{j(L)}^\alpha$	$\bar{P}_{j(R)}^\alpha$	$\bar{P}_{j(L)}^\alpha$	$\bar{P}_{j(R)}^\alpha$	$\bar{P}_{j(L)}^\alpha$	$\bar{P}_{j(R)}^\alpha$
1	0.6063945	0.6315314	0.6087487	0.6289154	0.6111551	0.6263127	0.6136056	0.6237259	0.6160933	0.6211579	0.6186124	0.6186124
2	0.2972732	0.3123386	0.2989390	0.3109342	0.3005526	0.3095165	0.3021222	0.3080829	0.3036547	0.3066305	0.3051558	0.3051558
3	0.0118338	0.01232287	0.0118794	0.0122719	0.0119262	0.0122212	0.0119738	0.0121708	0.0120222	0.01212076	0.0120712	0.0120712
4	0.0081875	0.0085259	0.0082191	0.0084906	0.0082514	0.0084555	0.0082844	0.0084207	0.0083179	0.0083861	0.0083518	0.0083518
5	0.0023391	0.0024359	0.0023482	0.0024259	0.0023574	0.0024158	0.0023669	0.0024059	0.0023765	0.0023959	0.0023862	0.0023862
6	0.0018195	0.0018947	0.0018265	0.0018868	0.0018337	0.0018790	0.0018410	0.0018713	0.0018485	0.0018636	0.0018559	0.0018559
7	0.0025191	0.0026233	0.0025288	0.0026125	0.0025388	0.0026016	0.0025489	0.0025909	0.0025593	0.0025803	0.0025697	0.0025697
8	0.0057988	0.0060941	0.0058315	0.0060666	0.0058632	0.0060389	0.0058941	0.0060109	0.0059824	0.0059824	0.0059535	0.0059535
9	0.0040119	0.0042164	0.0040347	0.0041974	0.0040566	0.0041782	0.004077	0.0041588	0.0040987	0.0041391	0.0041191	0.0041191
10	0.0011464	0.0012047	0.0011529	0.0011993	0.0011591	0.0011938	0.0011652	0.0011883	0.0011711	0.0011826	0.0011769	0.0011769
11	0.0008914	0.0009369	0.0008965	0.0009327	0.0009014	0.0009284	0.0009061	0.0009241	0.0009107	0.00091978	0.0009153	0.0009153
12	0.0012346	0.0012974	0.0012415	0.0012915	0.0012483	0.0012856	0.0012548	0.0012796	0.0012612	0.0012736	0.0012675	0.0012675
13	0.0005499	0.0005778	0.0005530	0.0005752	0.0005560	0.0005726	0.0005589	0.0005699	0.0005618	0.0005673	0.0005645	0.0005645

j	$\bar{P}_{j\beta}$ for $\beta = 0$		$\bar{P}_{j\beta}$ for $\beta = 0.2$		$\bar{P}_{j\beta}$ for $\beta = 0.4$		$\bar{P}_{j\beta}$ for $\beta = 0.6$		$\bar{P}_{j\beta}$ for $\beta = 0.8$		$\bar{P}_{j\beta}$ for $\beta = 1$	
	$\bar{P}_{j\beta(L)}$	$\bar{P}_{j\beta(R)}$	$\bar{P}_{j\beta(L)}$	$\bar{P}_{j\beta(R)}$	$\bar{P}_{j\beta(L)}$	$\bar{P}_{j\beta(R)}$	$\bar{P}_{j\beta(L)}$	$\bar{P}_{j\beta(R)}$	$\bar{P}_{j\beta(L)}$	$\bar{P}_{j\beta(R)}$	$\bar{P}_{j\beta(L)}$	$\bar{P}_{j\beta(R)}$
1	0.6035804	0.6349551	0.6063950	0.6316373	0.6093271	0.6283387	0.6123527	0.6250639	0.6154528	0.6218193	0.6186124	0.6186124
2	0.2948236	0.3141797	0.2970815	0.3124241	0.2992221	0.3106494	0.3012694	0.3088507	0.3032423	0.3070221	0.3051558	0.3051558
3	0.0117794	0.0123896	0.0118339	0.0123249	0.0118907	0.0122606	0.0119495	0.0121968	0.0120097	0.0121336	0.0120712	0.0120712
4	0.0081499	0.0085721	0.0081876	0.0085273	0.0082269	0.0084828	0.0082676	0.0084387	0.0083092	0.0083950	0.0083518	0.0083518
5	0.0023284	0.0024491	0.0023392	0.0024363	0.0023504	0.0024236	0.0023621	0.0024110	0.0023739	0.0023985	0.0023862	0.0023862
6	0.0018112	0.0019049	0.0018196	0.0018949	0.0018283	0.0018851	0.0018729	0.0018753	0.0018466	0.0018656	0.0018559	0.0018559
7	0.0025075	0.0026375	0.0025191	0.0026238	0.0025312	0.0026101	0.0025438	0.0025965	0.0025566	0.0025830	0.0025697	0.0025697
8	0.0057504	0.0061301	0.0057949	0.0060958	0.0058371	0.0060611	0.0058773	0.0060259	0.005916	0.0059901	0.0059535	0.0059535
9	0.0039785	0.0042413	0.004009	0.0042175	0.0040385	0.0041935	0.0040663	0.0041692	0.0040931	0.0041444	0.0041191	0.0041191
10	0.0011369	0.0012118	0.0011457	0.0012050	0.0011539	0.0011982	0.0011619	0.0011912	0.0011695	0.0011842	0.0011769	0.0011769
11	0.0008839	0.00094249	0.0008908	0.0009372	0.0008974	0.0009391	0.0009036	0.0009264	0.0009095	0.0009209	0.0009153	0.0009153
12	0.0012244	0.0013050	0.0012338	0.0012977	0.0012427	0.0012903	0.0012513	0.0012829	0.0012595	0.0012752	0.0012675	0.0012675
13	0.0005454	0.0005812	0.0005496	0.0005779	0.0005560	0.0005726	0.0005573	0.0005714	0.0005609	0.0005679	0.0005645	0.0005645

Table 2. System availability at t = 365 days

$\alpha, \beta \downarrow$	$\tilde{Av}^{\alpha}_{(L)}$	$\tilde{Av}^{\alpha}_{(R)}$	$\tilde{Av}_{\beta(L)}$	$\tilde{Av}_{\beta(R)}$
0	0.9036677	0.9438700	0.9237682	0.9237682
0.1	0.9056777	0.9418598	0.9212317	0.9263048
0.2	0.9076877	0.9398496	0.9186951	0.9288415
0.3	0.9096977	0.9378394	0.9161586	0.9313781
0.4	0.9117077	0.9358292	0.9136221	0.9339147
0.5	0.9137178	0.9338190	0.9110856	0.9364514
0.6	0.9157278	0.9318089	0.9085492	0.9389881
0.7	0.9177379	0.9297987	0.9060128	0.9415247
0.8	0.9197480	0.9277885	0.9034765	0.9440614
0.9	0.9217581	0.9257784	0.9009402	0.9465981
1.0	0.9237682	0.9237682	0.8984040	0.9491348

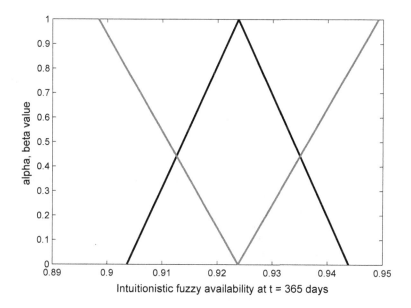

Fig. 2. Intuitionistic fuzzy availability (membership and non-membership functions are shown by black and red lines respectively) (Color figure online)

6   Conclusion

In the present article, authors have analyzed the reliability analysis of Butter oil processing plant using intuitionistic fuzzy differential equations. Based on the summary given in the tabular form, system analyst can predict the behavior of the system in more consistent manner. This methodology will assist the plant managers in design modifications to reduce the failures and to help in maintenance decision making.

Acknowledgement. The authors are grateful to the reviewers for their valuable comments and suggestions. The first author (Neha Singhal) acknowledges the University Grants Commission (UGC), New Delhi India, for financial support to carry out this work.

References

1. Dhillon, B.S., Singh, C.: Engineering Reliability: New Techniques and Applications. Wiley, New York (1981)
2. Knezevic, J., Odoom, E.R.: Reliability modeling of repairable systems using petri nets and fuzzy Lambda-Tau methodology. Reliab. Eng. Syst. Saf. **73**(1), 1–17 (2001)
3. Sharma, S.P., Garg, H.: Behavioral analysis of a urea decomposition system in a fertilizer plant. Int. J. Ind. Syst. Eng. **8**(3), 271–297 (2011)
4. Zadeh, L.A.: Fuzzy sets. Inf. Control **8**, 338–353 (1965)
5. Atanassov, K.T.: Intuitionistic fuzzy sets. In: VII ITKR Session, Sofia, 20–23 June 1983
6. Atanassov, K.T.: Intuitionistic fuzzy sets. Fuzzy Sets Syst. **20**(1), 87–96 (1986)
7. Atanasov, K.T.: Intuitionistic Fuzzy Sets: Theory and Applications. Springer Physica-Verlag, Heidelberg (1999)
8. De, S.K., Biswas, R., Roy, A.R.: An application of intuitionistic fuzzy sets in medical diagnosis. Fuzzy Sets Syst. **117**(2), 209–213 (2001)
9. Komal, C.D., Lee, S.: Fuzzy reliability analysis of dual-fuel steam turbine propulsion system in LNG carries considering data uncertainty. J. Nat. Sci. Eng. **23**, 148–164 (2015)
10. Garg, H.: An approach for analyzing the reliability of industrial system using fuzzy Kolmogorovs differential equations. Arab. J. Sci. Eng. **40**(3), 975–987 (2015)
11. Kaleva, O.: Fuzzy differential equations. Fuzzy Sets Syst. **24**(3), 301–317 (1987)
12. Buckley, J.J., Feuring, T.: Fuzzy initial problem for Nth-order linear differential equations. Fuzzy Sets Syst. **121**, 247–255 (2001)
13. Buckley, J.J., Feuring, T., Hayashi, Y.: Linear systems of first order differential equations: fuzzy differential equations. Soft Comput. **6**, 415–421 (2002)
14. Garg, H.: A novel approach for solving fuzzy differential equations using Runge-Kutta and Biogeography-based optimization. J. Intell. Fuzzy Syst. **30**, 2417–2429 (2016)
15. Ettoussi, R., Melliani, S., Elomari, M., Chadli, L.S.: Solution of intuitionistic fuzzy differential equations by successive approximations method. Not. Intuitionistic Fuzzy Sets **21**(2), 51–62 (2015)
16. Singhal, N., Sharma, S.P.: Solution of system of first order linear differential equations in intuitionistic fuzzy environment. Not. Intuitionistic Fuzzy Sets **22**(3), 70–79 (2016)
17. Gupta, P., Lal, A.K., Sharma, R.K., Singh, J.: Numerical analysis of reliability and availability of the serial processes in butter-oil processing plant. Int. J. Qual. Reliab. Manag. **22**(3), 303–316 (2005)

Applying Fuzzy Probabilistic PROMETHEE on a Multi-Criteria Decision Problem

Susmita Bandyopadhyay[1]([⊠]) and Indraneel Mandal[2]

[1] Department of Business Administration, The University of Burdwan,
Burdwan 713104, West Bengal, India
bandyopadhyaysusmita2010@gmail.com,
sbandyopadhyay@mba.buruniv.ac.in
[2] Asia Pacific Institute of Management, New Delhi, India
indraneel75@gmail.com

Abstract. This paper proposes fuzzy probabilistic PROMETHEE outranking multi-criteria decision making technique and applies the proposed technique to a real-life multi-criteria case study. In many situations, selection of particular alternatives based on certain criteria found out to be uncertain. In such conflicting situations, probabilistic treatment for the alternatives is required. Since the determination of values of such probabilities is difficult and unrealistic as well, thus this paper proposes fuzzy probability values for the alternatives. The fuzzy probabilistic PROMETHEE outranking technique is applied for the purpose. Numerical example through a case shows the applicability of the proposed technique.

Keywords: Fuzzy probability · Triangular Fuzzy Number · PROMETHEE · Multi-Criteria Decision Analysis technique

1 Introduction

Decision making in practical problem generally consists of multiple criteria. Thus the existing literature shows significant number of Multi-Criteria Decision Analysis (MCDA) techniques in order to solve such problems. A few of such techniques include Analytical Hierarchy Process (AHP), TOPSIS, PROMETHEE, MACBETH, ANP and so on [1]. These techniques have been applied in numerical applications. Some of those research studies the works of Bandyopadhyay and Bhattacharya [2], Kumar et al. [3], Ho et al. [4], Galankashi et al. [5], Lima et al. [6], Shemshadi et al. [7] and so on. Besides, there are hybrid methods as proposed in the literature. for example, Bilisik et al. [8] combined weighted satisfaction score with correlation coefficient; Feng et al. [9] combined multi-objective programming model ad Tabu search algorithm; Scott et al. [10] combined AHP with Quality Function Deployment; Rezaei and Davoodi [11] combined Integer Programming with Genetic Algorithm; Zeydan et al. [12] combined fuzzy AHP with fuzzy TOPSIS. In spite of such vast number of applications, there are still numerous aspects of MCDA techniques which need attention from the researchers and practitioners of the respective fields of study. Some of those aspects include the need of a method to compare the results of MCDA techniques, accounting

© Springer Nature Singapore Pte Ltd. 2017
K. Deep et al. (eds.), *Proceedings of Sixth International Conference on Soft Computing for Problem Solving*, Advances in Intelligent Systems and Computing 546, DOI 10.1007/978-981-10-3322-3_33

for the cases where the consideration of alternatives and/or criteria is uncertain and so on. This paper addresses the problem of uncertainty in the consideration of the alternatives in a MCDA problem. PROMETHEE multi-criteria technique has been applied for the purpose. The technique applied in this paper is briefly described through numerical example. The book of Ishizaka and Nemery [1] can be consulted for the concept of the basic PROMETHEE. The fuzzy probabilistic PROMETHEE as applied in this paper is described in the following section though a case study.

2   Fuzzy Probabilistic PROMETHEE

The following case has been used for this study. A fast food chain plans to open up a new branch in one of the following 3 areas – an open market area, an industrial area, a city center and a densely populated residential area. The area will be selected based on the rental cost, proximity to raw vegetables and meat, number of similar food shops in the area, average number of random customers visiting such shop in the locality. The company has 4 managers who are in conflict in selecting these criteria for particular alternatives and each has his own reason to exclude one or more alternatives for certain criteria. Thus an analyst is hired at cheap rate who selects the best location considering the uncertainty of selecting the alternative based on uncertain criteria. Thus, the criteria and alternatives as considered in this paper are - C1: Rental cost, C2: Proximity to raw vegetables and meat, C3: Number of similar food shops in the area, C4: Average number of random customers visiting such shop in the locality; A1: An open market area, A2: An industrial area, A3: A city center, A4: A densely populated residential area. The four managers (decision makers) assign various preference values in various linguistic terms. The linguistic terms for the criteria and the alternatives as considered in this paper are shown in Tables 1 and 2 respectively.

Table 1. Linguistic terms for criteria

Linguistic terms	Fuzzy number
Little important (LI)	(0.0, 0.10, 0.20)
Moderately important (MI)	(0.10, 0.20, 0.40)
Important (I)	(0.20, 0.40, 0.60)
Very important (VI)	(0.40, 0.60, 0.80)
Absolute importance (AI)	(0.60, 0.80, 1.00)

Table 2. Linguistic terms for alternatives

Linguistic terms	Fuzzy number
Very little important (VLI)	(0.0, 0.0, 0.16)
Little important (LI)	(0.0, 0.16, 0.32)
Moderately important (MI)	(0.16, 0.32, 0.48)
Important (I)	(0.32, 0.48, 0.64)
Very important (VI)	(0.48, 0.64, 0.80)
Extremely important (EI)	(0.64, 0.80, 1.00)

Tables 3 and 4 represent the linguistic preference values from the decision makers for the alternatives and the criteria respectively. To deal with the uncertainty in consideration of the alternatives by the decision maker, uniform fuzzy probability values have been assumed and are provided in Table 5. Thus Tables 1, 2, 3, 4 and 5 serves as input to the PROMETHEE technique.

Table 3. Preferences of decision makers for alternatives

DM1	C1	C2	C3	C4
A1	VI	VI	EI	EI
A2	MI	EI	EI	EI
A3	VI	VI	I	MI
A4	EI	LI	MI	EI

DM2				
A1	VI	MI	LI	LI
A2	EI	EI	I	I
A3	EI	VLI	MI	MI
A4	EI	VI	MI	I

DM3				
A1	MI	VI	VI	EI
A2	I	MI	EI	EI
A3	EI	EI	EI	VI
A4	VI	EI	MI	MI

DM4				
A1	VLI	MI	EI	VI
A2	I	LI	MI	EI
A3	VI	EI	I	EI
A4	EI	I	MI	VI

Table 4. Preferences of decision makers for criteria

	C1	C2	C3	C4
DM1	AI	AI	I	LI
DM2	VI	MI	LI	AI
DM3	VI	AI	MI	I
DM4	AI	VI	VI	LI

At first, the weights of the criteria are calculated just by averaging the fuzzy numbers representing the linguistic preferences from the decision makers. For example, for criterion C1, the linguistic preference values from the four decision makers are AI (0.60, 0.80, 1.00), VI (0.40, 0.60, 0.80), VI, AI. The average of these values are calculated as: $(0.60 + 0.40 + 0.40 + 0.60)/4, (0.80 + 0.60 + 0.60 + 0.80)/4, (1.00 + 0.80 + 0.80 + 1.00)/4 = (0.50, 0.70, 0.90)$. The weights of the criteria are shown in Table 6.

Table 5. Uniform fuzzy numbers for the alternatives from the decision makers

DM1	C1	C2	C3	C4
A1	70, 80	72, 80	80, 90	90, 100
A2	60, 70	60, 65	82, 92	95, 100
A3	80, 90	90, 95	80, 90	80, 90
A4	90, 100	70, 80	72, 78	85, 95

DM2				
A1	75, 85	70, 80	80, 90	70, 80
A2	90, 100	73, 83	80, 90	75, 85
A3	85, 95	70, 80	90, 100	72, 80
A4	85, 95	70, 80	90, 100	82, 90

DM3				
A1	85, 95	95, 100	75, 85	70, 80
A2	90, 100	95, 100	82, 92	72, 80
A3	70, 80	80, 88	80, 90	82, 90
A4	75, 85	82, 87	70, 80	92, 100

DM4				
A1	70, 80	70, 80	95, 97	90, 100
A2	90, 100	70, 80	85, 92	80, 90
A3	80, 90	72, 80	85, 95	70, 80
A4	70, 80	82, 85	70, 80	60, 70

Table 6. Weights of criteria

Criteria	Weight
C1	0.5, 0.7, 0.9
C2	0.425, 0.6, 0.8
C3	0.175, 0.325, 0.5
C4	0.2, 0.35, 0.5

Next, the linguistic preferences for the alternatives in Table 3 are multiplied with the fuzzy probability values as provided in Table 5 to include the uncertainty in considering the alternatives. For example, for decision maker DM1, for criterion C1 and for alternative A1, the linguistic preference is VI (0.48, 0.64, 0.80) (by Table 3) and the respective uniform fuzzy probability number is (70, 80). The mean representation of this fuzzy number is $(70\% + 80\%)/2 = 75\%$. Thus the fuzzy probabilistic preference is calculated as: $0.48 \times 0.75, 0.64 \times 0.75, 0.80 \times 0.75 = 0.36, 0.48, 0.60$. Similarly the other fuzzy probabilistic preferences are calculated. Now these values are multiplied with the fuzzy criteria values as provided in Table 6, in order to get weighted fuzzy probabilistic preferences. For example, the fuzzy probabilistic preference value for DM1, A1, C1 is (0.36, 0.48, 0.60) and the weight of C1 is (0.5, 0.7, 0.9) (Table 6). The weighted fuzzy probabilistic preferences are calculated as: $0.36 \times 0.5, 0.48 \times 0.7, 0.60 \times 0.9 = 0.18, 0.336, 0.54$. The other weighted fuzzy probabilistic preferences (rounded up to 4 decimal places) are also calculated similarly and the results are shown in Table 7.

Table 7. Weighted fuzzy probabilistic preferences

	C1	C2	C3	C4
A1	0.11, 0.22, 0.4032	0.1140, 0.2386, 0.4216	0.0675, 0.1703, 0.3403	0.0772, 0.1827, 0.3375
A2	0.135, 0.2707, 0.4718	0.1132, 0.2363, 0.4220	0.0672, 0.1698, 0.3395	0.0853, 0.1969, 0.3596
A3	0.1633, 0.3137, 0.5305	0.1254, 0.249, 0.4420	0.0637, 0.1637, 0.3292	0.0774, 0.1823, 0.3358
A4	0.1828, 0.3432, 0.5723	0.1251, 0.2495, 0.4424	0.0550, 0.1467, 0.3003	0.0750, 0.1781, 0.3291

PROMETHEE is now applied on Table 7. At first, the preference function values (Table 8) are calculated followed by preference index values (Table 9) and outranking flows (Table 10). Table 10 also shows the final ranking of the alternatives. Thus Table 10 indicates that the alternative A3 is the best alternative (with rank 1), that is, the company can open its new branch at the city center.

Table 8. Preference function values

	C1	C2	C3	C4
A1, A2	−0.024, −0.04667, −0.0686	0.000831, 0.002293, −0.00044	0.000296, 0.000462, 0.000733	−0.00814, −0.01416, −0.02206
A1, A3	−0.05229, −0.08973, −0.1273	−0.0114, −0.01043, −0.02043	0.003791, 0.006644, 0.011095	−0.00019, 0.000449, 0.001717
A1, A4	−0.0718, −0.11918, −0.16909	−0.01113, −0.01096, −0.02082	0.012484, 0.023582, 0.03997	0.002239, 0.004628, 0.00843
A2, A1	0.024, 0.046667, 0.0686	−0.00083, −0.00229, 0.000444	−0.0003, −0.00046, −0.00073	0.008142, 0.014156, 0.022056
A2, A3	−0.02829, −0.04307, −0.0587	−0.01223, −0.01272, −0.01999	0.003496, 0.006182, 0.010362	0.007949, 0.014604, 0.023773
A2, A4	−0.0478, −0.07251, −0.10049	−0.01196, −0.01326, −0.02038	0.012188, 0.02312, 0.039237	0.010381, 0.018784, 0.030485
A3, A1	0.052286, 0.089733, 0.1273	0.011403, 0.01043, 0.020432	−0.00379, −0.00664, −0.0111	0.000193, −0.00045, −0.00172
A3, A2	0.028286, 0.043067, 0.0587	0.012235, 0.012724, 0.019987	−0.0035, −0.00618, −0.01036	−0.00795, −0.0146, −0.02377

(continued)

Table 8. (*continued*)

	C1	C2	C3	C4
A3, A4	−0.01952, −0.02945, −0.04179	0.000271, −0.00053, −0.00039	0.008692, 0.016938, 0.028875	0.002432, 0.004179, 0.006712
A4, A1	0.071805, −0.224, 0.169089	0.011132, 0.010963, 0.020821	−0.01248, −0.02358, −0.03997	−0.00224, −0.00463, −0.00843
A4, A2	0.047805, −0.27067, 0.100489	0.011963, 0.013256, 0.020376	−0.01219, −0.02312, −0.03924	−0.01038, −0.01878, −0.03049
A4, A3	0.019519, −0.31373, 0.041789	−0.00027, 0.000533, 0.000389	−0.00869, −0.01694, −0.02887	−0.00243, −0.00418, −0.00671

Table 9. Preference index values

A1, A2	−0.01, −0.02709, −0.05474
A1, A3	−0.02346, −0.0549, −0.10651
A1, A4	−0.02997, −0.06682, −0.1257
A2, A1	0.009995, 0.027093, 0.054736
A2, A3	−0.01346, −0.02781, −0.05177
A2, A4	−0.01997, −0.03972, −0.07096
A3, A1	0.023459, 0.054904, 0.106508
A3, A2	0.013463, 0.027811, 0.051773
A3, A4	−0.00651, −0.01191, − 0.01919
A4, A1	0.029968, −0.13909, 0.125697
A4, A2	0.019973, −0.16618, 0.070961
A4, A3	0.00651, −0.194, 0.019189

Table 10. Outranking flows and rank of alternatives

	PHI+	PHI−	PHI	Mean	Rank
A1	−0.02114, −0.0496, −0.09565	0.021141, −0.01903, 0.095647	−0.04228, −0.03057, −0.19129	−0.08805	4
A2	−0.00781, −0.01348, −0.02267	0.007814, −0.05516, 0.022666	−0.01563, 0.041674, −0.04533	−0.00643	3
A3	0.010138, 0.023601, 0.046364	−0.01014, −0.09224, −0.04636	0.020275, 0.115838, 0.092728	0.07628	1
A4	0.018817, −0.16642, 0.071949	−0.01882, −0.03948, −0.07195	0.037634, −0.12694, 0.143898	0.018198	2

References

1. Ishizaka, A., Nemery, P.: Multi-Criteria Decision Analysis: Methods and Software. Wiley, US (2013)
2. Bandyopadhyay, S., Bhattacharya, R.: Finding optimum neighbor for routing based on multi-criteria, multi-agent and fuzzy approach. J. Intell. Manufact. **26**(1), 25–42 (2015)
3. Kumar, A., Jain, V., Kumar, S.: A comprehensive environment friendly approach for supplier selection. Omega **42**(1), 109–123 (2014)
4. Ho, L.H., Feng, S.Y., Lee, Y.C., Yen, T.M.: Using modified IPA to evaluate supplier's performance: multiple regression analysis and DEMATEL approach. Expert Syst. Appl. **39** (8), 7102–7109 (2012)
5. Galankashi, M.R., Chegeni, A., Soleimanynanadegany, A., Memari, A., Anjomshoae, A., Helmi, S.A., Dargi, A.: Prioritizing green supplier selection criteria using fuzzy analytical network process. In: 12th Global Conference on Sustainable Manufacturing, Procedia CIRP, vol. 26, pp. 689–694 (2015)
6. Lima Jr., F.R., Osiro, L., Carpinetti, L.C.R.: A comparison between fuzzy AHP and fuzzy TOPSIS methods to supplier selection. Appl. Soft Comput. **21**, 194–209 (2014)
7. Shemshadi, A., Shiraji, H., Toreihi, M., Tarokh, M.J.: A fuzzy VIKOR method for supplier selection based on entropy measure for objective weighting. Expert Syst. Appl. **38**, 12160–12167 (2011)
8. Bilisik, M.E., Çağlar, N., Bilisik, Ö.N.A.: A comparative performance analyze model and supplier positioning in performance maps for supplier selection and evaluation. In: 8th International Strategic Management Conference on Procedia – Social; and Behavioral Sciences, vol. 58, pp. 1434–1442 (2012)
9. Feng, B., Fan, Z.-P., Li, Y.: A decision method for supplier selection in multi-service outsourcing. Int. J. Prod. Econ. **132**, 240–250 (2011)
10. Scott, J., Ho, W., Dey, P.K., Talluri, S.: A decision support system for supplier selection and order allocation in stochastic, multi-stakeholder and multi-criteria environments. Int. J. Prod. Econ. **166**, 226–237 (2015)
11. Rezaei, J., Davoodi, M.: A deterministic multi-item inventory model with supplier selection and imperfect quality. Appl. Math. Model. **32**, 2106–2116 (2008)
12. Zeydan, M., Colpan Cüneyt, Ç.: A combined methodology for supplier selection and performance evaluation. Expert Syst. Appl. **38**, 2741–2751 (2011)

Erratum to: An Analysis of Decision Theoretic Kernalized Rough Intuitionistic Fuzzy C-Means

Ryan Serrao[1(✉)], B.K. Tripathy[1], and A. Jayaram Reddy[2]

[1] SCOPE, VIT University, Vellore 632014, Tamil Nadu, India
{ryangodwin.serrao2013, tripathybk}@vit.ac.in
[2] SITE, VIT University, 632014 Vellore, Tamil Nadu, India
ajayaramreddy@vit.ac.in

Erratum to:
**Chapter "An Analysis of Decision Theoretic Kernalized
Rough Intuitionistic Fuzzy C-Means" in:
K. Deep et al. (eds.):** *Proceedings of Sixth International
Conference on Soft Computing for Problem Solving,*
**Advances in Intelligent Systems and Computing,
DOI: 10.1007/978-981-10-3322-3_28**

The original version of the book was inadvertently published with an incorrect author name "J. Jayaram Reddy" which has been corrected to "A. Jayaram Reddy".

The updated original online version for this chapter can be found at
DOI: 10.1007/978-981-10-3322-3_28

© Springer Nature Singapore Pte Ltd. 2017
K. Deep et al. (eds.), *Proceedings of Sixth International Conference
on Soft Computing for Problem Solving*, Advances in Intelligent Systems
and Computing 546, DOI 10.1007/978-981-10-3322-3_34

Author Index

© Springer Nature Singapore Pte Ltd. 2017
K. Deep et al. (eds.), *Proceedings of Sixth International Conference on Soft Computing for Problem Solving*, Advances in Intelligent Systems and Computing 546, DOI 10.1007/978-981-10-3322-3

Printed in the United States
By Bookmasters